Wide Bandgap Based Devices

Wide Bandgap Based Devices: Design, Fabrication and Applications

Editor

Farid Medjdoub

MDPI • Basel • Beijing • Wuhan • Barcelona • Belgrade • Manchester • Tokyo • Cluj • Tianjin

Editor
Farid Medjdoub
IEMN (Institute of Electronics, Microelectronics and Nanotechnology),
CNRS (Centre National de Recherche Scientifique)
France

Editorial Office
MDPI
St. Alban-Anlage 66
4052 Basel, Switzerland

This is a reprint of articles from the Special Issue published online in the open access journal *Micromachines* (ISSN 2072-666X) (available at: https://www.mdpi.com/journal/micromachines/special_issues/Wide_Bandgap_Devices).

For citation purposes, cite each article independently as indicated on the article page online and as indicated below:

LastName, A.A.; LastName, B.B.; LastName, C.C. Article Title. *Journal Name* **Year**, *Volume Number*, Page Range.

ISBN 978-3-0365-0566-4 (Hbk)
ISBN 978-3-0365-0567-1 (PDF)

Contents

About the Editor

Farid Medjdoub is a CNRS Senior Scientist and has led a team focused on wide bandgap materials and devices at IEMN in France since 2014. He received his Ph.D. in Electrical Engineering from the University of Lille in 2004. Then, he moved to the University of Ulm in Germany where he served as a Research Associate before joining IMEC as a Senior Scientist in 2008. Multiple state-of-the-art results have been realized in the frame of his work. Among others, a world record thermal stability up to 1000 °C for a field effect transistor, the best combination of cut-off frequency/breakdown voltage or highest lateral GaN-on-silicon breakdown voltage using a local substrate removal have been achieved.

His research interests are in the design, fabrication, characterization, and simulation of innovative wide bandgap devices. He is author and co-author of more than 170 papers in this field. He holds several patents derived from his research.

Editorial

Editorial for the Special Issue on Wide Bandgap Based Devices: Design, Fabrication and Applications

Farid Medjdoub

IEMN (Institute of Electronics, Microelectronics and Nanotechnology), CNRS (Centre National de Recherche Scientifique), Avenue Poincaré, 59650 Villeneuve d'Ascq, France; farid.medjdoub@univ-lille.fr

Received: 12 January 2021; Accepted: 14 January 2021; Published: 15 January 2021

Emerging wide bandgap (WBG) semiconductors hold the potential to advance the global industry in the same way that, more than 50 years ago, the invention of the silicon (Si) chip enabled the modern computer era. SiC- and GaN-based devices are starting to become more commercially available. Smaller, faster and more efficient than counterpart Si-based components, these WBG devices also offer a greater expected reliability in tougher operating conditions. Furthermore, in this frame, a new class of microelectronic-grade semiconducting materials that have an even larger bandgap than the previously established wide-bandgap semiconductors, such as GaN and SiC, have been created, and are; thus, referred to as "ultra-wide-bandgap" materials. These materials, which include AlGaN, AlN, diamond and BN oxide-based, offer theoretically superior properties, including a higher critical breakdown field, higher temperature operation and potentially higher radiation tolerance. These attributes, in turn, make it possible to use revolutionary new devices for extreme environments, such as high-efficiency power transistors, because of the improved Baliga Figure of Merit, ultra-high voltage pulsed power switches, high efficiency UV-LEDs, laser diodes and RF electronics.

There are 20 papers published in this Special Issue focusing on Wide Bandgap-Based Devices: Design, Fabrication and Applications. Three papers [1–3] deal with RF power electronics for future 5G applications and other high-speed high-power applications. Nine of the papers, [4–12], explore various designs of wide bandgap high power devices. The remaining papers cover various applications based on wide bandgaps, such as ZnO Nanorods for High Photon Extraction Efficiency of GaN-Based Photonic Emitter [13], InGaZnO Thin-Film Transistors [14], Wide Band Gap WO_3 Thin Film [15], Silver Nanorings [16,17] and InGaN Laser Diode [18–20].

In particular, on RF GaN devices, Kuchta et al. [1] proposed a GaN-based power amplifier design with a reduced level of transmittance distortions. Lee et al. [2] demonstrated a compact 20 W GaN internally matched power amplifier for 2.5 to 6 GHz jammer systems that uses a high dielectric constant substrate, single-layer capacitors, and shunt/series resistors for low-Q matching and low-frequency stabilization. Lin et al. [3] showed a high output power density of 8.2 W/mm in the Ka band by integrating a thick copper metallization.

Concerning GaN power devices, Wu et al. [4] investigated a double AlGaN barrier design toward enhancement-mode characteristics. Ma et al. [5] presented a digitally controlled 2 kVA three-phase shunt APF system using GaN. Tajalli et al. [6] studied the origin of vertical leakage and breakdown in GaN-on-Si epitaxial structures by carrying out a buffer decomposition. The contribution of each buffer layer related to vertical leakage and breakdown voltage could be identified. Sun et al. [7] proposes a new approach to realize normally-off GaN HEMTs using TCAD. The concept is based on the transposition of the gate channel orientation from a long horizontal one to a short vertical one. Mao et al. [8] introduced a portion of the p-polySi/p-SiC heterojunction on the collector side of an IGBT to reduce the turn-off loss without sacrificing other characteristics of the device. Kim et al. [9] implemented a SiC micro-heater chip as a novel thermal evaluation device for next-generation power modules and to evaluate the heat resistant performance. Keum et al. [10] investigated the time-dependent dielectric breakdown (TDDB) characteristics of normally-off AlGaN/GaN gate-recessed MISHEMTs submitted to proton irradiation. Abid et al. [11] presented the fabrication of AlN-based thin and thick channel AlGaN/GaN heterostructures that have been regrown by molecular beam epitaxy on AlN/sapphire. A remarkable breakdown field of 5 MV/cm has been observed for short contact distances, which is far beyond the theoretical limit of the GaN-based material system. Sandupatla et al. [12] used vertical GaN-on-GaN Schottky diodes as α- particle radiation sensors. They reported the highest reverse breakdown voltage of −2400 V from Schottky barrier diodes on a freestanding GaN substrate with 30 μm drift layer.

Besides, Lee et al. [13] demonstrated self-aligned hierarchical ZnO nanorod nanosheet arrays on a conventional photonic emitter with a wavelength of 430 nm with an improved optical output power. Zhang et al. [14] improved the electrical performance and bias-stress stability of amorphous InGaZnO thin-film transistors using buried-channel devices with multiple-stacked channel layers. Liu et al. [15] developed a tungsten trioxide (WO3) wide band gap using ammonium tungstate to obtain a high electrochromic modulation ability roughly 40% at 700 nm wavelength. Li et al. [16] optimized silver nanoring for transparent flexible electrodes applied to wide bandgap devices. Y. Wang et al. [17] proposed, for the first time, a novel GaN-based heterostructure Gunn diode, which turns out to be an excellent solid-state source for terahertz oscillators. W. Wang et al. [18] carried out a theoretical investigation the optical field distribution and electrical property improvements of the InGaN laser diode with an emission wavelength around 416 nm. Device optimization is favorable for the achievement of low threshold current and high output power lasers. Deng et al. [19] describes an optimization of InGaN/GaN distributed feedback laser diodes to enhance the efficiency. Finally, Luo et al. [20] propose a design based on a p-type composition-graded $Al_xGa_{1-x}N$ electron blocking layer to improve the output power of GaN-based VCSEL.

I would like to take this opportunity to thank all the authors for submitting their papers to this Special Issue. I would also like to thank all the reviewers for dedicating their time and helping to improve the quality of the submitted papers.

Conflicts of Interest: The author declares no conflict of interest.

References

1. Kuchta, D.; Gryglewski, D.; Wojtasiak, W. A GaN HEMT Amplifier Design for Phased Array Radars and 5G New Radios. *Micromachines* **2020**, *11*, 398. [CrossRef] [PubMed]
2. Min-Lee, P.; Kim, S.; Hong, S.; Kim, D. Compact 20-W GaN Internally Matched Power Amplifier for 2.5 GHz to 6 GHz Jammer Systems. *Micromachines* **2020**, *11*, 375.
3. Lin, Y.C.; Chen, S.H.; Lee, P.H.; Lai, K.H.; Huang, T.J.; Chang, Y.E.; Hsu, H. Gallium Nitride (GaN) High-Electron-Mobility Transistors with Thick Copper Metallization Featuring a Power Density of 8.2 W/mm for Ka-Band Applications. *Micromachines* **2020**, *11*, 222. [CrossRef] [PubMed]
4. Wu, T.; Tang, S.; Jiang, H. Investigation of Recessed Gate AlGaN/GaN MIS-HEMTs with Double AlGaN Barrier Designs toward an Enhancement-Mode Characteristic. *Micromachines* **2020**, *11*, 163. [CrossRef] [PubMed]
5. Ma, C.; Gu, Z. Design and Implementation of a GaN-Based Three-Phase Active Power Filter. *Micromachines* **2020**, *11*, 134. [CrossRef] [PubMed]
6. Tajalli, A.; Borga, M.; Meneghini, M.; de Santi, C.; Benazzi, D.; Besendörfer, S.; Püsche, R.; Derluyn, J.; Degroote, S.; Germain, M.; et al. Vertical Leakage in GaN-on-Si Stacks Investigated by a Buffer Decomposition Experiment. *Micromachines* **2020**, *11*, 101. [CrossRef] [PubMed]
7. Sun, Z.; Huang, H.; Sun, N.; Tao, P.; Zhao, C.; Liang, Y.C. A Novel GaN Metal-Insulator-Semiconductor High Electron Mobility Transistor Featuring Vertical Gate Structure. *Micromachines* **2019**, *10*, 848. [CrossRef] [PubMed]
8. Mao, H.; Wang, Y.; Wu, X.; Su, F. Simulation Study of 4H-SiC Trench Insulated Gate Bipolar Transistor with Low Turn-Off Loss. *Micromachines* **2019**, *10*, 815. [CrossRef] [PubMed]
9. Kim, D.; Yamamoto, Y.; Nagao, S.; Wakasugi, N.; Chen, C.; Suganuma, K. Measurement of Heat Dissipation and Thermal-Stability of Power Modules on DBC Substrates with Various Ceramics by SiC Micro-Heater Chip System and Ag Sinter Joining. *Micromachines* **2019**, *10*, 745. [CrossRef] [PubMed]

10. Keum, D.; Kim, H. Proton Irradiation Effects on the Time-Dependent Dielectric Breakdown Characteristics of Normally-Off AlGaN/GaN Gate-Recessed Metal-Insulator-Semiconductor Heterostructure Field Effect Transistors. *Micromachines* **2019**, *10*, 723. [CrossRef] [PubMed]
11. Abid, I.; Kabouche, R.; Bougerol, C.; Pernot, J.; Masante, C.; Comyn, R.; Cordier, Y.; Medjdoub, F. High Lateral Breakdown Voltage in Thin Channel AlGaN/GaN High Electron Mobility Transistors on AlN/Sapphire Templates. *Micromachines* **2019**, *10*, 690. [CrossRef] [PubMed]
12. Sandupatla, A.; Arulkumaran, S.; Ing, N.G.; Nitta, S.; Kennedy, J.; Amano, H. Vertical GaN-on-GaN Schottky Diodes as α-Particle Radiation Sensors. *Micromachines* **2020**, *11*, 519. [CrossRef] [PubMed]
13. Lee, W.; Kwon, S.; Choi, H.; Im, K.; Lee, H.; Oh, S.; Kim, K. Self-Aligned Hierarchical ZnO Nanorod/NiO Nanosheet Arrays for High Photon Extraction Efficiency of GaN-Based Photonic Emitter. *Micromachines* **2020**, *11*, 346. [CrossRef] [PubMed]
14. Zhang, Y.; Xie, H.; Dong, C. Electrical Performance and Bias-Stress Stability of Amorphous InGaZnO Thin-Film Transistors with Buried-Channel Layers. *Micromachines* **2019**, *10*, 779. [CrossRef] [PubMed]
15. Liu, J.; Zhang, G.; Guo, K.; Guo, D.; Shi, M.; Ning, H.; Qiu, T.; Chen, J.; Fu, X.; Yao, R.; et al. Effect of the Ammonium Tungsten Precursor Solution with the Modification of Glycerol on Wide Band Gap WO_3 Thin Film and Its Electrochromic Properties. *Micromachines* **2020**, *11*, 311. [CrossRef] [PubMed]
16. Li, Z.; Guo, D.; Xiao, P.; Chen, J.; Ning, H.; Wang, Y.; Zhang, X.; Fu, X.; Yao, R.; Peng, J. Silver Nanorings Fabricated by Glycerol-Based Cosolvent Polyol Method. *Micromachines* **2020**, *11*, 236. [CrossRef] [PubMed]
17. Wang, Y.; Li, L.; Ao, J.; Hao, Y. Physical-Based Simulation of the GaN-Based Grooved-Anode Planar Gunn Diode. *Micromachines* **2020**, *11*, 97. [CrossRef] [PubMed]
18. Wang, W.; Xie, W.; Deng, Z.; Liao, M. Improving Output Power of InGaN Laser Diode Using Asymmetric $In_{0.15}Ga_{0.85}N/In_{0.02}Ga_{0.98}N$ Multiple Quantum Wells. *Micromachines* **2019**, *10*, 875. [CrossRef] [PubMed]
19. Deng, Z.; Li, J.; Liao, M.; Xie, W.; Luo, S. InGaN/GaN Distributed Feedback Laser Diodes with Surface Gratings and Sidewall Gratings. *Micromachines* **2019**, *10*, 699. [CrossRef] [PubMed]
20. Luo, H.; Li, J.; Li, M. Improved Output Power of GaN-based VCSEL with Band-Engineered Electron Blocking Layer. *Micromachines* **2019**, *10*, 694. [CrossRef]

Article

A GaN HEMT Amplifier Design for Phased Array Radars and 5G New Radios

Dawid Kuchta *, Daniel Gryglewski and Wojciech Wojtasiak

Institute of Radioelectronics and Multimedia Technology, Warsaw University of Technology, Nowowiejska 15/19, 00-662 Warsaw, Poland; dgrygle@ire.pw.edu.pl (D.G.); wwojtas@ire.pw.edu.pl (W.W.)
* Correspondence: dkuchta@ire.pw.edu.pl; Tel.: +48-22-234-5886

Received: 29 February 2020; Accepted: 10 April 2020; Published: 10 April 2020

Abstract: Power amplifiers applied in modern active electronically scanned array (AESA) radars and 5G radios should have similar features, especially in terms of phase distortion, which dramatically affects the spectral regrowth and, moreover, they are difficult to be compensated by predistortion algorithms. This paper presents a GaN-based power amplifier design with a reduced level of transmittance distortions, varying in time, without significantly worsening other key features such as output power, efficiency and gain. The test amplifier with GaN-on-Si high electron mobility transistors (HEMT) NPT2018 from MACOM provides more than 17 W of output power at the 62% PAE over a 1.0 GHz to 1.1 GHz frequency range. By applying a proposed design approach, it was possible to decrease phase changes on test pulses from 0.5° to 0.2° and amplitude variation from 0.8 dB to 0.2 dB during the pulse width of 40 μs and 40% duty cycle.

Keywords: power amplifier; GaN 5G; high electron mobility transistors (HEMT); new radio; RF front-end; AESA radars; transmittance; distortions; optimization

1. Introduction

Radar systems, mainly 3D Active Electronically Scanned Array (AESA), strongly supported by the latest achievements in information technology, bring new challenges to the designers of transmit/receive (T/R) modules, especially High-Power Amplifiers (HPAs) based on solid-state devices [1,2]. In addition, there are currently rapid advances in high-speed wireless technology, such as 5G [3–9]. In particular, the power amplifiers, as a key element of RF transmitters, directly and significantly affect the operation quality of modern wireless communication systems and new generation radars [2,3,9]. The requirements concerning linearity and efficiency of HPAs are confronted with the needs of both systems for higher output power and improved the efficiency of heat management. In case of AESA, due to very complex beamforming techniques used, the strong emphasis is put on amplitude and phase constancy of the amplifier's transmittance during the RF pulse and pulse-to-pulse [10], as in pulse the signal is increasingly more often modulated not only in frequency but also in amplitude and phase [11]. The same is true for the 5G and Long-Term Evolution Advanced (LTE-A) network systems which are using quadrature amplitude modulation (QAM) of higher orders and orthogonal frequency division multiplexing (OFDM) methods [11–13]. Both QAM and OFDM are particularly sensitive to transmittance changes generated mainly by output stages of base station transmitter power amplifiers [8–13]. Currently, such amplifiers must be linearized to meet wireless transmission standards defined e.g., by following parameters: in-band error vector magnitude (EVM), adjacent channel power ratio (ACPR), or the shape of spectrum mask [14–16]. There are many techniques for the amplifier linearity improvement such as e.g., analog feedforward, digital pre-distortion [6], dynamic biasing [7], envelope tracking [10], or Chireix's outphasing method [17]. However, only a few of them are suitable for linearization of amplifiers operating with wideband spectrally efficient signals and high peak-to-average power ratio (PAPR) like LTE or 5G [5,9,18]. The common linearization technique applied in broadband transmitters of contemporary wireless systems is the baseband digital pre-distortion technique (DPD) preceded by the shaping of waveform crest factor (CF) [6]. All these methods have restrictions on applicability and require additional external hardware and/or software implementation consuming system resources. To reduce the system resource consumption, the nonlinearities of amplifiers should be as small as possible. There is a new challenge facing amplifier's designers in developing new amplifier solutions for modern applications, like AESA, LTE, 5G as well as next-generation systems [2–4]. Our research just goes in this direction to examine amplifier transmittance changes as a function of load and source impedance of a transistor. Since the transmittance changes are not only a symptom of transistor nonlinearity but they are also a response to the temperature changes inside a transistor due to the complexity of amplified signals. It is known that the thermal effects in the transistor active layer are essentially responsible for the amplitude distortion while variations of internal time delays inside the transistor are the main reason for a phase distortion during quick signal envelope changes in time, e.g., in pulse [19]. Therefore, it is necessary to use large-signal electro-thermal models [20]. The paper presents the dependencies of changes in the transmittance phase and magnitude on the GaN high electron mobility transistors (HEMT) load impedance. The design strategy based on the derived relationships and numerical simulations are confirmed by sophisticated time-domain measurements.

Our goal was to develop a temperature-dependent HPA modeling technique under operating conditions with signals of a variable envelope with built-in a function of transistor load impedance optimization for minimal transmittance distortions. For this purpose, the typical power amplifier structure was analyzed to find origins and relationships of changes in amplifier transmittance during radar pulse as well as in the defined time window in case of broadband wireless communication signals. The minimization of HPA transmittance changes will facilitate the use of correction methods in baseband, such as DPD and radar calibration.

In the paper, the proposed method is particularly focused on GaN HEMTs but it can be generalized to other types of transistors. There are few publications regarding GaN-on-Si HEMTs despite their price is twice as low as GaN-on-SiC. Thus, for the purpose of this work GaN-on-Si HEMT was chosen to examine its performance.

2. Power Amplifier Analysis

In order to identify sources of transmittance variations, the complete amplifier structure with the GaN HEMT was measured and analyzed [21,22]. On this basis, the relationship between the transistor load impedance and amplifier transmittance changes have been derived. We also want to show what the changes in the transmittance phase and magnitude depend on. The amplifier model schematic circuit to be analyzed is shown in Figure 1. For this purpose, it is sufficient that we use a small-signal model. The model represented by the equivalent circuit shown in Figure 1 was described by a set of parameters dependent on current and voltage values at the transistor operating points i.e., its parameters were extracted at the properly selected transistor operating conditions.

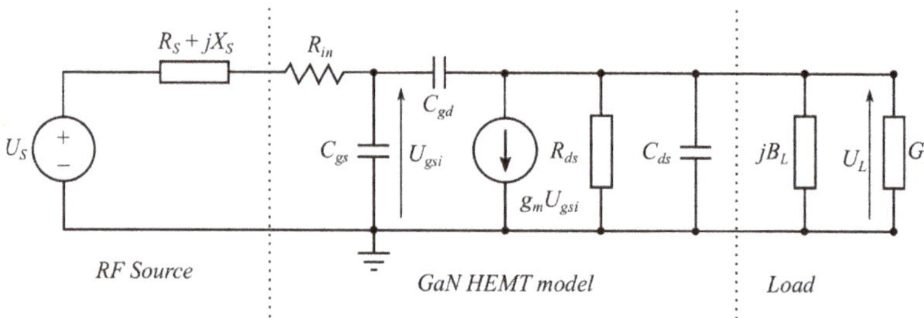

Figure 1. Power amplifier model with GaN high electron mobility transistors (HEMT).

It consists of a typical GaN HEMT transistor model with parameters determined at the average drain current I_{DA} corresponding to a saturated output power. Other transistor model components invisible in Figure 1 are included in the source impedance $Z_s = R_s + jX_s$, $(X_s > 0)$ and the load admittance $Y_L = G_L + jB_L$. The source impedance is connected in series with the RF signal source modeled as an ideal voltage source U_S. For clarity of our analysis, Miller theory is applied but there is a problem with determining an internal voltage gain K_{unit} [23]. In general, the gain K_{unit} should be calculated with account of all elements of the equivalent circuit shown in Figure 1. However, for low frequencies and roughly estimation for higher frequencies the internal voltage gain K_{unit} can be simplified in the first approximation to the following form:

$$K_{uint} = \frac{U_L}{U_{gsi}} = -\frac{g_m}{G'_L} \tag{1}$$

under the following assumption:

$$R_S + R_{in} > \frac{1}{\omega\left[C_{gs} + \left(1 + \frac{g_m}{G'_L}\right)C_{gd}\right]} \ or \ X_S \ll \frac{1}{\omega\left[C_{gs} + \left(1 + \frac{g_m}{G'_L}\right)C_{gd}\right]}, \tag{2}$$

where:

$$G'_L = G_L + \frac{1}{R_{ds}}. \tag{3}$$

The simplified HEMT model schematic circuit takes the form as shown in Figure 2.

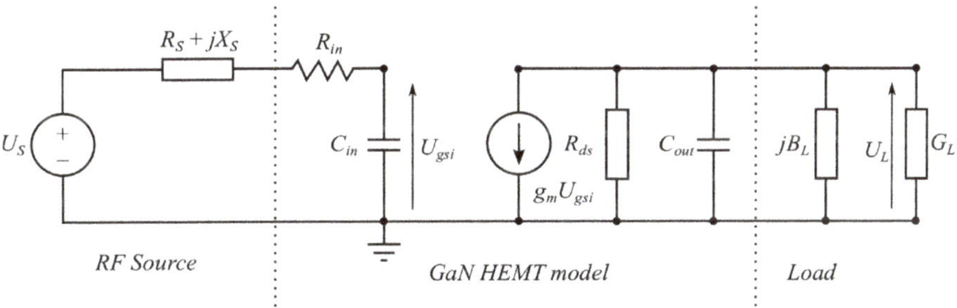

Figure 2. Amplifier model simplified by the Miller theorem.

The Miller theorem expresses only equivalents of C_{gd}; C_{gs} and C_{ds} are parallel to these equivalents, thus they can be added giving C_{in} and C_{out}. Hence the input and output equivalent Miller's capacitances are given as:

$$C_{in} = C_{gs} + C_{dg}\left(1 + \frac{g_m}{G'_L}\right) \tag{4}$$

$$C_{out} = C_{ds} + C_{dg}\left(1 + \frac{G'_L}{g_m}\right). \tag{5}$$

When the amplifier is operated close to saturated output power, the ratio G'_L/g_m is much smaller than unity as given in Table 1. Under this assumption the Equation (5) can be simplified to the following form:

$$C_{out} \approx C_{ds} + C_{dg}. \tag{6}$$

Parameters of three L-band GaN HEMTs, provided by the manufacturer are given in Table 1. They were calculated at the average drain current I_{DA} for output power close to P_{sat} and implemented in Keysight software ADS.

Table 1. Selected parameters for three different GaN HEMTs calculated at the average current for corresponding to P_{sat}.

Parameter	NPTB0004A	NPT2018	NPT2022
I_{DA}	0.34 A	0.6 A	4 A
P_{Sat}	5 W	16 W	120 W
G'_L	16 mS	14 ms	101 ms
g_m	670 mS	584 ms	3.88 S
G'_L/g_m	0.024	0.025	0.026
C_{gs}	6.9 pF	7.8 pF	42.9 pF
C_{gd}	0.27 pF	0.291 pF	1.66 pF
C_{ds}	0.62 pF	0.56 pF	9.9 pF
U_{DS}	28 V	50 V	48 V

When the GaN HEMT is operated as the current source the output capacitance is almost constant depending on the drain-to-source voltage U_{DS} and RF signal amplitudes in a wide range [8]. For the sake of clarity, it was assumed that all frequency harmonics are shorted as for ideal class A and AB [24]. To eliminate the imaginary part of output admittance in the transistor model the load susceptance B_L should be as follows:

$$B_L = -\omega C_{out}, \tag{7}$$

which leads to the amplifier model schematic diagram as shown in Figure 3.

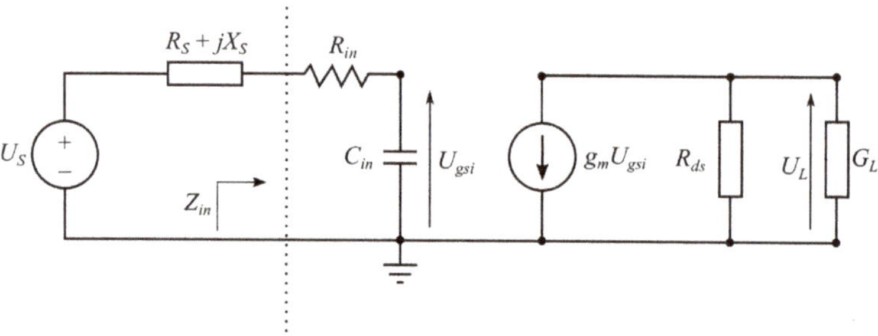

Figure 3. Simplified power amplifier model with GaN HEMT.

Applying the Kirchhoff voltage law to the model from Figure 3, the transmittance was derived. The transmittance phase and magnitude are expressed by (8) and (9). These formulas reveal details of the transmittance dependence on transistor parameters and parameters of the matching networks.

$$arg\left(\frac{U_L}{U_S}\right) = arg\left(j\frac{g_m}{G'_L \omega C_{in}} \frac{1}{R_S + R_{in} + j\left(X_S - \frac{1}{\omega C_{in}}\right)}\right) \tag{8}$$

$$\left|\frac{V_L}{V_S}\right| = \frac{g_m}{G'_L \omega C_{in}} \frac{1}{\sqrt{(R_S + R_{in})^2 + \left(X_S - \frac{1}{\omega C_{in}}\right)^2}}. \tag{9}$$

Equations (8) and (9) show that changes in the amplifier transmittance magnitude are influenced by both transistor parameters as well as the structure and parameters of matching networks.

Obviously, the transistor parameters in Equations (8) and (9) depend on temperature [19,25,26]. This explains the impact of temperature on the transmittance changes during pulse transfer as well as under the large amplitude variations of the signal with higher PAPR. In both cases, due to the high signal amplitude, the dynamic power dissipated in the transistor also strongly varies in time. As a response the temperature inside the active layer of the transistor is changed which proves the first thesis of this work.

Moreover, Equations (8) and (9) show that the transmittance phase increases monotonically with the change of load conductance G_L. Test results obtained in [27] confirm a validity of the relations (8), (9). To increase the accuracy of transmittance model the output capacitance C_{out} was taken into account. The phase and magnitude are given by (10) and (11) accordingly.

$$arg\left(\frac{V_L}{V_S}\right) = arg\left(j\frac{g_m}{\left(G'_L + j(B_L + \omega C_{out})\right)\omega C_{in}}\frac{1}{R_S + R_{in} + j\left(X_S - \frac{1}{\omega C_{in}}\right)}\right) \tag{10}$$

$$\left|\frac{V_L}{V_S}\right| = \frac{g_m}{\omega C_{in}}\left(\begin{array}{c}\left((R_S + R_{in})G'_L - \left(X_S - \frac{1}{\omega C_{in}}\right)(B_L + \omega C_{out})\right)^2 \\ +\left((R_S + R_{in})(B_L + \omega C_{out}) + \left(X_S - \frac{1}{\omega C_{in}}\right)G'_L\right)^2\end{array}\right)^{-\frac{1}{2}}. \tag{11}$$

The Equations (8)–(11) clearly show the dependence of the transmittance on the load impedance Z_L, which can be determined during the amplifier design.

In conclusion, using Equations (8) and (9) it is possible to facilitate initial steps of power amplifier design for the minimal transmittance changes. Using (8)–(11) formulas we can estimate the level of reduction of the transmittance changes by tuning Z_S and Z_L impedances. In our case, these values are 0.254° and 0.4 dB, for phase and magnitude, respectively, and are consistent with the simulation results which are 0.3° and 0.6 dB. Although this is a qualitative analysis, it is quite well in agreement with quantitative simulations using advanced software.

3. Measurement Setup and Amplifier Modelling

As a part of the research, measurements of waveforms of power amplifiers were performed. These measurements were performed using the Keysight DSAV334A Infiniium V-Series digital signal analyzer (DSA) (Keysight, Santa Rosa, CA, USA) and Keysight N5172B EXG X-Series vector signal generator with the option of generating training pulses. The purpose of such measurements was to examine the amplitude and phase distortions caused by the amplifier. The test program assumed the use of four pulse trains with a carrier frequency f_0, which are illustrated in Figure 4.

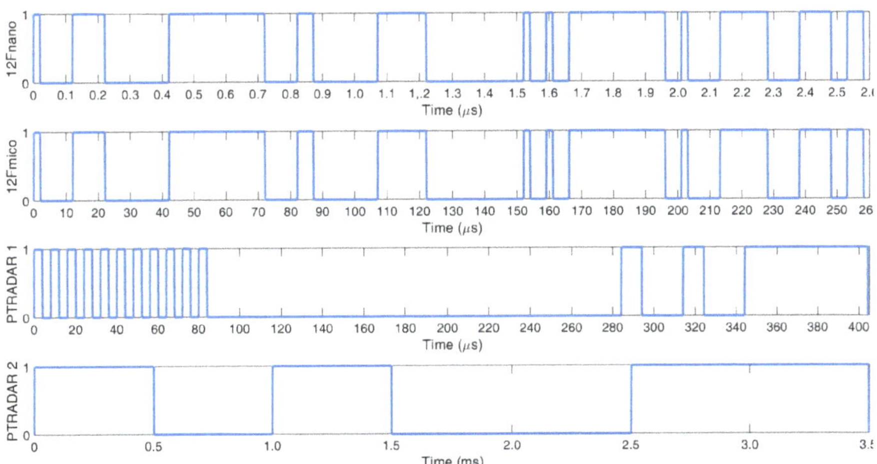

Figure 4. Train pulses used for power amplifier measurements.

The measurement consisted of recording in the time domain, previously designed, four train pulses, using DSA. To detect changes in the signal handled by the amplifier under test, the training signal was recorded before and after passing through the amplifier, working with a power close to P_{1dB} region. For this reason, the signal from the generator was split into two paths using HP11667A power divider (HP, Palo Alto, CA, USA). One of the paths was directly connected to DSA while the second one was connected to the amplifier input. The output of the amplifier was connected to the second DSA channel via the HP778D directional coupler. To protect the measurement instruments in both paths proper attenuators were used (the impact of attenuation on the obtained results was checked). The measurement setup is shown in Figure 5.

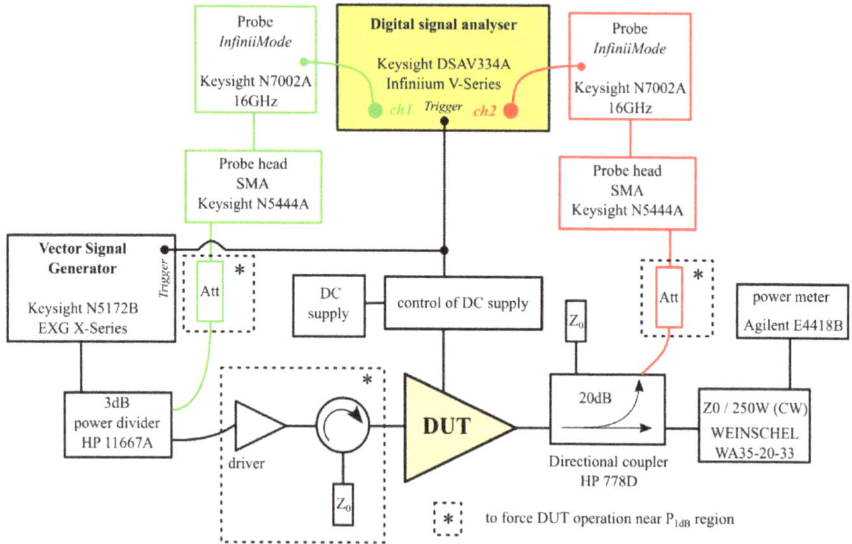

Figure 5. Measurement setup used for transmittance characterization with pulse trains.

The sample measurement results are shown in Figure 6. In the upper part of Figure 6 the input waveform is presented, in the middle, the output waveform coming directly from the amplifier under test and in the lower part the difference between these waveforms is calculated as a change of the transmittance phase during the pulse duration.

Based on the measurements, derived formulas, and using the Envelope simulation technique available in Keysight ADS software, we developed the power amplifier design method for minimization of the transmittance changes. To develop an algorithm for determining optimum load impedance, a test amplifier with 14W GaN HEMT NTP2018 from MACOM (Lowell, MA, USA) was designed using the large-signal transistor model provided by MACOM. The test amplifier was designed according to *Cripps* design methodology to achieve a trade-off between maximum output power and high efficiency [24]. The MACOM model is closed, and enables only the temperature characteristics under thermal steady-state conditions to be calculated. Therefore, for the purpose of this work, we developed our own large-signal model based on the Angelov [28]. Our model includes an extensive thermal part representing by 5-6 serially connected (R_{thi}, C_{thi}) parallel cells. The values of Angelov model parameters were fitted to obtain simulations consistent with the original non-linear MACOM model. The model thermal parameters were determined by fitting RC-ladder network to the thermal impedance $Z_{th}(t)$ of the transistor with was measured by the modified $DeltaU_{gs}$ method [29,30]. To determine load impedances the load-pull method was applied in ADS Software. The high compliance of the simulations and measurements of the amplifier was achieved.

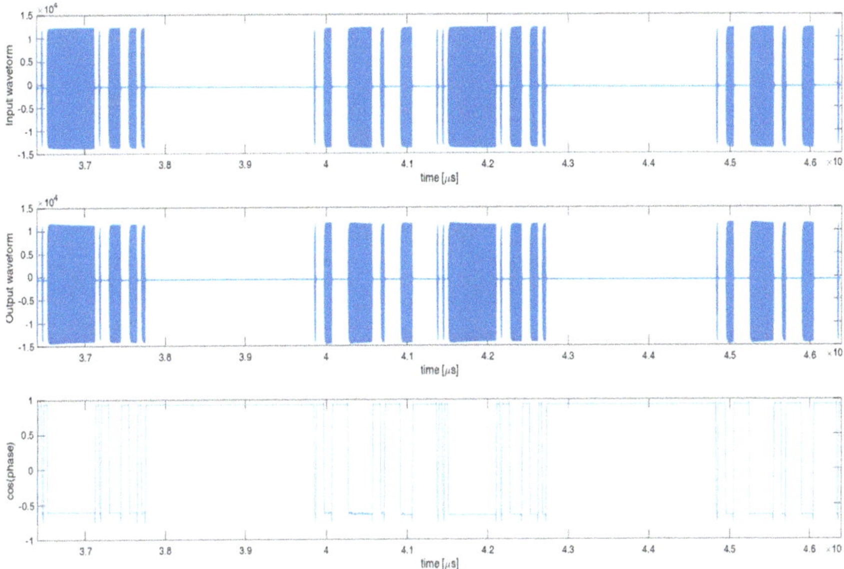

Figure 6. Sample measurement results: input waveform, output waveform and phase difference between waveforms above.

An appropriate use of the Envelope simulation [31] together with the large-signal electro-thermal model allows for simulating and modeling of amplifier's transmittance changes. The simulation provides great opportunities to optimize the structure of the amplifiers and facilitates the process of finding the optimal load impedance by viewing time waveforms. There are no studies in the literature that combine the electro-thermal model with the envelope simulation to study changes in the parameters of the transmitted signal. However, so far, a load impedance optimization of microwave amplifiers has been performed only under quasi-static conditions using simulations in the frequency domain. This made it impossible to analyze changes, in the signal during operation caused by widely understood memory effects at the stage of amplifier design. Moreover, optimization of amplifiers' parameters (e.g., ACPR) generally was based on the appropriate selection of load impedance, during load-pull measurements, in order to determine its value for which the optimized parameters reach a satisfactory range [15,32–34]. This approach is unenforceable in case of transmittance changes that are dynamic over time due to the phase measurement, requiring an accurate calibration of the reference track with a phase shifter that has to be perfectly synchronized with the load-pull tuners.

4. Test Amplifier

To show the proposed design methodology, step by step, the test amplifier with GaN HEMT NTP2018 was designed. The amplifier achieves more than 17 W of output power over a 1 GHz to 1.1 GHz frequency range, while MACOM in datasheet provides the information about 14 W. Assembly drawing and photo of the fabricated amplifier is shown in Figures 7 and 8, respectively. The amplifier was fabricated on Rogers RO4003C laminate ($\varepsilon_r = 3.55$, h = 0.020′′, T = 1oz.).

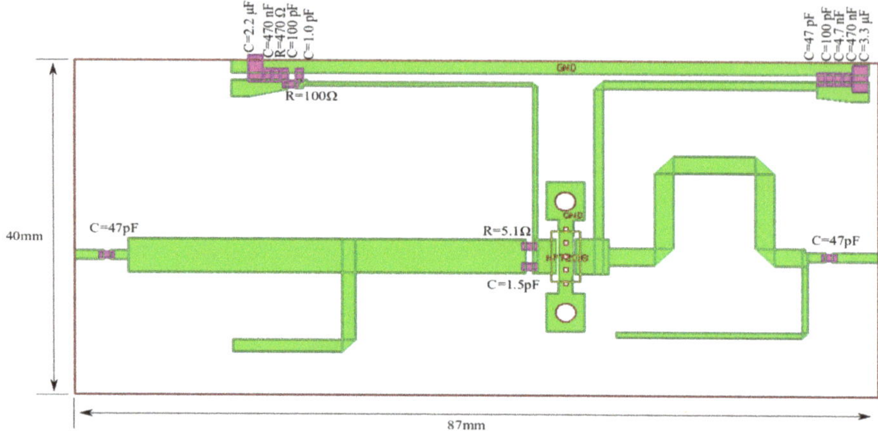

Figure 7. Assembly drawing of the test PA with GaN HEMT NTP2018.

Figure 8. Photography of the test PA with GaN HEMT NPT2018.

To generate a sequence of training pulses of a given power, as a stimulus an internally matched simulation port was used. This enabled us to simulate the pulses transferred by the amplifier in the time domain. The average pulse power $P_{AVout}(t)$ over a time interval Δt was calculated according to the following formula [35]:

$$P_{AVout}(t) = \frac{1}{\Delta t} \int_0^{\Delta t} PDF_{P_0}(t) P_{out}(t) dt, \tag{12}$$

where PDF denotes Probability density function (or the probability distribution function) of the output power. The simulation ADS schematic is shown in Figure 9.

The simulation was performed for two pulses with a duration of 40 µs and a period of 100 µs at the carrier frequency f = 1.05 GHz being the center frequency of the amplifier working band. In order to simulate the amplifier operating with output power close to P1dB the excitation power was set to 25 dBm in pulse. The test amplifier was characterized by the measurement setup shown in Figure 5 with the same parameters as given above. GaN HEMT NPT 2018 was biased at the quiescent operating point U_{DS} = 50 V and I_{DQ} = 75 mA. The simulation and measurement results of pulses at the amplifier output for frequency f = 1.05 GHz are shown in Figure 10.

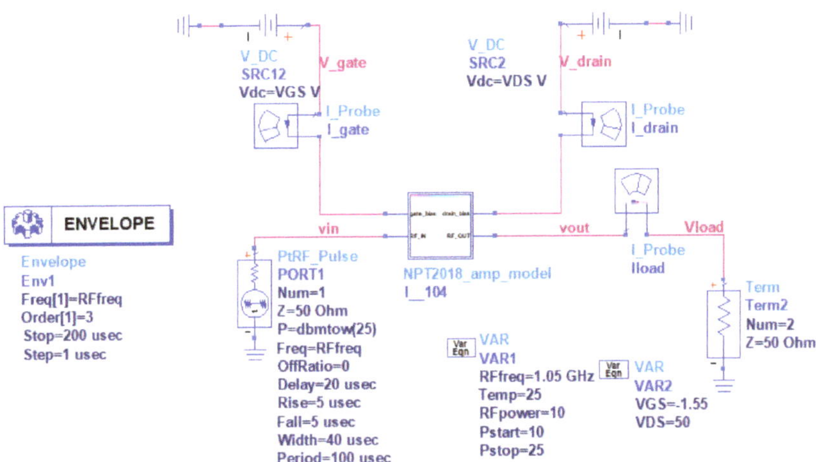

Figure 9. Simulation ADS schematic for modeling of PA transmittance.

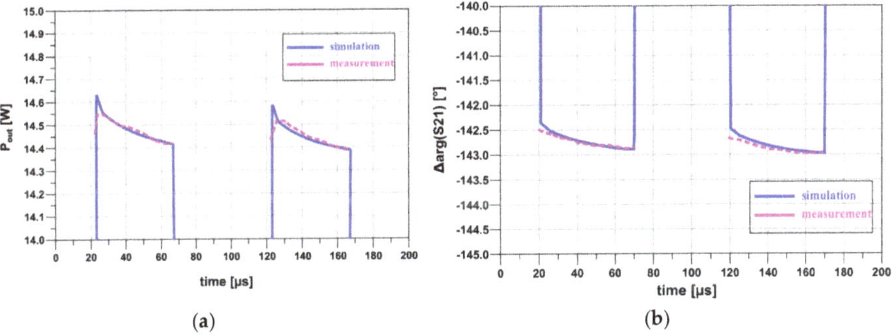

Figure 10. Transmittance changes of the amplifier: (**a**) amplitude changes, (**b**) phase changes during pulse for carrier frequency f = 1.05 GHz.

The presented approach allows for simulating the amplitude and phase changes of the amplifier transmittance during pulse in a wide range of transistor load impedance. It is the basis to develop an algorithm for optimizing of the source and load impedances for minimal transmittance distortions.

To find the optimum impedance of Z_{LT}, the Envelope simulation was used. It allows for modeling the amplifiers with very different waveforms and observing the waveforms at the amplifier output depending on the input power levels. With access to large-signal electro-thermal models and envelope simulation, it is possible to calculate the phase and amplitude changes caused by the amplifier during the pulse transfer. The range of phase and amplitude variations depends on the load impedance. For the optimum impedance Z_{LT} this range is the smallest. The impedance was adjusted using load-pull tuners. The ADS schematic used for the load impedance optimization is shown in Figure 11.

The starting point for searching for impedance Z_{LT} is the load impedance Z_L which is a compromise between maximum output power and maximum power-added efficiency (PAE) impedance. The impedance search area is narrowed down to impedances that meet the following conditions:

$$\begin{aligned} P_{out} &\geq 0.7P_{outmax} \\ PAE &\geq 0.8PAE_{max} \end{aligned} \quad (13)$$

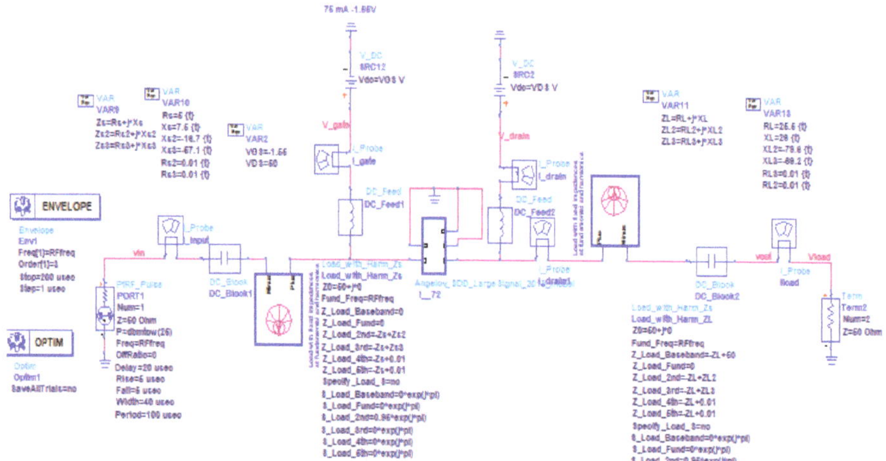

Figure 11. ADS schematic for source and load impedance optimization.

It is assumed in the optimization that all harmonics are shorted. The values of Z_S and Z_L source and load impedances seen by transistor before and after optimization are shown in Table 2.

Table 2. Source and load impedances after and before optimization.

	Original	After Optimization
$Z_L(f_0)$ [Ω]	26.4 + 26.6j	31.8 + 35.7j
$Z_S(f_0)$ [Ω]	5.6 + 12.0j	5.2 + 17.5j

Assembly drawing of amplifier optimized for minimum transmittance changes is shown in Figure 12.

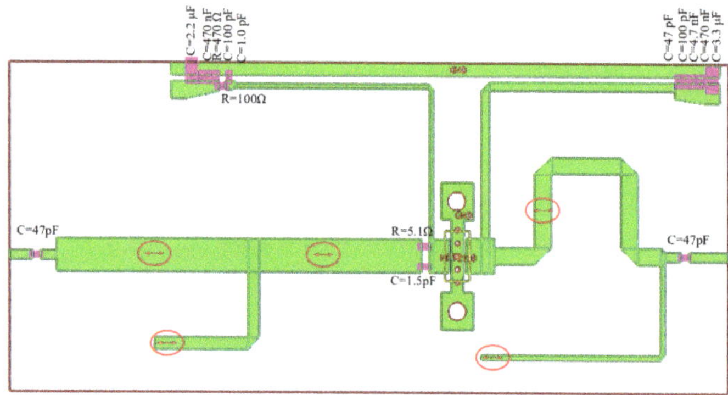

Figure 12. Assembly drawing of the PA with GaN HEMT NTP2018 optimized for minimum transmittance changes, modifications of the matching networks in relation to the previous version of the PA are marked in red.

Pulses shapes as visible result of transmittance variations after and before optimization are shown in Figure 13.

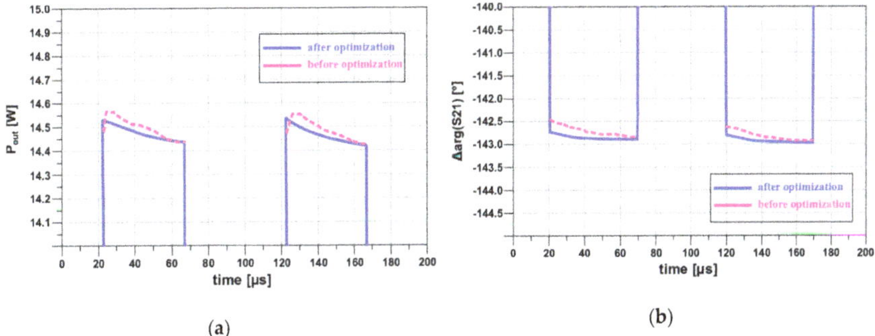

(a)

(b)

Figure 13. Transmittance changes of the amplifier after and before optimization: (**a**) amplitude changes and (**b**) phase changes during pulse for carrier frequency f = 1.05 GHz.

Transmittance variations before and after optimization with obtained output power and PAE of the test amplifier are presented in Table 3.

Table 3. Transmittance changes before and after optimization.

	$\Delta arg(S_{21})[°]$	$\Delta \lvert S_{21} \rvert [dB]$	$P_{out}[W]$	$PAE[\%]$
Before optimization	0.5	0.8	17 W	61
Optimization applied	0.2	0.2	14 W	54

The simplified guideline for the proposed design method is presented in Figure 14.

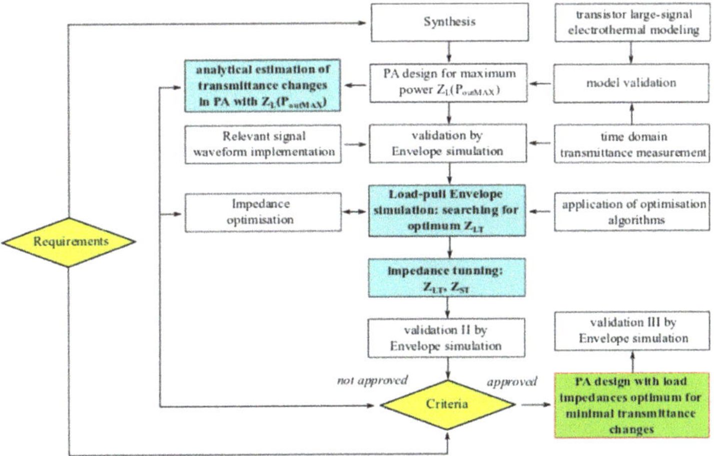

Figure 14. Simplified guidelines for proposed design methodology.

5. Conclusions

The paper presents the method to improve the amplifier transmittance flatness during pulse as well as other time-varying parameters of amplified signals without deterioration of other significant amplifier parameters such as output power, PAE and gain. The method comprises the Envelope simulations and sophisticated measurement. We simulated the transmittance phase and amplitude in time, e.g., during the pulse. The results of such simulations were consistent with the measurement results. The formulas for the amplifier transmittance were derived. These formulas enable the power amplifiers with minimum transmittance phase changes to be easier designed. For the transparency of this work, presented results relate to simple pulses, but the calculation can be done for a more complex radio signal, like 5G. We are currently working on the application of the presented method to this kind of signal. By applying new design approach, it was possible to improve phase changes on test pulses from 0.5° to 0.2° and decrease amplitude variation from 0.8 dB to 0.2 dB during the pulse width of 40µs and 40% duty cycle with the 17 W of output power and PAE more than 62%. Though we used GaN-on-Si HEMT, the results are very promising, and we are currently testing and modeling amplifiers with GaN-on-SiC HEMTs.

Author Contributions: D.K. designed, and performed characterization, analysis and optimization of the microwave amplifiers as well as measurements of PAs. W.W. performed the design and optimization of GAN HEMT PAs as well as performed modeling of GaN HEMTs and. D.G. performed large-signal electro-thermal modeling of HEMTs as well as performed analysis verification. All authors have read and agreed to the published version of the manuscript.

Funding: The research was supported by the National Centre for Research and Development WidePOWER Project, Contract No. 1/346922/4/NCBR/2017.

Conflicts of Interest: The authors declare no conflict of interest.

References

1. Walker, J.L.B. *Handbook of RF and Microwave Power Amplifiers*; Cambridge University Press: Cambridge, UK, 2012.
2. Kuchta, D.; Wojtasiak, W. GaN HEMT Power Amplifer for Radar Waveforms. *Proc. SPIE Radioelectron. Syst.* **2018**, *10715*, 1–6.
3. Asbeck, P.M. Will Doherty Continue to Rule for 5G? IEEE MTT-S International Microwave Symposium (IMS): San Francisco, CA, USA, 2016; pp. 1–4.
4. Hueber, G.; Niknejad, A.M. *Millimeter-Wave Circuits for 5G and Radar*; Hueber, G., Niknejad, A.M., Eds.; Cambridge University Press: Cambridge, UK, 2019.
5. Shafi, M.; Molisch, A.F.; Smith, P.J.; Haustein, T.; Zhu, P.; De Silva, P.; Tufvesson, F.; Benjebbour, A.; Wunder, G. 5G: A tutorial overview of standards, trials, challenges, deployment, and practice. *IEEE J. Sel. Areas Commun.* **2017**, *35*, 1201–1221. [CrossRef]
6. Katz, A.; Wood, J.; Chokola, D. The evolution of PA linearization: From classic feedforward and feedback through analog and digital predistortion. *IEEE Microw. Mag.* **2016**, *2*, 32–40. [CrossRef]
7. Pelaz, J.; Collantes, J.-M.; Otegi, N.; Anakabe, A.; Collins, G. Experimental Control and Design of Low-Frequency Bias Networks for Dynamically Biased Amplifiiers. *IEEE Trans. Microw. Theory Tech.* **2015**, *63*, 1923–1936. [CrossRef]
8. Onoe, S. Evolution of 5G mobile technology toward 1 2020 and beyond. In Proceedings of the 2016 IEEE International Solid-State Circuits Conference (ISSCC), San Francisco, CA, USA, 31 January–4 Feberuary 2016.
9. Lien, S.-Y.; Shieh, S.-L.; Huang, Y.; Su, B.; Hsu, Y.-L.; Wei, H.-Y. 5G new radio: Waveform, frame structure, multiple access, and initial access. *IEEE Commun. Mag.* **2017**, *55*, 64–71. [CrossRef]
10. Kim, B.; Kim, J.; Kim, D.; Son, J.; Cho, Y.; Kim, J.; Park, B. Push the envelope: Design concepts for envelope-tracking power amplifiers. *IEEE Microw. Mag.* **2013**, *14*, 68–81. [CrossRef]
11. Uysal, F.; Yeary, M.; Goodman, N.; Rincon, R.F.; Osmanoglu, B. Waveform design for wideband beampattern and beamforming. In Proceedings of the 2015 IEEE Radar Conference, Arlington, VA, USA, 10–15 May 2015; pp. 1062–1066.
12. Roh, W.; Seol, J.-Y.; Park, J.; Lee, B.; Lee, J.; Kim, Y.; Cho, J.; Cheun, K.; Aryanfar, F. Millimeter-wave beamforming as an enabling technology for 5G cellular communications: Theoretical feasibility and prototype results. *IEEE Commun. Mag.* **2014**, *52*, 106–113. [CrossRef]
13. Vook, F.W.; Ghosh, A.; Thomas, T.A. MIMO and beamforming solutions for 5G technology. In Proceedings of the 2014 IEEE MTT-S International Microwave Symposium (IMS2014), Tampa, FL, USA, 1–6 June 2014; pp. 1–4.

14. Dunn, Z.; Yeary, M.; Fulton, C. Frequency-dependent power amplifier modeling and correction for distortion in wideband radar transmissions. In Proceedings of the 2014 IEEE 57th International Midwest Symposium on Circuits and Systems (MWSCAS), College Station, TX, USA, 3–6 August 2014; pp. 61–64.
15. Fellows, M.; Baylis, C.; Martin, J.; Cohen, L.; Marks, R.J. Direct algorithm for the Pareto load-pull optimization of power-added efficiency and adjacent channel power ratio. *IET Radar Sonar Navig.* **2014**, *8*, 1280–1287. [CrossRef]
16. Baylis, C.; Fellows, M.; Barkate, J.; Tsatsoulas, A.; Rezayat, S.; Lamers, L.; Marks, R.J.; Cohen, L. Circuit optimization algorithms for real-time spectrum sharing between radar and communications. In Proceedings of the 2016 IEEE Radar Conference (RadarConf), Philadelphia, PA, USA, 2–6 May 2016; pp. 1–4.
17. Chireix, H. High power outphasing modulation. *Proc. Inst. Radio Eng.* **1935**, *23*, 1370–1392. [CrossRef]
18. Wood, J. *Behavioral Modeling and Linearization of RF Power Amplifiers*; Artech House: Norwood, MA, USA, 2014.
19. Gryglewski, D.; Wojtasiak, W. Temperature-Dependent Modeling of High-Power MESFET Using Thermal FDTD Method. In Proceedings of the IEEE MTT-S International Microwave Symposium Digest 2001, Phoenix, AZ, USA, 20–26 May 2001; pp. 411–414.
20. Remsburg, R. *Thermal Design of Electronic Equipment*; CRC Press: Boca Raton, FL, USA, 2001.
21. Golio, J.M.; Golio, M. RF and Microwave Passive and Active Technologies. In *RF and Microwave Handbook*, 2nd ed.; CRC Press: Boca Raton, FL, USA, 2007.
22. Komiak, J.J. GaN HEMT: Dominant Force in High-Frequency Solid-State Power Amplifiers. *IEEE Microw. Mag.* **2015**, *16*, 97–105. [CrossRef]
23. Miller, J.M. Dependence of the input impedance of a three-electrode vacuum tube upon the load in the plate circuit. *Sci. Pap. Bur. Stand.* **1920**, *15*, 367–385. [CrossRef]
24. Steve, C. *Cripps Advanced Techniques in RF Power Amplifier Design*; Artech House: Norwood, MA, USA, 2002.
25. Crupi, G.; Raffo, A.; Avolio, G.; Schreurs, D.M.M.-P.; Vannini, G.; Cassemi, A. Temperature Influence on GaN HEMT Equivalent Circuit. *IEEE Microw. Wirel. Compon. Lett.* **2016**, *26*, 813–815. [CrossRef]
26. Holman, J.P. *Heat Transfer*, 10th ed.; McGraw-Hill: London, UK, 2010.
27. Nunes, L.C.; Cabral, P.M.; Pedro, J.C. AM/AM and AM/PM Distortion Generation Mechanisms in Si LDMOS and GaN HEMT Based RF Power Amplifiers. *IEEE Trans. Microw. Theory Tech.* **2014**, *62*, 799–809. [CrossRef]
28. Angelov, I.; Thorsell, M.; Andersson, K.; Rorsman, N.; Kuwata, E.; Ohtsuka, H.; Yamanaka, K. *On the Large-Signal Modeling of High Power AlGaN/GaN HEMTs*; 2012 IEEE/MTT-S International Microwave Symposium Digest: Montreal, QC, USA, 2012.
29. Gryglewski, D. The GaN HEMT Thermal Characterization. In Proceedings of the 8th Wide Bandgap Semiconductors and Components Workshop, Harwell, UK, 12–13 September 2016.
30. Crupia, G.; Avolio, G.; Raffo, A.; Barmuta, P.; Schreurs, D.M.M.-P.; Caddemi, A.; Vannini, G. Investigation on the thermal behavior of microwave GaN HEMTs. *Solid-State Electron.* **2011**, *64*, 28–33. [CrossRef]
31. Ngoya, E.; Larcheveque, R. Envelop transient analysis: A new method for the transient and steady state analysis of microwave communication circuits and systems. In Proceedings of the 1996 IEEE MTT-S International Microwave Symposium Digest, San Francisco, CA, USA, 17–21 June 1996; Volume 3, pp. 1365–1368.
32. Dunn, Z.; Yeary, M.; Fulton, C.; Goodman, N. Wideband digital predistortion of solid-state radar amplifiers. *IEEE Trans. Aerosp. Electron. Syst.* **2016**, *52*, 2452–2466. [CrossRef]
33. Martin, J.; Moldovan, M.; Baylls, C.; Marks, R.J.; Cohen, L.; de Graaf, J. Radar chirp waveform selection and circuit optimization using ACPR load-pull measurements. In Proceedings of the WAMICON 2012 IEEE Wireless & Microwave Technology Conference, Cocoa Beach, FL, 15–17 April 2012; pp. 1–4.
34. Apel, T. RF Power Device Characterization Accomplished Through Direct and Indirect Techniques. In *Microwave Systems News*; EW Communications: Palo Alto, CA, USA, 1984.
35. Kazimierczuk, M. *RF Power Amplifiers*, 2nd ed.; John Wiley & Sons: Hoboken, NJ, USA, 2015.

Article

Compact 20-W GaN Internally Matched Power Amplifier for 2.5 GHz to 6 GHz Jammer Systems

Min-Pyo Lee, Seil Kim, Sung-June Hong and Dong-Wook Kim *

Department of Radio and Information Communications Engineering, Chungnam National University, Daejeon 34134, Korea; dignitymp20@naver.com (M.-P.L.); ksl4896@naver.com (S.K.); hsj_1006@naver.com (S.-J.H.)

* Correspondence: dwkim21c@cnu.ac.kr; Tel.: +82-42-821-6887

Received: 25 February 2020; Accepted: 1 April 2020; Published: 2 April 2020

Abstract: In this paper, we demonstrate a compact 20-W GaN internally matched power amplifier for 2.5 to 6 GHz jammer systems which uses a high dielectric constant substrate, single-layer capacitors, and shunt/series resistors for low-Q matching and low-frequency stabilization. A GaN high-electron-mobility transistor (HEMT) CGH60030D bare die from Wolfspeed was used as an active device, and input/output matching circuits were implemented on two different substrates using a thin-film process, relative dielectric constants of which were 9.8 and 40, respectively. A series resistor of 2.1 Ω was chosen to minimize the high-frequency loss and obtain a flat gain response. For the output matching circuit, double λ/4 shorted stubs were used to supply the drain current and reduce the output impedance variation of the transistor between the low-frequency and high-frequency regions, which also made wideband matching feasible. Single-layer capacitors effectively helped reduce the size of the matching circuit. The fabricated GaN internally matched power amplifier showed a linear gain of about 10.2 dB, and had an output power of 43.3–43.9 dBm (21.4–24.5 W), a power-added efficiency of 33.4–49.7% and a power gain of 6.2–8.3 dB at the continuous-wave output power condition, from 2.5 to 6 GHz.

Keywords: wideband; GaN; HEMT; power amplifier; jammer system

1. Introduction

GaN high-electron-mobility transistor (HEMT) wideband power amplifiers have been studied for multi-mode communication systems, electronic warfare systems, and other frequency-agile systems that required high-power operation over a wide frequency range [1–10]. While the transistor gain decreases with the frequency, a wideband power amplifier requires a flat gain performance in the interested bandwidth. In addition, the inherent reactance of the transistor, mainly due to its drain-to-source capacitance, limits the frequency bandwidth of the power amplifier as the Bode–Fano gain-bandwidth product describes [11,12]. To overcome this limitation, a multiple inductor-capacitor (LC) ladder configuration may be a proper choice, but it is area-inefficient and degrades the cost competitiveness, especially for expensive GaN monolithic microwave integrated circuits.

In this paper, we present a compact 20-W internally matched power amplifier for 2.5 to 6 GHz electronic warfare jammer applications which uses a GaN HEMT device and can be integrated into a small metal package. The novelty of this work originates in a practical combination of a series resistor, a shunt resistor-capacitor (RC)sub-circuit, a high dielectric constant matching substrate, and single-layer capacitors to maximize the size reduction effect. Input and output matching circuits for the power amplifier are implemented on two different thin-film substrates with the relative dielectric constants of 9.8 and 40, and, as a part of the matching circuit, single-layer capacitors and thin-film shunt/series resistors are utilized to achieve the circuit stabilization and flat gain, in addition to its compact size.

2. Power Amplifier Design

2.1. Device Description

The GaN HEMT bare die with the 0.4 μm gate length (CGH60030D, Wolfspeed, Inc., Research Triangle Park, NC, USA) that is provided by Wolfspeed has a typical power density of 4.5 W/mm and a maximum output power of 30 W at the drain pad and is applicable up to 6 GHz. Considering the loss of the matching circuit itself and the impedance mismatch loss, CGH60030D is a proper choice for the output power of more than 20 W from 2.5 to 6 GHz. Figure 1 and Table 1 show the photograph and device parameters of CGH60030D [13]. The transistor bare die consists of two 10 × 360 μm cells and has four pads on the gate and drain sides. Because the large-signal model provided by the manufacturer has only two ports for the gate and drain pads, it is difficult to fully include the bonding wire effect and the phase balance on each pad. In our work, we use a modified large-signal model combining the transistor unit cell model, which is based on the manufacturer's foundry process, with three-dimensional electromagnetic simulation models of the gate/drain pads and bonding wires [14,15]. Figure 2 shows the modified large-signal transistor model, the port reference planes of which are on the middle of its gate and drain pads to consider practical wire bonding effects effectively.

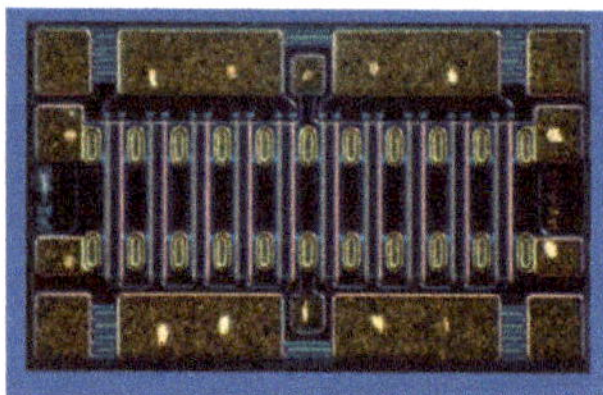

Figure 1. Photograph of a GaN high-electron-mobility transistor (HEMT) bare die (CGH60030D) from Wolfspeed.

Table 1. Device parameters of CGH60030D.

Parameters	Specifications
operating frequency	DC—6 GHz
saturated output power	30 W
power-added efficiency	65% at 4 GHz
small-signal gain	15 dB at 4 GHz
operating voltage	28 V
size	920 μm × 1660 μm

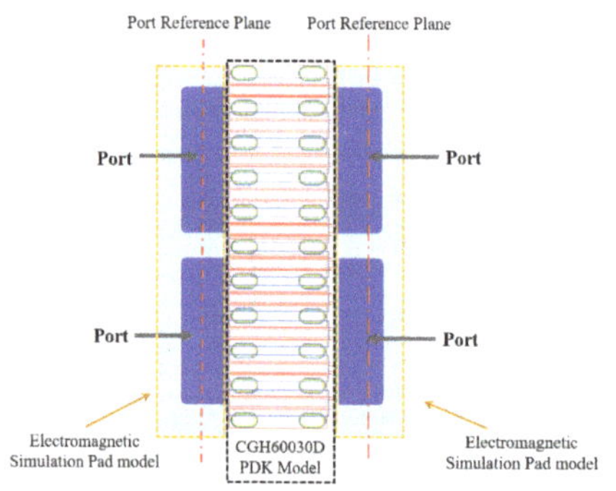

Figure 2. Modified large-signal transistor model with a foundry process design kit model and electromagnetic simulation pad models.

2.2. Input and Output Matching Circuit Design

We simulated the GaN HEMT under the bias conditions of V_{DS} = 28 V and I_{DS} = 250 mA, and predicted small-signal input impedances at 2.5 GHz and 6 GHz as $Z_{S,2.5\,GHz}$ = 1.39 + j2.11 Ω and $Z_{S,6\,GHz}$ = 0.89 + j0.11 Ω. Because the transistor was unstable below 7.9 GHz, we performed source-pull and load-pull simulations after stabilizing the transistor with shunt and series resistors. With the drain pad effect eliminated, the transistor was simulated up to the third harmonic by a load-pull tuner [16,17]. The simulation results showed that the optimum load impedances at 2.5 GHz and 6 GHz were $Z_{L,2.5\,GHz}$ = 8.00 + j4.91 Ω and $Z_{L,6\,GHz}$ = 5.41 + j3.86 Ω, respectively.

Figure 3 shows a schematic circuit diagram of the GaN HEMT with the stabilization circuit. To secure the low-frequency stability of the transistor and apply the gate bias voltage through the resistor, the shunt circuit of the resistor R_{BIAS} and the capacitor C_{Bypass} was inserted between the transistor and the input matching circuit. The element values of the shunt circuit were determined to be C_{Bypass} = 39 pF and R_{BIAS} = 36 Ω to simultaneously achieve stability and low loss.

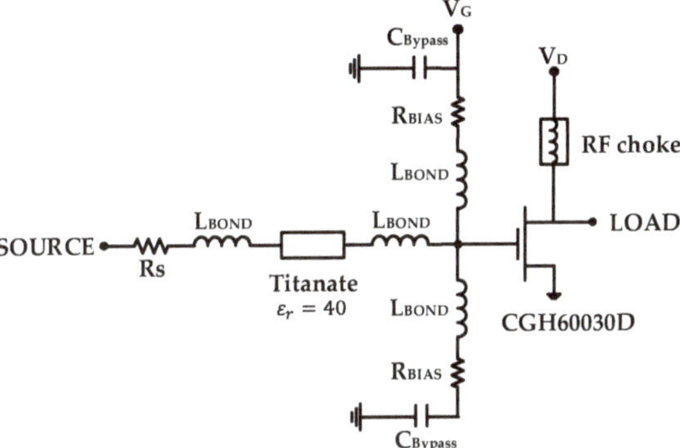

Figure 3. GaN HEMT with the stabilization circuit of the series and shunt resistors.

Because the high-frequency gain of the transistor is typically lower than its low-frequency gain, the high-frequency loss of the stabilization circuit should be lower than the circuit's low-frequency loss. However, because the input impedance of the transistor is very low, low-loss impedance matching is very difficult in a wide frequency range only with the simple matching circuit of a few LC elements. To compromise the low loss and compact size, we used a titanate substrate with the high relative dielectric constant of 40, and applied a series resistor of R_S = 2.1 Ω which made the stability factor k more than 1 and a flat gain in the interested bandwidth by increasing the low-frequency loss and reducing the high-frequency loss.

Figure 4 shows the variation of the stability factor k, maximum available gain (G_{max}), and input impedance before and after the insertion of the series resistor R_S. As the dotted impedance trace on the Smith chart in Figure 4 implies, the titanate substrate and bonding wires spread the input impedance trace of the transistor greatly in both of the low-frequency region and the high-frequency region, which maintains the low-frequency impedance as very low, and moves the high-frequency impedance to a high-value region. Therefore, a small series resistor greatly reduces the low-frequency gain, and, by contrast, does not affect the high-frequency gain because of the frequency-dependent voltage-dividing ratio.

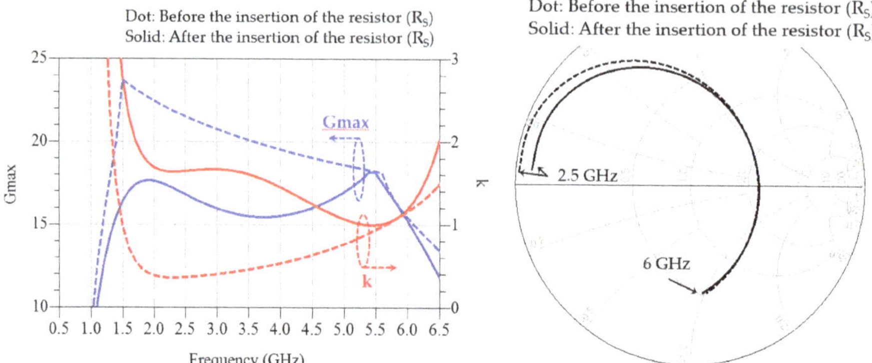

Figure 4. Variation of the stability factor k, maximum available gain (G_{max}), and input impedance after the insertion of the series resistor R_S, with the shunt resistor attached.

Figure 5 shows the output matching circuit that is tuned to the optimum load impedance. Double λ/4 shorted stubs were used for a DC drain current supply and had an LC-parallel resonator effect at the center frequency, thus reducing the variation of the transistor's output impedance trace and facilitating wideband impedance-matching within a low-Q region on the Smith chart [18,19]. Figure 6 shows the variation of the output impedance trace before and after the insertion of the λ/4 shorted stubs. The output impedance trace was extracted as we saw the drain of the transistor at the bias line position. A single-layer capacitor followed the λ/4 shorted stubs, which resulted in the size reduction of the output matching circuit, although multiple thin-film substrates complicated the assembly process a little. A cascaded low-pass LC network was used for fine tuning of the desired output impedance, and was implemented on an alumina substrate.

Figure 7 shows a schematic circuit diagram of the designed power amplifier, exemplary input/output impedance matching traces at 4 GHz, and the designed load impedance trace seen from the drain of the transistor. The input impedance was not well matched because of the very low input impedance of the transistor. On the other hand, the output impedance showed a reasonably good matching. To improve the input matching condition, we should use lossy matching or multiple LC matching techniques, but the former degrades the gain performance, and the latter increases the circuit size. An insufficient gain margin makes it difficult to choose lossy matching, and the large circuit size makes compact integration into a small standard metal package infeasible. Therefore, the input matching properly compromised with the size of the input matching circuit because the input mismatch could be improved with the use of the balanced amplifier configuration [20–22].

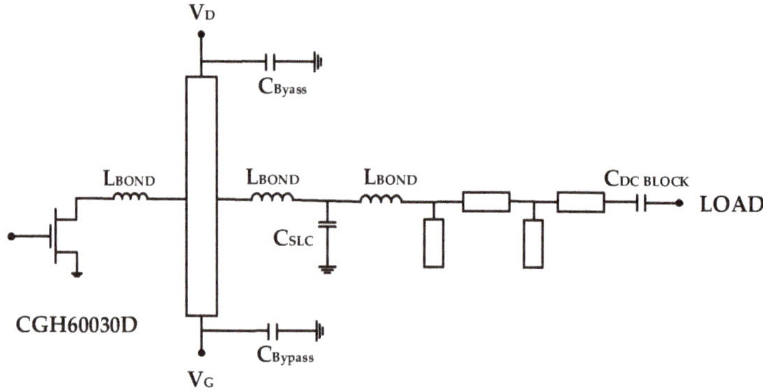

Figure 5. Output matching circuit for the optimum load impedance.

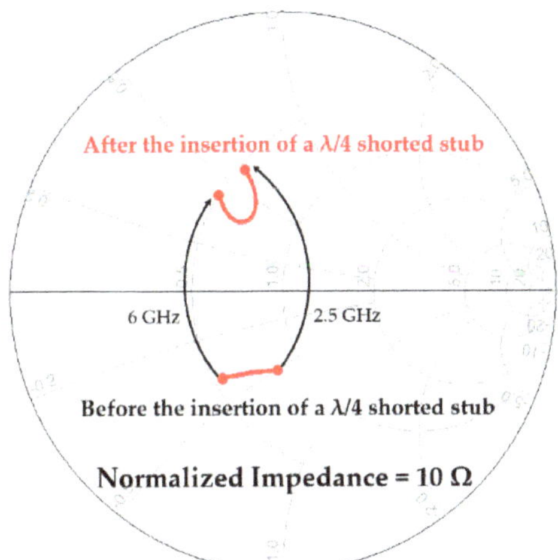

Figure 6. Variation of the transistor's output impedance before and after the insertion of λ/4 shorted stubs.

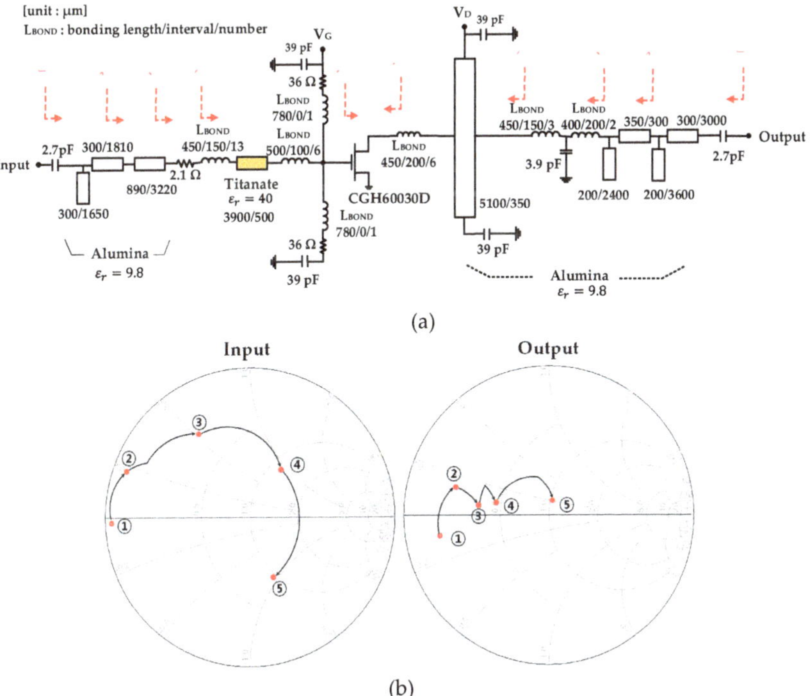

Figure 7. *Cont.*

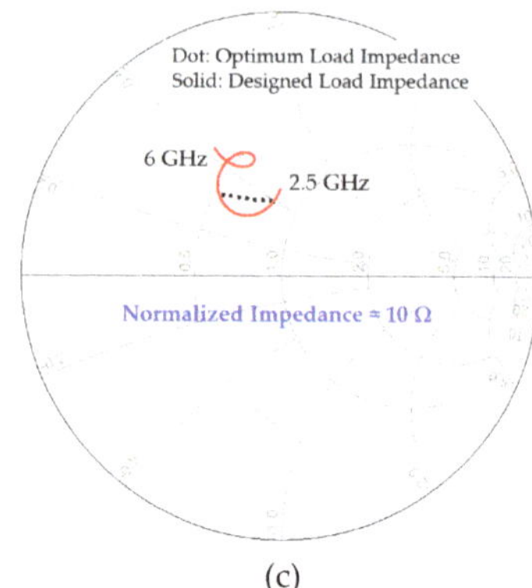

(c)

Figure 7. Schematic circuit diagram of the designed power amplifier (**a**), input/output matching impedance traces at 4 GHz (**b**), and the designed load impedance trace seen from the drain of the transistor (**c**).

3. Fabrication and Measurement

3.1. Power Amplifier Fabrication

Figure 8 shows a fabricated internally matched power amplifier. Input and output matching circuits were implemented on alumina substrates and a titanate substrate using a thin-film process [23]. The GaN HEMT bare die was attached onto a CPC (Cu/Mo70Cu/Cu) carrier with high thermal conductivity using an AuSn (80/20) eutectic process. The complete circuit occupied 9.9 mm × 6.8 mm on the carrier. The GaN HEMT bare die, single-layer capacitors, input/output matching circuits, and microstrip feed lines on RO4003C (Rogers Corporation, Chandler, AZ, USA) for microwave testing were interconnected by 1 mil Au wedge bonding wires.

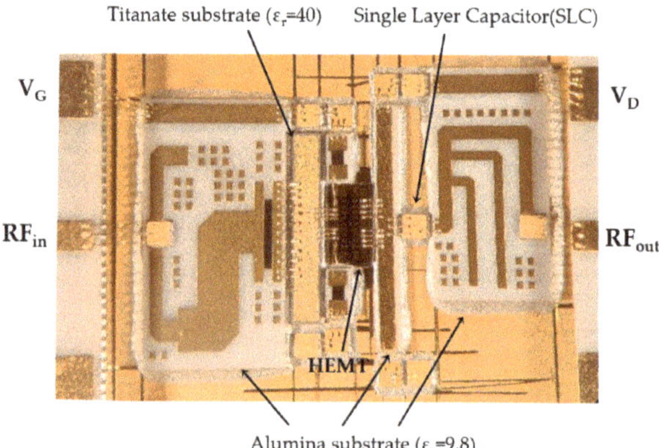

Figure 8. Fabricated internally matched power amplifier using a GaN HEMT bare die, input/output thin-film matching substrates, and single-layer capacitors (circuit area = 9.9 mm × 6.8 mm).

3.2. Power Amplifier Measurement

The fabricated power amplifier was measured on a heat-sinking jig under the bias conditions of V_{DS} = 28 V and I_{DS} = 250 mA. The measured S-parameter data were compared with the simulated data in Figure 9. The measured results showed a linear gain of more than 10.2 dB and a return loss of more than 2.3 dB from 2.5 to 6 GHz. The low-frequency gain decreased slightly, and the roll-off of the high-frequency gain moved to about 6.5 GHz due to the minute change of the implemented load impedance after the device assembly.

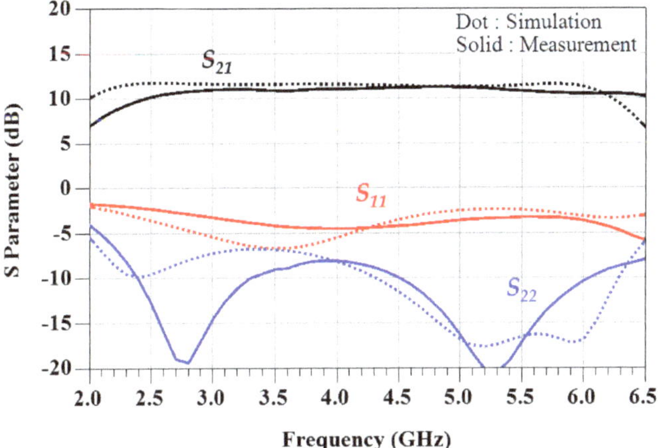

Figure 9. Simulated and measured S-parameter results (simulation: dotted lines, measurement: solid lines).

Figure 10 shows the designed load impedance trace and implemented load impedance trace estimated after the device assembly on Smith charts, together with the power gain contours and the output power contours at 2.5 GHz and 6 GHz. As shown in Figure 10, the small shift of the implemented load impedance from the designed load impedance slightly increased the high-frequency gain.

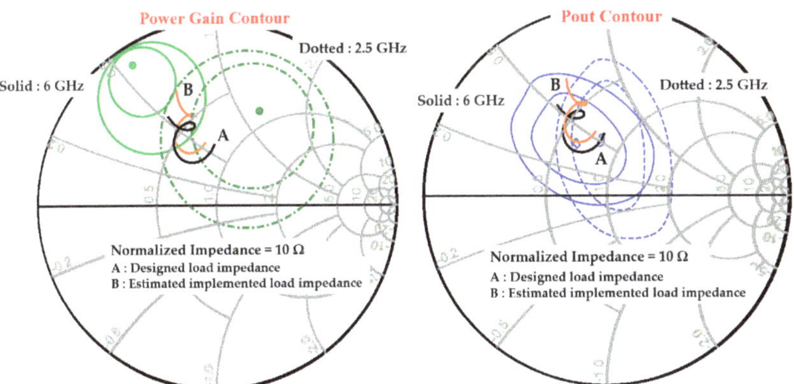

Figure 10. Designed and implemented load impedance traces with the power gain contours and the output power contours at 2.5 GHz and 6 GHz.

Figure 11 shows the output power, power gain, and power-added efficiency (PAE) of the fabricated power amplifier across the input power at 5 GHz where the maximum PAE was measured. The power gain was decreased from 11.1 to 7.6 dB with the varying input power, which corresponded to 3.5 dB power compression. The saturated output power was 43.9 dBm, and the power-added efficiency was 49.7% at the power saturation condition.

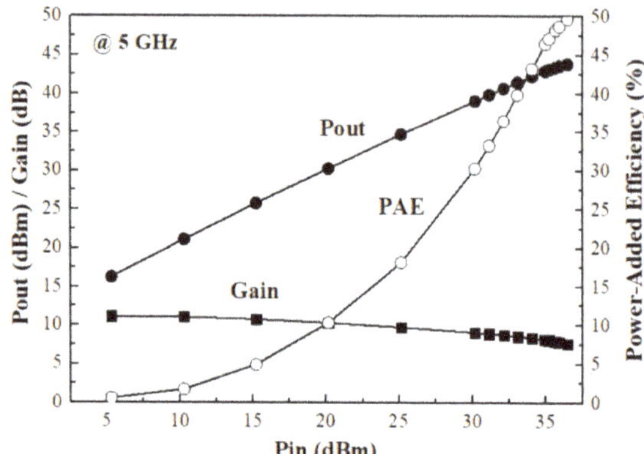

Figure 11. Output power, power gain, and power-added efficiency (PAE) across the input power at 5 GHz.

Figure 12 shows the measured continuous-wave output power performance of the power amplifier under the bias conditions of V_{ds} = 28 V and I_{ds} = 250 mA. The measurement was done from 2.5 to 6 GHz with a 0.5 GHz step frequency. The measured results showed that the amplifier had an output power of 43.3–43.9 dBm, a power-added efficiency (PAE) of 33.4–49.7%, and a power gain of 6.2–8.3 dB. The maximum deviation of the output power was 0.7 dB at 4.5 GHz, and the power-added efficiency (PAE) was slightly degraded due to some mismatch of the designed load impedance and the implemented load impedance after the device assembly. Our simulations estimated an insertion loss of 0.3–0.52 dB for the output matching circuit itself across 2–6 GHz [24]. The power gain decreased at both ends of the designed bandwidth with the maximum 1.5 dB, which was caused by slightly earlier compression with the increase in the input power, as predicted from the trace comparison of the designed load impedance and the implemented load impedance on the output power contours in Figure 10.

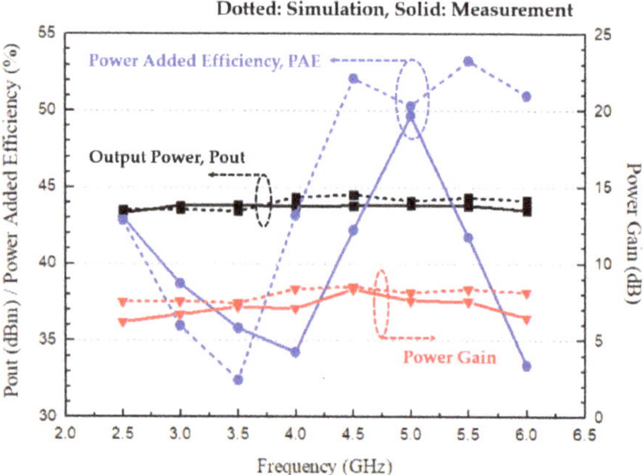

Figure 12. Measured continuous-wave output power performance of the fabricated power amplifier at V_{DS} = 28 V and I_{DS} = 250 mA in the frequency range of 2.5–6 GHz under the power saturation condition.

Table 2 compares our measured power performance with other published results, showing that our work is very competitive in terms of the bandwidth, output power, and power-added efficiency, despite the compact size.

Table 2. Comparison of our work and previously published wideband GaN HEMT power amplifiers.

References	Frequency [GHz]	Power Gain [dB]	P_{out} [dBm]	PAE [%]	Technology	Drain Voltage [V]	Size [mm^2]
Ref. [25]	1.1–2.7	9.5–11.5	43–45	59–72	PCB	28	120 × 50
Ref. [26]	2–6	8–9	43.5–44.5	40–47	PCB	30	est. 68 × 28
Ref. [27]	1.7–3	9.8–10.7	43.8–44.4	57.2–71.1 [1]	PCB	28	est. 75 × 35
Ref. [28]	0.3–2.3	≥10	40–43.5	58–69	PCB	28	59 × 50
Ref. [29]	0.6–3.8	9–14	40–42	46–75	PCB	28	69 × 40
Ref. [30]	2–4	≥9.8	44	37–52 [1]	Quasi-MMIC	50	≤420
This work	2.5–6	6.2–8.3	43.3–43.9	33.4–49.7	Thin film	28	9.9 × 6.8

[1]: drain efficiency; est.: estimated.

4. Conclusions

We presented the compact 20-W GaN internally matched power amplifier operating from 2.5 to 6 GHz that utilizes a GaN HEMT bare die, single-layer capacitors, and thin-film substrates for input and output matching circuits. The fabricated power amplifier achieved stable operation and flat gain performance by using shunt and series resistors. Under the continuous-wave output power condition, it showed the output power of 43.3–43.9 dBm (21.4–24.5 W), the PAE of 33.4–49.7% and the power gain of 6.2–8.3 dB in the frequency range of 2.5–6 GHz. The developed power amplifier will be effectively used in wireless and military applications that require high output power over a wide frequency range.

Author Contributions: Conceptualization, D.-W.K.; design, M.-P.L. and S.K.; simulation, M.-P.L. and S.K.; measurement, M.-P.L., S.K., and S.-J.H.; writing, D.-W.K.; supervision and project administration, D.-W.K. All authors have read and agreed to the published version of the manuscript.

Funding: This work was partly supported by the Technology Development Program of MSS (S2581160) and the NRF grant of MSIT (2018R1D1A1B07049609).

Conflicts of Interest: The authors declare no conflict of interest.

References

1. Runton, D.W.; Trabert, B.; Shealy, J.B.; Vetury, R. History of GaN: High-power RF gallium nitride (GaN) from infancy to manufacturable process and beyond. *IEEE Microw. Mag.* **2013**, *14*, 82–93. [CrossRef]
2. Gonzalez-Garrido, M.A.; Grajal, J.; Cubilla, P.; Cetronio, A.; Lanzieri, C.; Uren, M. 2–6 GHz GaN MMIC power amplifiers for electronic warfare applications. In Proceedings of the 2008 European Microwave Integrated Circuit Conference, IEEE, Amsterdam, The Netherlands, 27–28 October 2008; pp. 83–86.
3. Saad, P.; Fager, C.; Cao, H.; Zirath, H.; Andersson, K. Design of a highly efficienct 2–4 GHz octave bandwidth GaN-HEMT power amplifier. *IEEE Trans. Microw. Theory Tech.* **2010**, *58*, 1677–1685. [CrossRef]
4. Dennler, P.; Quay, R.; Ambacher, O. Novel semi-reactively-matched multistage broadband power amplifier architecture for monolithic ICs in GaN technology. In Proceedings of the 2013 IEEE MTT-S International Microwave Symposium Digest (MTT), IEEE, Seattle, WA, USA, 2–7 June 2013; pp. 1–4.
5. Giofre, R.; Colantonio, P.; Giannini, F. 1–6 GHz ultrawideband 4 W single-ended GaN power amplifier. *Microw. Opt. Tech. Lett.* **2014**, *56*, 215–217. [CrossRef]
6. Kim, D.W. An output matching technique for a GaN distributed power amplifier MMIC using tapered drain shunt capacitors. *IEEE Microw. Wirel. Compon. Lett.* **2015**, *25*, 603–605.
7. Bae, K.T.; Kim, D.W. 2-6 GHz GaN distributed power amplifier MMIC with tapered gate-series/drain-shunt capacitors. *Int. J. RF Microw. Comput. Aided Eng.* **2016**, *26*, 456–465. [CrossRef]
8. Wilcox, G.; Andrews, M. TriQuint Delivers High Power–Wideband GaN Technology. Available online: https://www.rfmw.com/data/triquint%20gan.pdf (accessed on 5 July 2019).
9. Integrated Electronic Warfare System Own Enemy's Battlespace. Available online: https://www.qualcomm.com/media/documents/files/spectrum-for-4g-and-5g.pdf (accessed on 20 January 2020).
10. Huawei Technologies. 5G Spectrum-Huawei. 2017. Available online: https://www-file.huawei.com/-/media/CORPORATE/PDF/public-policy/public_policy_position_5g_spectrum.pdf (accessed on 8 January 2020).
11. Pozar, D.M. *Microwave Engineering*; Wiley: Hoboken, NJ, USA, 2006.

12. Virdee, B.S.; Virdee, A.S.; Banyamin, B.Y. *Broadband Microwave Amplifiers*; Artech House: Norwood, MA, USA, 2004.
13. Wolfspeed, GaN HEMT CGH60030D. Available online: https://www.wolfspeed.com/cgh60030d (accessed on 1 June 2018).
14. Keysight Technologies. PathWave EM Design (EMPro). Available online: https://www.keysight.com/kr/ko/assets/7018-02343/brochures/5990-4819.pdf (accessed on 18 May 2018).
15. Kim, S.I.; Lee, M.P.; Hong, S.J.; Kim, D.W. Ku-band 50 W GaN HEMT power amplifier using asymmetric power combining of transistor cells. *Micromachines* **2018**, *9*, 619. [CrossRef] [PubMed]
16. Yeom, K.W. *Microwave Circuit Design: A Practical Approach Using ADS*; Prentice Hall: Upper Saddle River, NJ, USA, 2015.
17. Ghannouchi, F.M.; Hashmi, M.S. *Load-Pull Techniques with Applications to Power Amplifier Design*; Springer: Berlin/Heidelberg, Germany, 2013.
18. Itoh, Y.; Nii, M.; Kohno, Y.; Mochizuki, M.; Takagi, T. A 4 to 25 GHz 0.5 W monolithic lossy match amplifier. In Proceedings of the IEEE MTT-S International Microwave Symposium Digest, San Diego, CA, USA, 23–27 May 1994; pp. 257–260.
19. Lee, S.K.; Bae, K.T.; Kim, D.W. 2~6 GHz GaN HEMT power amplifier MMIC with bridged-T all-pass filters and output reactance compensation shorted stubs. *J. Semicond. Technol. Sci.* **2016**, *16*, 312–318. [CrossRef]
20. Bahl, I.J. *Fundamentals of RF and Microwave Transistor Amplifiers*; Wiley: Hoboken, NJ, USA, 2009.
21. Berrached, C.; Bouw, D.; Camiade, M.; Barataud, D. Wideband high efficiency high power GaN amplifiers using MIC and quasi-MMIC technologies. In Proceedings of the 2013 European Microwave Conference, IEEE, Nuremberg, Germany, 6–10 October 2013; pp. 1395–1398.
22. Le, Q.H.; Zimmer, G. Wideband high efficiency 50 W GaN-HEMT balanced power amplifier. In Proceedings of the 2018 48th European Microwave Conference (EuMC), IEEE, Madrid, Spain, 23–27 September 2018; pp. 348–351.
23. Compex Corporation. Available online: http://www.compexcorp.com/index (accessed on 10 June 2017).
24. Besser, L.; Gilmore, R. *Practical RF Circuit Design for Modern Wireless Systems—Volume 1 Passive Circuits and Systems*; Artech House: Norwood, MA, USA, 2003.
25. Arnous, M.T.; Barbin, S.E.; Boeck, G. Design of multi-octave highly efficient 20 Watt harmonically tuned power amplifier. In Proceedings of the 2016 21st International Conference on Microwave, Radar and Wireless Communications (MIKON), Krakow, Poland, 9–11 May 2016; pp. 1–4.
26. Nia, H.T.; Hosseini, S.H.J.; Nayyeri, V. Design and fabrication of 2-6 GHz 25 W and 35 W power amplifiers. In Proceedings of the 2018 9th International Symposium on Telecommunications (IST2018), Tehran, Iran, 17–19 December 2018; pp. 647–651.
27. Sayed, A.S.; Ahmed, H.N. Wideband high efficiency power amplifier design using precise high frequency GaN HEMT parasitics modeling/compensation. In Proceedings of the 2019 IEEE Topical Conference on RF/Microwave Power Amplifiers for Radio and Wireless Applications (PAWR), Orlando, FL, USA, 20–23 January 2019; pp. 1–4.
28. Zhuang, Y.; Zhou, J.; Huang, Y.; Rabbi, K.; Rayit, R. Design of ultra broadband highly efficient GaN power amplifier using voronoi diagram. In Proceedings of the ARMMS RF and Microwave Society Conference, Thame, Oxfordshire, UK, 23–24 April 2018; Volume 1, pp. 1–6.
29. Rubio, J.J.M.; Camarchia, V.; Quaglia, R.; Malaver, E.F.A.; Pirola, M. A 0.6-3.8 GHz GaN power amplifier designed through a simple strategy. *IEEE Microw. Wirel. Compon. Lett.* **2016**, *26*, 446–448. [CrossRef]
30. Berrached, C.; Bouw, D.; Camiade, M.; El-Akhdar, K.; Barataud, D.; Neveux, G. Wideband, high-efficiency, high-power GaN amplifiers, using MIC and quasi-MMIC technologies, in the 1-4 GHz range. *Int. J. Microw. Wirel. Technol.* **2014**, *7*, 1–12. [CrossRef]

Article

Self-Aligned Hierarchical ZnO Nanorod/NiO Nanosheet Arrays for High Photon Extraction Efficiency of GaN-Based Photonic Emitter

Won-Seok Lee [1], Soon-Hwan Kwon [1], Hee-Jung Choi [1], Kwang-Gyun Im [2], Hannah Lee [1], Semi Oh [3,*] and Kyoung-Kook Kim [1,2,*]

1. Department of Advanced Convergence Technology, Research Institute of Advanced Convergence Technology, Korea Polytechnic University, Gyeonggi-do 15073, Korea; dldnjstjr37@kpu.ac.kr (W.-S.L.); canwkd21@kpu.ac.kr (S.-H.K.); gmlwjd0889@kpu.ac.kr (H.-J.C.); leehn331@kpu.ac.kr (H.L.)
2. Department of Nano & Semiconductor Engineering, Korea Polytechnic University, Gyeonggi-do 15073, Korea; rhkdrbs87@kpu.ac.kr
3. Department of Electrical Engineering and Computer Science, University of Michigan, Ann Arbor, MI 48109, USA

* Correspondence: semio@umich.edu (S.O.); kim.kk@kpu.ac.kr (K.-K.K.)

Received: 29 February 2020; Accepted: 25 March 2020; Published: 26 March 2020

Abstract: Advancements in nanotechnology have facilitated the increased use of ZnO nanostructures. In particular, hierarchical and core–shell nanostructures, providing a graded refractive index change, have recently been applied to enhance the photon extraction efficiency of photonic emitters. In this study, we demonstrate self-aligned hierarchical ZnO nanorod (ZNR)/NiO nanosheet arrays on a conventional photonic emitter (C-emitter) with a wavelength of 430 nm. These hierarchical nanostructures were synthesized through a two-step hydrothermal process at low temperature, and their optical output power was approximately 17% higher than that of ZNR arrays on a C-emitter and two times higher than that of a C-emitter. These results are due to the graded index change in refractive index from the GaN layer inside the device toward the outside as well as decreases in the total internal reflection and Fresnel reflection of the photonic emitter.

Keywords: self-align; hierarchical nanostructures; ZnO nanorod/NiO nanosheet; photon extraction efficiency; photonic emitter

1. Introduction

The advancements in nanotechnology have facilitated the increased use of ZnO nanostructures that, for example, are widely utilized in photonic devices because of their peculiar chemical and physical properties [1–5]. Different dimensions from zero to three-dimensional ZnO nanostructures have been synthesized using various precursors. These nanostructures are particularly important for realizing many applications, such as electronic devices, catalysis, and biomedical and sensing usage, especially, visible ultraviolet optical devices [6–11].

Especially, one-dimensional (1D) ZnO nanostructures can be widely used in photonic emitters and photodetectors because of their easy refractive index control, transparency in the visible light range, high photoreactivity, and light waveguide properties [12–14]. According to effective medium approximation (EMA), the effective refractive index (n_{eff}) of ZnO (ZnO film: n = 2.1 at visible wavelengths) decreases when it is converted into nanostructures [15–18].

For achieving a higher photon extraction efficiency (PEE), a material with a refractive index lower than that of ITO (2.1 at visible wavelengths) is required to reduce the total internal reflection (TIR) in a conventional photonic emitter (C-emitter) and increase its outward light emission, that is, in air

(n = 1). Although the ZnO nanostructures can partially mitigate the abrupt change of refractive indices between p-type GaN and air, TIR and Fresnel reflection losses occur at the ZnO/air interface [19,20].

Therefore, alternative materials and structures are required for effective photon extraction from a GaN-based photonic emitter to the outside by matching the refractive indices and for realizing exceptional photon emission from surface nanostructures.

Recently, hierarchical and core–shell nanostructures that provide graded refractive index changes have been applied to achieve high PEE in photonic emitters [11,21–25]. However, for the realization of hierarchical nanostructures, a separate seed layer deposition, high cost vacuum systems, and complicate fabrication processes are required [24,26,27]. To solve these problems, rapid manufacturing techniques are required.

In this study, we demonstrate self-aligned hierarchical ZnO nanorod (ZNR)/NiO nanosheet (NNS) arrays to realize the high PEE of a GaN-based C-emitter. These hierarchical nanostructures are synthesized through a two-step hydrothermal process at low temperatures. The optical output power of the as-obtained C-emitter is approximately 17% and it is two times higher than that of the C-emitter with ZNRs and C-emitter without nanostructures. This increase can be ascribed to a graded change in the refractive index between the GaN layer and the device exterior, as well as a decrease in the TIR and Fresnel reflection of the photonic emitter.

2. Materials and Methods

2.1. Device Fabrication

Epilayers were grown on a sapphire substrate by metal–organic chemical vapor deposition. The photonic emitter (chip size: 350 × 350 μm^2) with a 430 nm wavelength consisted of a 0.12 µm-thick p-type GaN:Mg ($n = 3 \times 10^{17}$ cm^{-3}) layer, a 0.08 µm active layer, a 2.5 µm-thick undoped GaN layer, and a 4.0 µm-thick n-type GaN:Si ($n = 5 \times 10^{18}$ cm^{-3}) layer on the sapphire substrate. All emitter samples were ultrasonically degreased with acetone, methanol, deionized (DI) water, and a mixture of sulfuric acid and hydrogen peroxide (3:1) for 5 min in each step to remove organic and inorganic contaminants. Then, to fabricate the n-electrode, the epilayers were partially etched until the n-type GaN layer was exposed. The 200 nm thick ITO layer was deposited using an electron-beam evaporator on the remaining parts of the p-type GaN layer and annealed at 600 °C in O_2 atmosphere for 1 min using the rapid thermal annealing. The Ti/Al (50/200 nm) layers were deposited as an n-electrode. Finally, the Cr/Al (30/200 nm) layers were deposited on the p- and n-electrodes and annealed at 300 °C for 1 min.

2.2. ZNRs Synthesis

A ZnO seed layer was formed on the selectively deposited ITO by a simple dipping process as follows. First, 105 mM zinc acetate ($Zn(C_2H_3O_2)_2$) dissolved in DI water was synthesized at 90 °C for 1 h. Then, ZNRs were grown using 37.5 mM zinc nitrate hexahydrate ($Zn(NO_3)_2$*$6H_2O$) and 75 mM hexamethylenetetramine ($C_6H_{12}N_4$) dissolved in 300 mL of DI water at 90 °C for 6 h.

2.3. Hierarchical ZNR/NNS Arrays Synthesis

Nickel nitrate hexahydrate ($Ni(NO_3)_2$*$6H_2O$) (10.4 mg) was dissolved in DI water (50 mL) and stirred for 30 min. Then, this solution was used to synthesis NNSs on the ZNRs at 90 °C for 1 h.

2.4. Characterization

The structural shapes of the self-aligned ZNRs and hierarchical ZNR/NNS arrays were observed using a field emission-scanning electron microscope (FESEM, Hitachi S4300, Tokyo, Japan), and the hierarchical nanostructures were analyzed using an energy-dispersive spectroscopy (EDS) mounted on the FESEM. The electrical and optical properties were measured using a parameter analyzer (Keithley 2400, Tektronix, Beaverton, OR, USA), an optical power meter (Newport 1830C, Irvine, CA, USA), and an optical microscope (Velcam CVC5220, Chun Shin Electronics Inc., Taipei, Taiwan). Finite-difference time-domain (FDTD) simulation was conducted to compare the light output of the fabricated photonic emitter with that of a C-emitter.

3. Results and Discussion

The schematic diagram of the ZNR/NNS arrays synthesis on the C-emitter by following the proposed experimental procedures is shown in Figure 1.

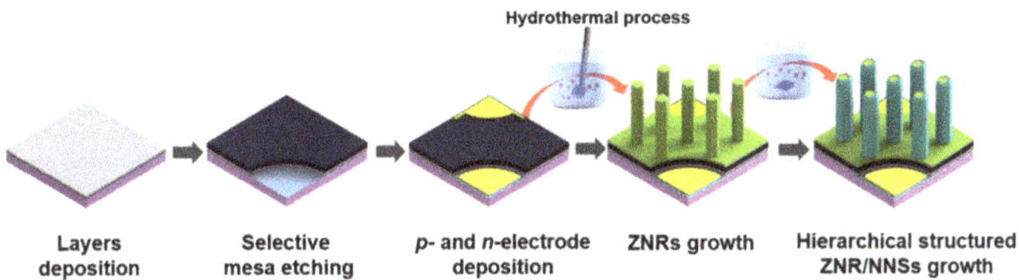

Figure 1. Fabrication process of C-emitter with hierarchical ZnO nanorod (ZNR)/NiO nanosheet (NNS) arrays.

The FESEM image in Figure 2a shows the ZNRs with an average diameter and length of 300 nm and 3.5 μm, respectively. To understand the growth mechanism of the ZNR/NNS arrays, the study of the morphology evolution of hierarchical ZNR/NNS arrays during the reaction time has been carried out, as shown in Figure 2b–e.

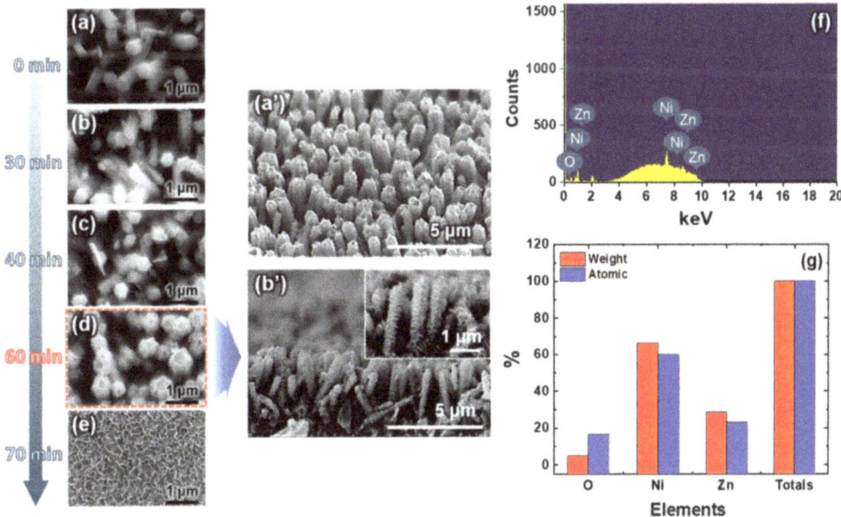

Figure 2. Field emission-scanning electron microscope (FESEM) images of (**a**) ZNRs and (**b**–**e**) hierarchical ZNR/NNS arrays grown for different times; (**a'**) tilt and (**b'**) cross-sectional FESEM images of (**d**). (**f**) an energy-dispersive spectroscopy (EDS) and (**g**) atomic composition of the hierarchical ZNR/NNS arrays.

Since the system tends to minimize the overall surface energy, the ZNRs grew preferentially along the [0001] direction [28,29]. Then, Ni-based nanoparticles (NPs) nucleated on the surface of the ZNRs to form active sites, which minimized the interfacial energy barrier to promote the subsequent growth of Ni-based NPs. The merge of these NPs reduced the overall energy by decreasing the surface energy, which was beneficial for adjacent Ni-based NPs to spontaneously self-organize together. The self-organized NPs shared a common crystallographic orientation and formed a planar interface, as shown in Figure 2b. When the reaction proceeded for 30 min, these NPs self-assembled to form large nanosheets (NSs) and finally generated the ZNR/NNS arrays. The above hypothesis is supported by examining the morphologies of NiO NSs at different growth stages by controlling the reaction time, as shown in Figure 2c,d. After exceeding a growth time of 60 min, the NSs transformed into nanowalls (Figure 2e). Therefore, we selected 60 min as the optimum growth time for obtaining stable ZNR/NNS arrays. Figure 2a',b' show the tilted and cross-sectional FESEM images of the well-aligned NNS arrays grown for 60 min. The results of an EDS analysis along with the atomic and weight values of the hierarchical ZNR/NNS arrays are shown in Figure 2f. The atomic contents of oxygen, nickel, and zinc were 16.6%, 60.09%, and 23.32%, respectively. The lower percentage of oxygen was ascribed to the fact that oxygen is relatively lighter than nickel and zinc [30]. These results confirmed the successful synthesis of the ZnO and NiO nanostructures.

To evaluate the properties of the fabricated photonic emitters, we measured their current–voltage curves and optical output intensity. Under the injection current of 20 mA, the threshold voltages of the C-emitters without nanostructures, with ZNRs, and with hierarchical ZNR/NNS arrays ranged between 3.00 and 3.03 V, as shown in Figure 3a. This indicates that the ZnO nanostructures did not affect the electrical properties of the emitters because they were grown using a low-temperature growth process. At the injection current of 100 mA, the optical output power of the C-emitter with the hierarchical ZNR/NNS arrays was approximately 17% higher than that of the device with ZNRs and two times higher than that of the photonic emitter without any nanostructures (Figure 3b). The inset emission images in Figure 3c–e show that the hierarchical ZNR/NNS arrays on the C-emitter are brighter than the other emitters at the injection current of 0.05 mA. Compared to other studies on this topic, we realized the optimal refractive index between the ITO layer and air [31,32]. In addition, the higher optical output power can probably be attributed to the graded refractive index and the reduced Fresnel reflection.

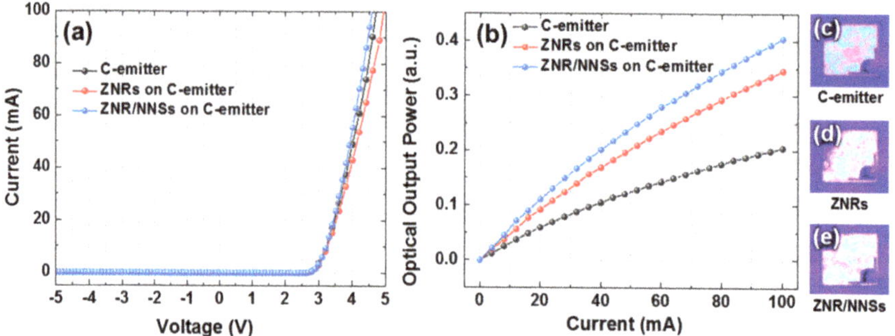

Figure 3. (**a**) Current–voltage curves, (**b**) current–optical output power curves, and (**c**–**e**) emission images (at an injection current of 0.05 mA) of the C-emitter without nanostructures, with ZNRs, and with hierarchical ZNR/NNS arrays.

Then, we performed FDTD simulations to compare the three C-emitter types. Figure 4a,b show schematic diagrams of the overall structures of the devices with the ZNR and the hierarchical ZNR/NNS.

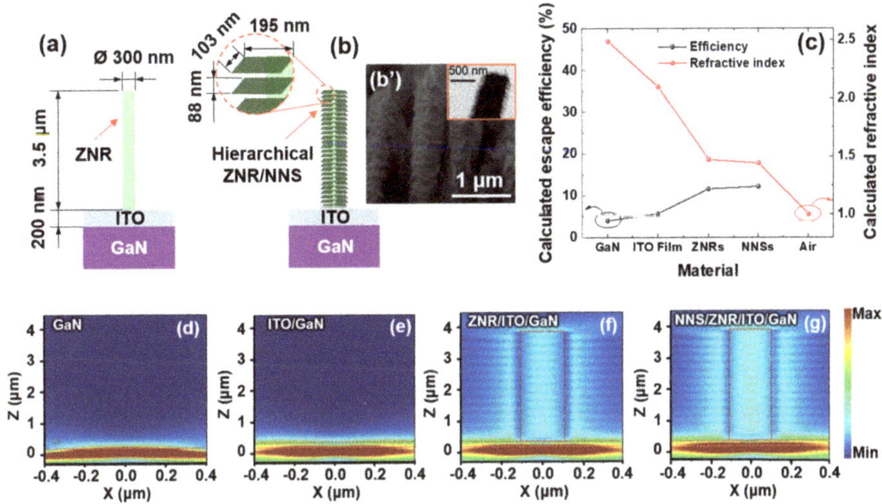

Figure 4. Layouts of the C-emitters with (**a**) ZNR and (**b**) hierarchical ZNR/NNS. (**b'**) FESEM and transmission electron microscope (inset) images of the hierarchical ZNR/NNS arrays. (**c**) Calculated refractive indices and escape efficiencies. (**d–g**) Electric field propagation for various C-emitters.

In these simulations, we used the following parameters: ZNR diameter and height of 300 nm and 3.5 μm, respectively, based on the corresponding FESEM image (Figure 2a); NNS width and height of 195 nm and 103 nm, respectively; and consistent interval of 88 nm between adjacent NNSs based on the FESEM and transmission electron microscope results (Figure 4b'). These structures were simulated on the 200 nm thick ITO layer deposited on the 430 nm photonic emitters. The refractive indices of various materials were calculated using EMA [15–17]:

$$n_{\text{eff}} = \left[n_{\text{ZnO}}^2 V_{\text{ZnO}} + n_{\text{Air}}^2 (1 - V_{\text{ZnO}})\right]^{\frac{1}{2}} \tag{1}$$

where n_{eff} is the effective refractive index of the ZNRs; n_{ZnO} and n_{Air} are the refractive indices of ZnO and air, respectively; and V_{ZnO} is the volume fraction of ZnO in the effective medium. The refractive indices and the volume fraction were determined from the FESEM image, as shown in Figure 2a. The average refractive index of the ZNRs was significantly lower than that of the ZnO film because of the inclusion of air in the effective medium. Therefore, according to EMA, ZNRs have a lower refractive index, even though the refractive index of the ZnO film was similar to that of the ITO film.

Furthermore, the refractive index of NNSs was lower than that of the ZNRs because the refractive index of NiO is lower (1.68) than that of the ZnO film, even though NiO has the lower air volume, as can be inferred from the FESEM image shown in Figure 2d. Therefore, we can confirm that the calculated refractive indices of the GaN, ITO, ZNRs, NNSs are 2.49, 2.1, 1.47, and 1.44, respectively.

The optimal refractive index of the antireflection layer between the ITO film and air can be computed using the formula: [33,34]

$$n_1 = \sqrt{n_0 n_s}. \tag{2}$$

The formula yielded a value of 1.449, which is very similar to the refractive index of the NNSs for a high antireflection effect.

The escape efficiencies of GaN, ITO, ZNRs, and NNSs (Figure 4c), based on the corresponding calculated refractive indices, can be given as follows [35]:

$$\frac{P_{escape}}{P_{source}} \approx \frac{1}{4}\frac{n_{air}^2}{n_{GaN}^2} \tag{3}$$

where n_{GaN} is the refractive index of the GaN layer. P_{escape} and P_{source} are the escape and source powers of the photonic emitter, respectively. According to the equation, the photonic emitter without the ITO electrode has an escape efficiency of only 4%. This efficiency can be increased to 12.2% by controlling the refractive index through NNSs growth. However, because this value represents the escape efficiency of photon extraction obtained by considering only the refractive index, it will decrease when considering the efficiency lost through GaN, ITO, and ZNRs. Moreover, the photon efficiency of the device based on various nanostructures is not discussed here, because this calculation considers only layered thin films.

Figure 4d–g illustrate the simulated electric field propagation for the various photonic emitters. As can be inferred from the simulation results shown in Figure 4d,e, the two emitters have similar propagation images. However, the C-emitter shows a higher field extraction toward the outside compared to the emitter without the ITO electrode because of Fresnel reflection. In the case of Figure 4f, although the ZNR on the C-emitter shows a remarkably higher propagation of electric field toward the outside because of the refractive index control and wave guide effect of ZNR, it is limited by the refractive index difference between the ZNR and air and the flat end of the ZNR. This can be explained based on the strong intensity of the electric field inside the ZNR because of interference of the reflected electric field. However, the simulation results of ZNR/NNS on the C-emitter, which are presented in Figure 4g, shows the stronger electric propagation image than that of ZNR on the C-emitter. This difference can be explained by the fact that the NNSs have a smoother graded effective refractive index change and lower TIR loss from the GaN-based photonic emitter to the outside.

4. Conclusions

In conclusion, we successfully grew self-aligned hierarchical ZNR/NNS arrays on a C-emitter through a low-temperature two-step hydrothermal process. Compared to the device with only ZNRs, the device with the hierarchical ZNR/NNS arrays exhibited optimal refractive indices (GaN: 2.49, ITO: 2.1, ZNR: 1.47, NNS: 1.44) and antireflection properties as a result of the graded refractive index and waveguide effect. Therefore, at the injection current of 100 mA, its output power was approximately 17% higher than that of the C-emitter with ZNRs and twice as high as that of the device without nanostructures; in addition, there was no degradation of electrical properties. The proposed nanostructures can be used to realize various nanotechnology applications, such as photonic emitters, gas sensors, supercapacitors, electrochromic devices, and solar cells.

Author Contributions: W.-S.L., S.-H.K., and K.-K.K. conceived and designed the experiments; H.-J.C. and S.O. conducted the simulation; W.-S.L., S.-H.K., K.-K.K., and H.L. performed the experiments; W.L., S.-H.K., and H.-J.C. analyzed the data; W.-S.L., S.O., K.-G.I. and K.-K.K. wrote the paper. All authors have read and agreed to the published version of the manuscript.

Funding: This research was partly supported by the MSIT(Ministry of Science and ICT), Korea, under the ITRC(Information Technology Research Center) support program(IITP-2020-2019-2018-0-01426) supervised by the IITP(Institute for Information & Communications Technology Planning & Evaluation, Korea Institute for Advancement of Technology (KIAT) grant funded by the Korea Government(MOTIE) (P0008458, The Competency Development Program for Industry Specialist), and the work reported in this paper was conducted during the sabbatical year of Korea Polytechnic University in 2019.

Conflicts of Interest: The authors declare no conflict of interest.

References

1. Lupan, O.; Pauporté, T.; Viana, B. Low-Voltage UV-Electroluminescence from ZnO-Nanowire Array/p-GaN Light-Emitting Diodes. *Adv. Mater.* **2010**, *22*, 3298–3302. [CrossRef] [PubMed]
2. Drobek, M.; Kim, J.-H.; Bechelany, M.; Vallicari, C.; Julbe, A.; Kim, S.S. MOF-Based Membrane Encapsulated ZnO Nanowires for Enhanced Gas Sensor Selectivity. *ACS Appl. Mater. Interfaces* **2016**, *8*, 8323–8328. [CrossRef] [PubMed]
3. Li, C.; Han, C.; Zhang, Y.; Zang, Z.; Wang, M.; Tanga, X.-S.; Du, J. Enhanced photoresponse of self-powered perovskite photodetector based on ZnO nanoparticles decorated $CsPbBr_3$ films. *Sol. Energy Mater. Sol. Cells* **2017**, *172*, 341–346. [CrossRef]
4. Huang, M.H.; Mao, S.; Feick, H.; Yan, H.; Wu, Y.; Kind, H.; Weber, E.; Russo, R.; Yang, P. Room-temperature ultraviolet nanowire nanolasers. *Science* **2001**, *292*, 1897–1899. [CrossRef]
5. Lee, H.K.; Kim, M.S.; Yu, J.S. Effect of AZO seed layer on electrochemical growth and optical properties of ZnO nanorod arrays on ITO glass. *Nanotechnology* **2011**, *22*, 445602. [CrossRef]
6. Özgür,, Ü.; Hofstetter, D.; Morkoç, H. ZnO Devices and Applications: A Review of Current Status and Future Prospects. *Proc. IEEE* **2010**, *98*, 1255–1268. [CrossRef]
7. Litton, C.W.; Reynolds, D.C.; Collins, T.C. *Zinc Oxide Materials for Electronic and Optoelectronic Device Applications*, 1st ed.; John Wiley & Sons, Ltd.: Chichester, UK, 2011; pp. 29–86.
8. Sun, Y.H.; Chen, L.; Bao, Y.; Zhang, Y.; Wang, J.; Fu, M.; Wu, J.; Ye, D. The Applications of Morphology Controlled ZnO in Catalysis. *Catalysts* **2016**, *6*, 188. [CrossRef]
9. Mirzaei, H.; Darroudi, M. Zinc oxide nanoparticles: Biological synthesis and biomedical applications. *Ceram. Int.* **2017**, *43*, 907–914. [CrossRef]
10. Ao, D.; Li, Z.; Fu, Y.Q.; Tang, Y.; Yan, S.; Zu, X.T. Heterostructured NiO/ZnO Nanorod Arrays with Significantly Enhanced H_2S Sensing Performance. *Nanomaterials* **2019**, *9*, 900. [CrossRef]
11. Oh, S.; Ha, K.; Kang, S.-H.; Yohn, G.-J.; Lee, H.-J.; Park, S.-J.; Kim, K.-K. Self-standing ZnO nanotube/SiO_2 core–shell arrays for high photon extraction efficiency in III-nitride emitter. *Nanotechnology* **2017**, *29*, 15301. [CrossRef]
12. Kim, K.-K.; Lee, S.-D.; Kim, H.; Park, J.-C.; Lee, S.-N.; Park, Y.; Park, S.-J.; Kim, S. Enhanced light extraction efficiency of GaN-based light-emitting diodes with ZnO nanorod arrays grown using aqueous solution. *Appl. Phys. Lett.* **2009**, *94*, 071118.
13. Park, Y.J.; Song, H.; Ko, K.B.; Ryu, B.D.; Cuong, T.V.; Hong, C.-H. Nanostructural Effect of ZnO on Light Extraction Efficiency of Near-Ultraviolet Light-Emitting Diodes. *J. Nanomater.* **2016**, *2016*, 1–6. [CrossRef]
14. Ong, C.B.; Ng, L.Y.; Mohammad, A. A review of ZnO nanoparticles as solar photocatalysts: Synthesis, mechanisms and applications. *Renew. Sustain. Energy Rev.* **2018**, *81*, 536–551. [CrossRef]
15. ElAnzeery, H.; El Daif, O.; Buffiere, M.; Oueslati, S.; Ben Messaoud, K.; Agten, D.; Brammertz, G.; Guindi, R.; Kniknie, B.; Meuris, M.; et al. Refractive index extraction and thickness optimization of $Cu_2ZnSnSe_4$ thin film solar cells. *Phys. Status Solidi (A)* **2015**, *212*, 1984–1990. [CrossRef]
16. Xi, J.-Q.; Schubert, M.F.; Kim, J.K.; Schubert, E.F.; Chen, M.; Lin, S.-Y.; Liu, W.; Smart, J.A.; Lončar, M. Optical thin-film materials with low refractive index for broadband elimination of Fresnel reflection. *Nat. Photon.* **2007**, *1*, 176–179. [CrossRef]
17. Chao, Y.-C.; Chen, C.-Y.; Lin, C.-A.; Dai, Y.-A.; He, J.-H. Antireflection effect of ZnO nanorod arrays. *J. Mater. Chem.* **2010**, *20*, 8134. [CrossRef]
18. Tsai, D.-S.; Lin, C.-A.; Lien, W.-C.; Chang, H.-C.; Wang, Y.-L.; He, J.-H. Ultra-High-Responsivity Broadband Detection of Si Metal–Semiconductor–Metal Schottky Photodetectors Improved by ZnO Nanorod Arrays. *ACS Nano* **2011**, *5*, 7748–7753. [CrossRef]
19. Lei, P.-H.; Yang, C.-D.; Huang, P.-C.; Yeh, S.-J. Enhancement of Light Extraction Efficiency for InGaN/GaN Light-Emitting Diodes Using Silver Nanoparticle Embedded ZnO Thin Films. *Micromachines* **2019**, *10*, 239. [CrossRef]
20. Leem, Y.-C.; Seo, O.; Jo, Y.-R.; Kim, J.H.; Chun, J.; Kim, B.-J.; Noh, D.Y.; Lim, W.; Kim, Y.-I.; Park, S.-J. Titanium oxide nanotube arrays for high light extraction efficiency of GaN-based vertical light-emitting diodes. *Nanoscale* **2016**, *8*, 10138–10144. [CrossRef]

21. Mao, P.; Mahapatra, A.K.; Chen, J.; Chen, M.; Wang, G.; Han, M. Fabrication of Polystyrene/ZnO Micronano Hierarchical Structure Applied for Light Extraction of Light-Emitting Devices. *ACS Appl. Mater. Interfaces* **2015**, *7*, 19179–19188. [CrossRef]
22. Ho, C.-H.; Hsiao, Y.-H.; Lien, D.-H.; Tsai, M.S.; Chang, D.; Lai, K.-Y.; Sun, C.-C.; He, J.-H. Enhanced light-extraction from hierarchical surfaces consisting of p-GaN microdomes and SiO_2 nanorods for GaN-based light-emitting diodes. *Appl. Phys. Lett.* **2013**, *103*, 161104. [CrossRef]
23. Leem, Y.; Park, J.S.; Kim, J.H.; Myoung, N.; Yim, S.-Y.; Jeong, S.; Lim, W. Light-Emitting Diodes: Light-Emitting Diodes with Hierarchical and Multifunctional Surface Structures for High Light Extraction and an Antifouling Effect (Small 2/2016). *Small* **2016**, *12*, 138. [CrossRef]
24. Park, M.J.; Kim, C.U.; Kang, S.B.; Won, S.H.; Kwak, J.S.; Kim, C.-M.; Choi, K.J. 3D Hierarchical Indium Tin Oxide Nanotrees for Enhancement of Light Extraction in GaN-Based Light-Emitting Diodes. *Adv. Opt. Mater.* **2016**, *5*, 1600684. [CrossRef]
25. Zhang, C.; Marvinney, C.E.; Xu, H.; Liu, W.Z.; Wang, C.L.; Zhang, L.X.; Wang, J.; Ma, J.G.; Liu, Y. Enhanced waveguide-type ultraviolet electroluminescence from ZnO/MgZnO core/shell nanorod array light-emitting diodes via coupling with Ag nanoparticles localized surface plasmons. *Nanoscale* **2015**, *7*, 1073–1080. [CrossRef]
26. Ko, S.H.; Lee, S.; Kang, H.W.; Nam, K.H.; Yeo, J.; Hong, S.J.; Grigoropoulos, C.P.; Sung, H.J. Nanoforest of Hydrothermally Grown Hierarchical ZnO Nanowires for a High Efficiency Dye-Sensitized Solar Cell. *Nano Lett.* **2011**, *11*, 666–671. [CrossRef]
27. Ren, X.; Sangle, A.; Zhang, S.; Yuan, S.; Zhao, Y.; Shi, L.; Hoye, R.L.Z.; Cho, S.; Li, D.; MacManus-Driscoll, J.L. Photoelectrochemical water splitting strongly enhanced in fast-grown ZnO nanotree and nanocluster structures. *J. Mater. Chem. A* **2016**, *4*, 10203–10211. [CrossRef]
28. Cai, Y.; Li, X.; Sun, P.; Wang, B.; Liu, F.; Cheng, P.; Du, S.; Lu, G. Ordered ZnO nanorod array film driven by ultrasonic spray pyrolysis and its optical properties. *Mater. Lett.* **2013**, *112*, 36–38. [CrossRef]
29. Li, D.; Zhang, Y.; Liu, D.; Yao, S.; Liu, F.; Wang, B.; Sun, P.; Gao, Y.; Chuai, X.; Lu, G. Hierarchical core/shell ZnO/NiO nanoheterojunctions synthesized by ultrasonic spray pyrolysis and their gas-sensing performance. *CrystEngComm* **2016**, *18*, 8101–8107. [CrossRef]
30. Wang, D.; Zhao, L.; Ma, H.; Zhang, H.; Guo, L.H. Quantative Analysis of Reactive Oxygen Species Photogenerated on Metal Oxide Nanoparticles and Their Bacteria Toxicity: The Role of Superoxide Radicals. *Environ. Sci. Technol.* **2017**, *51*, 10137–10145. [CrossRef]
31. Raut, H.; Ganesh, V.A.; Nair, A.S.; Ramakrishna, S. Anti-reflective coatings: A critical, in-depth review. *Energy Environ. Sci.* **2011**, *4*, 3779. [CrossRef]
32. Cho, C.-Y.; Kim, N.-Y.; Kang, J.-W.; Leem, Y.-C.; Hong, S.-H.; Lim, W.; Kim, S.-T.; Park, S.-J. Improved Light Extraction Efficiency in Blue Light-Emitting Diodes by SiO_2-Coated ZnO Nanorod Arrays. *Appl. Phys. Express* **2013**, *6*, 042102. [CrossRef]
33. Park, J.; Shin, D.S.; Kim, D.-H. Enhancement of light extraction in GaN-based light-emitting diodes by Al_2O_3-coated ZnO nanorod arrays. *J. Alloy. Compd.* **2014**, *611*, 157–160. [CrossRef]
34. Moreno, I.; Araiza, J.D.J.; Avendaño-Alejo, M. Thin-film spatial filters. *Opt. Lett.* **2005**, *30*, 914. [CrossRef] [PubMed]
35. Sun, W.; Shatalov, M.; Hu, X.; Yang, J.; Lunev, A.; Bilenko, Y.; Shur, M.S.; Gaska, R. *Milliwatt Power 245 nm Deep Ultraviolet Light-Emitting Diodes*; Institute of Electrical and Electronics Engineers (IEEE): Piscataway, NJ, USA, 2009; pp. 109–110.

Article

Effect of the Ammonium Tungsten Precursor Solution with the Modification of Glycerol on Wide Band Gap WO_3 Thin Film and Its Electrochromic Properties

Jinxiang Liu [1], Guanguang Zhang [1], Kaiyue Guo [1], Dong Guo [2], Muyang Shi [1], Honglong Ning [1,*], Tian Qiu [3], Junlong Chen [1], Xiao Fu [1], Rihui Yao [1,4,*] and Junbiao Peng [1]

1 Institute of Polymer Optoelectronic Materials and Devices, State Key Laboratory of Luminescent Materials and Devices, South China University of Technology, Guangzhou 510640, China; 201765340428@mail.scut.edu.cn (J.L.); msgg-zhang@mail.scut.edu.cn (G.Z.); 201836320090@mail.scut.edu.cn (K.G.); 201430320229@mail.scut.edu.cn (M.S.); msjlchen@gmail.com (J.C.); 201630343721@mail.scut.edu.cn (X.F.); psjbpeng@scut.edu.cn (J.P.)

2 School of Medical Instrument & Food Engineering, University of Shanghai for Science and Technology, No.516 Jungong Road, Shanghai 200093, China; guodong99@tsinghua.org.cn

3 Department of Intelligent Manufacturing, Wuyi University, Jiangmen 529000, China; timeqiu@hotmail.com

4 Guangdong Province Key Lab of Display Material and Technoloy, Sun Yat-sen University, Guangzhou 510275, China

* Correspondence: ninghl@scut.edu.cn (H.N.); yaorihui@scut.edu.cn (R.Y.); Tel.: +86-20-8711-4525 (H.N.)

Received: 11 February 2020; Accepted: 14 March 2020; Published: 16 March 2020

Abstract: Tungsten trioxide (WO_3) is a wide band gap semiconductor material, which is commonly not only used, but also investigated as a significant electrochromic layer in electrochromic devices. WO_3 films have been prepared by inorganic and sol-gel free ammonium tungstate ($(NH_4)_2WO_4$), with the modification of glycerol using the spin coating technique. The surface tension, the contact angle and the dynamic viscosity of the precursor solutions demonstrated that the sample solution with a 25% volume fraction of glycerol was optimal, which was equipped to facilitate the growth of WO_3 films. The thermal gravimetric and differential scanning calorimetry (TG-DSC) analysis represented that the optimal sample solution transformed into the WO_3 range from 220 °C to 300 °C, and the transformation of the phase structure of WO_3 was taken above 300 °C. Fourier transform infrared spectroscopy (FT-IR) spectra analysis indicated that the composition within the film was WO_3 above the 300 °C annealing temperature, and the component content of WO_3 was increased with the increase in the annealing temperature. The X-ray diffraction (XRD) pattern revealed that WO_3 films were available for the formation of the cubic and monoclinic crystal structure at 400 °C, and were preferential for growing monoclinic WO_3 when annealed at 500 °C. Atomic force microscope (AFM) images showed that WO_3 films prepared using ammonium tungstate with modification of the glycerol possessed less rough surface roughness in comparison with the sol-gel-prepared films. An ultraviolet spectrophotometer (UV) demonstrated that the sample solution which had been annealed at 400 °C obtained a high electrochromic modulation ability roughly 40% at 700 nm wavelength, as well as the optical band gap (E_g) of the WO_3 films ranged from 3.48 eV to 3.37 eV with the annealing temperature increasing.

Keywords: tungsten trioxide film; spin coating; optical band gap; morphology; electrochromism

1. Introduction

Currently, an increasing amount of attention has been concentrated on energy saving in buildings for the reason that the energy used for these is in excess of 30% of the consumption of the total energy in the word, as a matter of fact [1]. It is worth mentioning that it is the transition metal oxides that are the fascinating semiconducting materials which possess a variety of properties, applications and functions [2]. Tungsten trioxide (WO_3), the most widely studied and used electrochromic material, has been studied for many years since having been found by Deb in the 1960s, thanks to its attractive characteristics of high coloring efficiency, good optical modulation ability and decent chemical stability in the field of electrochromism [3–6]. It is remarkable that WO_3 with various crystalline structures by different synthesis methods is extensively appropriate for diverse device applications [7–9], such as the electrochromic smart window [10,11], rear-view automatic-dimming rearview mirror [12], gas sensor device [13–15], and military camouflage [16], etc. A large variety of techniques can be used to prepare WO_3 thin films, such as chemical vapor deposition (CVD) [17], physical vapor deposition (PVD) [18] and wet chemical deposition methods base on solution [19–21]. In fact, the sol-gel technique that is used to prepare WO_3 thin films is commonly considered feasible among wet chemical deposition methods due to a good few favorable and conspicuous advantages, such as large-area film formation, a repeatable process and inexpensive experimental facilities [22–24]. It is the formation of the gel network by polymerization that the principle of the sol-gel technique for preparing thin films depends on [3]. Nevertheless, it cannot be neglected that there is less satisfaction with thin films prepared by the sol-gel technique because of these following deficiencies, which would be limited in applications in electrochromism and thin film transistors, etc. Not only the stability of the precursor solution for the formation of gel structure is generally not adequate enough, and it would not gratify the industrial production as a significant parameter [25], but also the homogeneity throughout the sol-gel process would not be guaranteed effectively by producing a homogeneous precursor solution at room temperature [26]. Furthermore, there are practical limitations for the reason that thin films with microstructures and nanostructures prepared by the sol-gel technique are more fragile, so that they possibly cannot maintain well their structure in the course of the assembly process of electrochromic devices [27]. It is important to note that wrinkle cracks are observed on the surface of the thin films prepared by the sol-gel technique with the increase in the thickness of films [28,29].

To effectively solve this problem, WO_3 thin films can be prepared by the spin coating technique using pure-inorganic ammonium tungstate precursor solution in this present study, which would not obtain the gel structure. In this work, WO_3 thin films were successfully prepared using the above precursor solution by the spin coating technique. The glycerol, one kind of excellent solvent with high viscosity, has been utilized to magnify the adhesiveness of the pure-inorganic ammonium tungstate precursor solution, so as to sufficiently facilitate the growth of the WO_3 films. The sample solution performance, the film surface morphology, components, crystallization, optical properties and the electrochromic properties, were investigated and discussed through different characterization methods.

2. Materials and Methods

Tungsten trioxide (WO_3, <100 nm, 99.9% metals basis, Macklin Biochemical Co. Ltd, Shanghai, China) and ammonia hydroxide ($NH_3{\cdot}H_2O$, AR, 26%, Guangzhou Chemical Reagent Factory, Guangzhou, China) were mixed in a beaker to prepare an ammonium tungstate ($(NH_4)_2WO_4$) solution. Subsequently, five kinds of sample solution were prepared using glycerol ($C_3H_8O_3$, AR, Richjoint, Shanghai, China) for modification. According to the different volume fraction of glycerol, the volume fractions of glycerol were 0%, 12.5%, 25%, 37.5% and 50%, respectively, and the concentration of ammonium tungstate solution was approximately 5.5 mol/L. The precursor solutions were uniformly obtained after being ultrasonically oscillated, which were not only transparent, but also sol-gel free. Eventually, all of the WO_3 thin films were prepared, using the spin coating technique, onto indium tin oxide (ITO) plane glass substrates (2×2 cm^2), with spin coating parameters (3500 revolutions per minute for 60 s).

The WO_3 thin films were annealed at different temperatures in the air atmosphere: 200 °C, 250 °C, 300 °C, 350 °C, 400 °C, 450 °C and 500 °C, respectively.

The surface tension and the contact angle of the precursor solutions were measured by Attension Theta (Biolin Scientific, TL200, Gothenburg, Sweden). The dynamic viscosity measurements of the precursor solutions were implemented using a rotational rheometer (Thermo Fisher Scientific, HAAKE MARS 40, MA, USA). Thermal gravimetric analysis (TGA) and differential scanning calorimetry were characterized by a Differential Scanning Calorimeter (DSC, Differential Scanning Calorimeter, DSC214, NETZSCH Scientific Instruments Trading (Shanghai) Ltd., Selb, Germany). The thickness of all films was measured by a probe surface profiler (VeecoDektak150, Veeco, Somerset, NJ, USA). The crystalline structure of WO_3 thin films was characterized by X ray Diffraction using Cu Kα radiation (XRD, PANalytical Empyrean DY1577, PANalytical, Almelo, The Netherlands), and the XRD patterns were analyzed using the software Jade 6.0. As well as this, Fourier transform infrared spectroscopy (FT-IR) spectra of the solutions were recorded by an FT-IR spectrophotometer (SHIMADZU IR Prestige-21, SHIMADZU, Tokyo, Japan) with an ITO glass substrate acting as a blank. The surface morphology was observed by an atomic force microscopy (AFM, Being Nano-Instruments BY3000, Being Nano-Instruments, Beijing, China). The transmittance of the films at the initial state, colored state and bleached state were measured by an ultraviolet spectrophotometer (SHIMADZU UV2600, SHIMADZU, Tokyo, Japan), with air acting as a blank. The current of the electrochromic test and the relationship between the change of transmittance and the time were recorded by an electrochemical workstation (CH Instruments CHI600E, CH Instruments, Shanghai, China) and a micro-spectrometer (Morpho PG2000, Morpho, Shanghai, China), respectively.

3. Results and Discussion

The surface tension, the contact angle and the dynamic viscosity of all the precursor solutions are illustrated in Figure 1. Figure 1a shows that as the volume fraction of modified-ammonia tungstate glycerol increases, keeping the volume of the precursor solutions consistent, both of the surface tension and the contact angle decrease first, and then increase while obtaining a minimum in the sample solution with 25% volume fraction glycerol. As shown in Figure 1b, the dynamic viscosity of the solutions characterizes an increasing tendency with the increase in the volume of the glycerol. As defined, when the droplet has a tendency to spread out on the solid surface, not only the solid–liquid contact surface increases and the contact angle decreases, but also the surface tension decreases. In other words, the smaller the surface tension and the contact angle, the better the surface wettability of the solution on the substrate. Additionally, bigger dynamic viscosity could facilitate the growth of films. In comparison to dynamic viscosity, the surface tension and the contact angle exert a decisive part in selecting the optimal solution in this work. As a consequence, the ammonium tungstate solution with 25% volume fraction glycerol is used to grow WO_3 thin films in this work.

Figure 2 shows the thermal gravimetric and differential scanning calorimetry (TG-DSC) curves of the sample solution with 25% volume fraction glycerol and the thickness of the films annealed at different temperatures. The sample solution successively loses weight from room temperature to approximately 300 °C, as shown in the TG curve. The drastic weight loss can be as a result of the water evaporation range from room temperature to 140 °C [30]. It was the pyrolysis of glycerol and the further evaporation of water that a continuous weight loss of the solution range from 140 °C to 220 °C results from [31]. It can be seen that there are multiple weak endothermic peaks and exothermic peaks, and the weight of the sample solution reduces in the temperature between 220 °C and 300 °C, which is presumably ascribed to the formation of WO_3 from ammonium tungstate. The weight of the sample solution was not observed to change obviously above 300 °C. Besides, the thickness of the WO_3 films decreases drastically between 200 °C and 300 °C, as shown in Figure 2b. The drastic drop of thickness probably results from the water evaporation and the pyrolysis of glycerol. The further change in thickness would be analyzed through the FT-IR and XRD. Furthermore, the slowly endothermic process between 300 °C and 460 °C could be observed, which is possibly attributed to the transformation of phase structure of WO_3 from ammonium tungstate.

On the other hand, according to DSC analysis, it is concluded that the phase structure of WO_3 was also transformed along with exothermic process above 460 °C. The further phase structure transformation of WO_3 would be characterized and analyzed through XRD measurement.

The FT-IR spectra of the sample solution with a 25% volume fraction of glycerol are illustrated in Figure 3, which were annealed at 300 °C, 350 °C, 400 °C, 450 °C and 500 °C, respectively. It can be seen that there are a sharp peak at approximately 980 cm^{-1} and a weak peak at around 690 cm^{-1}, which respectively correspond to the characteristic W=O stretching vibration and W–O–W stretching vibration [32,33], revealing the successful formation of WO_3 from ammonium tungstate when WO_3 films were annealed above 300 °C, which is consistent with the result of the TG-DSC analysis. In addition, the sharp peak shifts to a higher wavenumber, and becomes sharper from 300 °C to 350 °C in indication of that the W=O bond length shortens and the W=O bond vibrational steric resistance increases. It is indicated that the film structure becomes dense from loose, and the thickness of film deceases at 350 °C. Besides, the decrease in the film thickness in the temperature stage of 400–500 °C is also attributed to the densification of the film in the case of the stable content of WO_3 in the film. As shown in Figure 3, both of the two peaks were intensified with the increase in the annealing temperature, which indicated that a mounting number of WO_3 was generated, and the WO_3 thin film was prone to be purer. According to FT-IR analysis, WO_3 is the dominant composition within the films instead of glycerol as the annealing temperature increases.

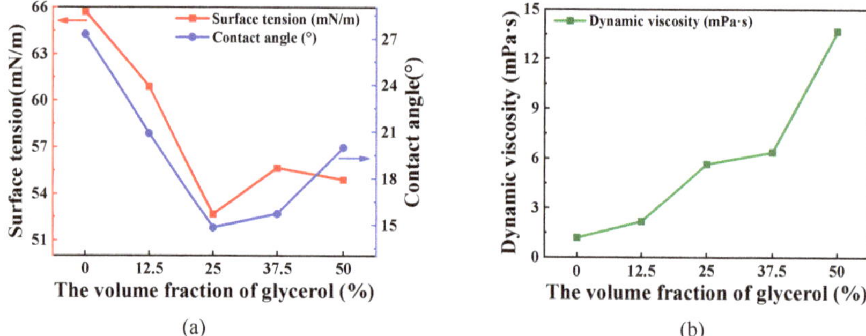

Figure 1. The surface tension and the contact angle and the dynamic viscosity of the all sample precursor solutions. (**a**) The surface tension and the contact angle. (**b**) The dynamic viscosity.

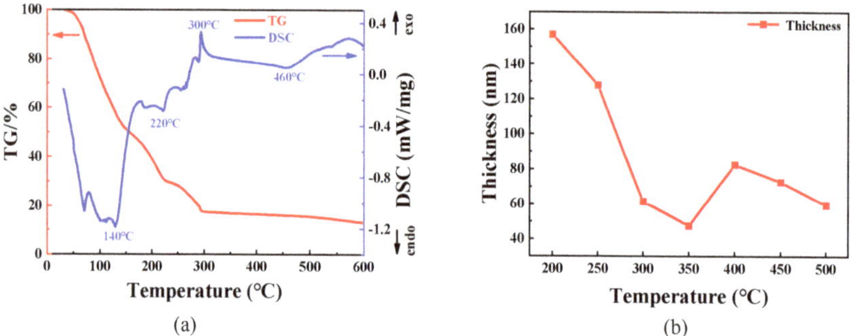

Figure 2. (**a**) The thermal gravimetric and differential scanning calorimetry (TG-DSC) curves of the sample solution with 25% volume fraction glycerol. The endothermic process and exothermic process are appointed as "endo" and "exo", respectively. (**b**) The thickness of the WO_3 films annealed at different temperatures.

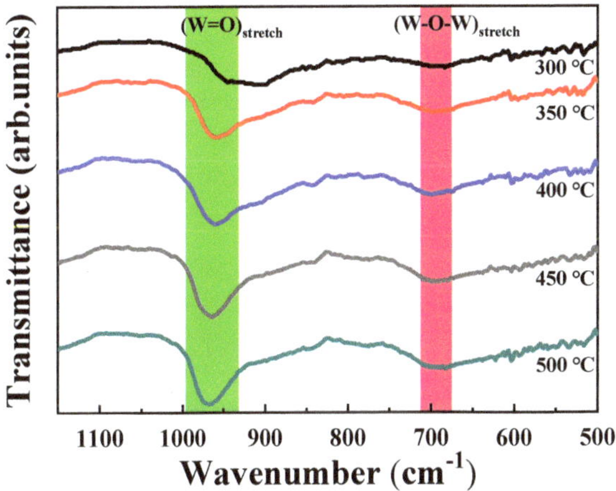

Figure 3. Fourier transform infrared (FT-IR) spectra of WO_3 films prepared using the sample solution with 25% volume fraction glycerol.

Figure 4a–f presents the AFM 2D and 3D images (5000 nm × 5000 nm) of the WO_3 thin films prepared using the sample solution with 25% volume fraction glycerol, which were annealed at different temperatures for an hour. Moreover, the surface roughness and the surface grain size of these films are illustrated in Figure 4g. As shown in Figure 4a,b, it is revealed that the WO_3 thin film which was annealed at 200 °C is almost smooth and homogeneous. There are few surface grains on the WO_3 thin films annealed at 200 °C for the reason that definite boundary would not be observed. According to the result of TG-DSC analysis, the film annealed at 200 °C did not transform into WO_3 yet. With the increase in the annealing temperature, a slightly rough surface is commenced to emerge on the WO_3 thin film, which can be observed in Figure 4c,d. There are a great many easily identifiable grains on the surface of the WO_3 film in Figure 4e,f, which may be attributed to the crystallization of WO_3 when the WO_3 film was annealed at 500 °C. This can be confirmed through the result of XRD analysis. The average surface grain size of the WO_3 thin films annealed at 200 °C, 350 °C and 500 °C are, respectively, 7.8 nm, 63.6 nm and 74.7 nm. The surface root mean square (RMS) roughness of the WO_3 films annealed at 200 °C, 350 °C and 500 °C are 0.38 nm, 1.01 nm and 1.68 nm, respectively. It is worth mentioning that the WO_3 films prepared in this work are less rough, with a surface roughness of less than 2 nm in comparison with the sol-gel-prepared WO_3 films, although the surface roughness of WO_3 films exists a slowly increasing tendency with annealing temperature [34–36]. In other words, the WO_3 films with less roughness prepared in this work can reduce scattering of the incident light, and are beneficial to the transmission of incident light, so that they are more qualified for application in an electrochromic device [34].

X-ray Diffraction (XRD) is an effective experimental technique to characterize the crystalline structure and the grain size of material. The XRD patterns of the WO_3 films prepared using the sample solution with 25% volume fraction glycerol onto ITO glass substrate, and annealed at different temperatures, are illustrated in Figure 5. The diffraction peaks were analyzed using Jade 6.0 and PDF#06-0416, PDF#41-0905 and PDF#30-1387. In the XRD of Figure 5, the diffraction peaks of the WO_3 thin films annealed at 200 °C and 300 °C are matched well with Indium Oxide (In_2O_3), according to PDF#06-0416, indicating acquirement of the amorphous structure of the WO_3 film [20,25,37]. It can be confirmed that there are some diffraction peaks of the WO_3 film annealed at 400 °C, except for the diffraction peaks of ITO glass substrate in Figure 5. In other words, it is indicated that the crystalline temperature for the WO_3 thin films is between 300 °C and 400 °C [2,38]. According to the broad diffraction peak located at approximately 24°, the formation of the WO_3 cubic crystal structure is revealed, and monoclinic WO_3 also exists at the same time [32]. It is suggested that the crystalline transformation in thin film, making the uplift of the film, results in the increase in thickness at 400 °C. Along with temperature being further increased, the WO_3 film was confirmed to be preferential growth to monoclinic crystalline structure, rather than cubic crystalline structure when the WO_3 film was annealed at 500 °C [16,39]. Furthermore, this can be in good agreement with the results of TG-DSC curves and AFM images.

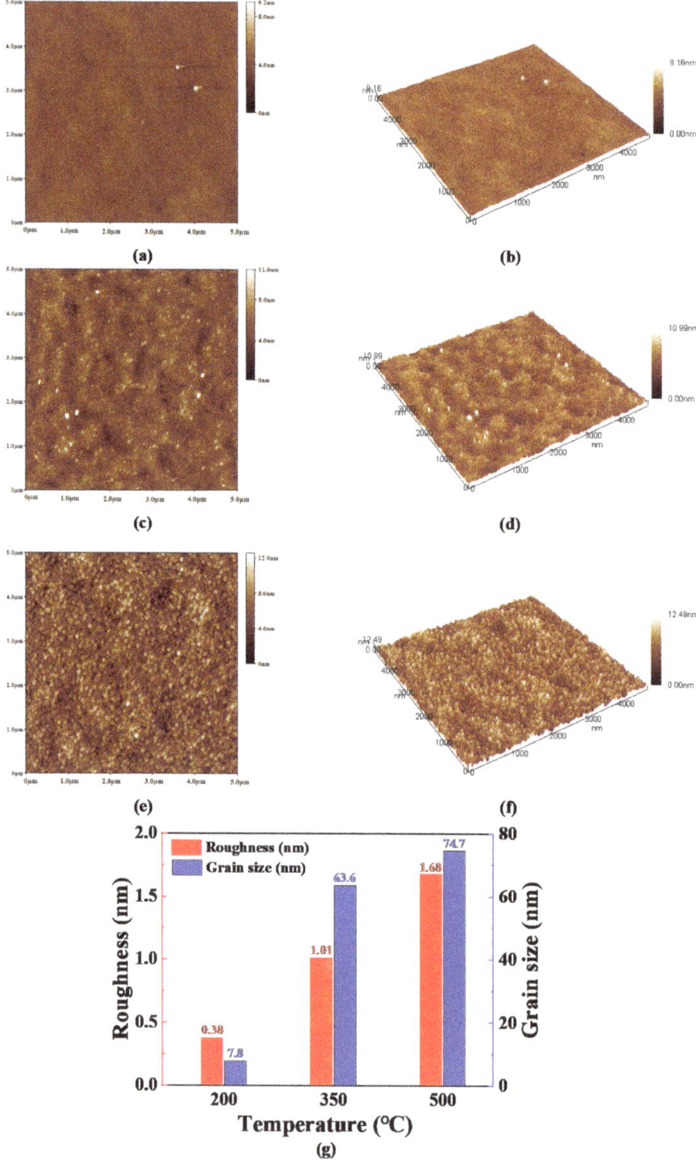

Figure 4. The atomic force microscope (AFM) 2D and 3D images of the WO_3 thin films prepared using the sample solution with 25% volume fraction glycerol annealed at different temperatures: (**a**,**b**) 200 °C, (**c**,**d**) 350 °C, (**e**,**f**) 500 °C, respectively. (**g**) The surface roughness and the surface grain size of these films.

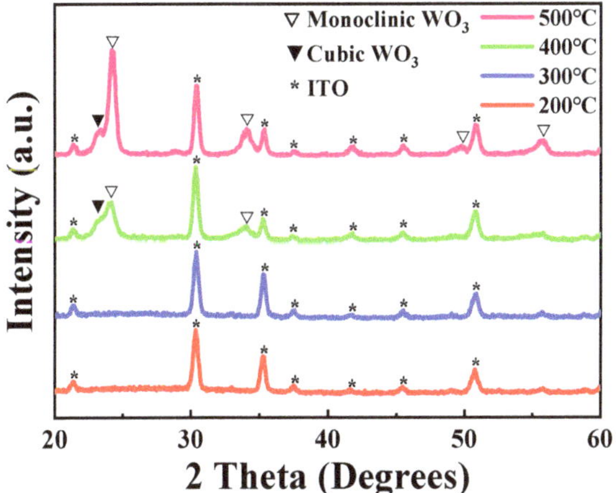

Figure 5. The X-ray Diffraction (XRD) patterns of the WO_3 films prepared using the sample solution with 25% volume fraction glycerol annealed at different temperatures: 200 °C, 300 °C, 400 °C and 500 °C, respectively.

Figure 6a–g display the optical transmittance spectra of the WO_3 films at the initial state, colored state and bleached state in the wavelength range from 400 nm to 800 nm, which were prepared using the sample solution with 25% volume fraction glycerol, and annealed at different temperatures, and the optical transmittance modulation ability (ΔT at 700 nm) curve of these films is illustrated in Figure 6h. The transmittance modulation ability (ΔT) can be defined by the following formula at 700 nm wavelength:

$$\Delta T = |T_c - T_b| \tag{1}$$

In this formula, T_c and T_b are the optical transmittance of the WO_3 films at a colored state and bleached state at the 700 nm wavelength, respectively. It is meant that the greater the ΔT, the better the optical transmittance modulation ability. As shown in Figure 6h, ΔT increased first from 200 °C to 300 °C, and subsequently decreased slightly at 350 °C. After ΔT achieved a maximum roughly 40% at 400 °C, it decreased again at 450 °C and 500 °C.

In the temperature stage of 200–300 °C, the precursor has not completely transformed. The quite thick film is attributed to a large quantity of amorphous carbon from the pyrolysis of glycerol. Besides, the amorphous carbon is the good binding receptor for Li^+ [40,41], thereby there is a competitive relationship between amorphous carbon and WO_3 in binding to Li^+. As the annealing temperature increases, the thickness of the film decreases, suggesting that the amorphous carbon is gradually oxidized, and the content of amorphous carbon decreases. WO_3 gradually plays a dominant role in the competition of binding to Li^+ with the continuous formation of tungsten trioxide. As a consequence, the ΔT of the film gradually increases range from 200 °C to 300 °C.

When the annealing temperature rises above 350 °C, the composition of the film is basically WO_3, and the effect of temperature on the mass of WO_3 is negligible, according to the result of the TG curve in Figure 2. The ΔT of the film decreases slightly at 350 °C, which is attributed to the following possible reason: In the FT-IR of Figure 3, when the annealing temperature increases from 300 °C to 350 °C, the W=O absorption band shifts to a higher wavenumber and becomes sharper, which indicates that the W=O bond length shortens, as well as the W=O bond vibrational steric resistance increases. In addition, the mass of WO_3 is basically constant, while the thickness of film decreases. It demonstrates that the film structure becomes dense from loose. The steric hindrance of Li^+ into and out the thin film increases, and the binding sites may decrease at the same time, which is not conducive to electrochromism [42–44]. Therefore, there is a slight decrease in the ΔT at 350 °C.

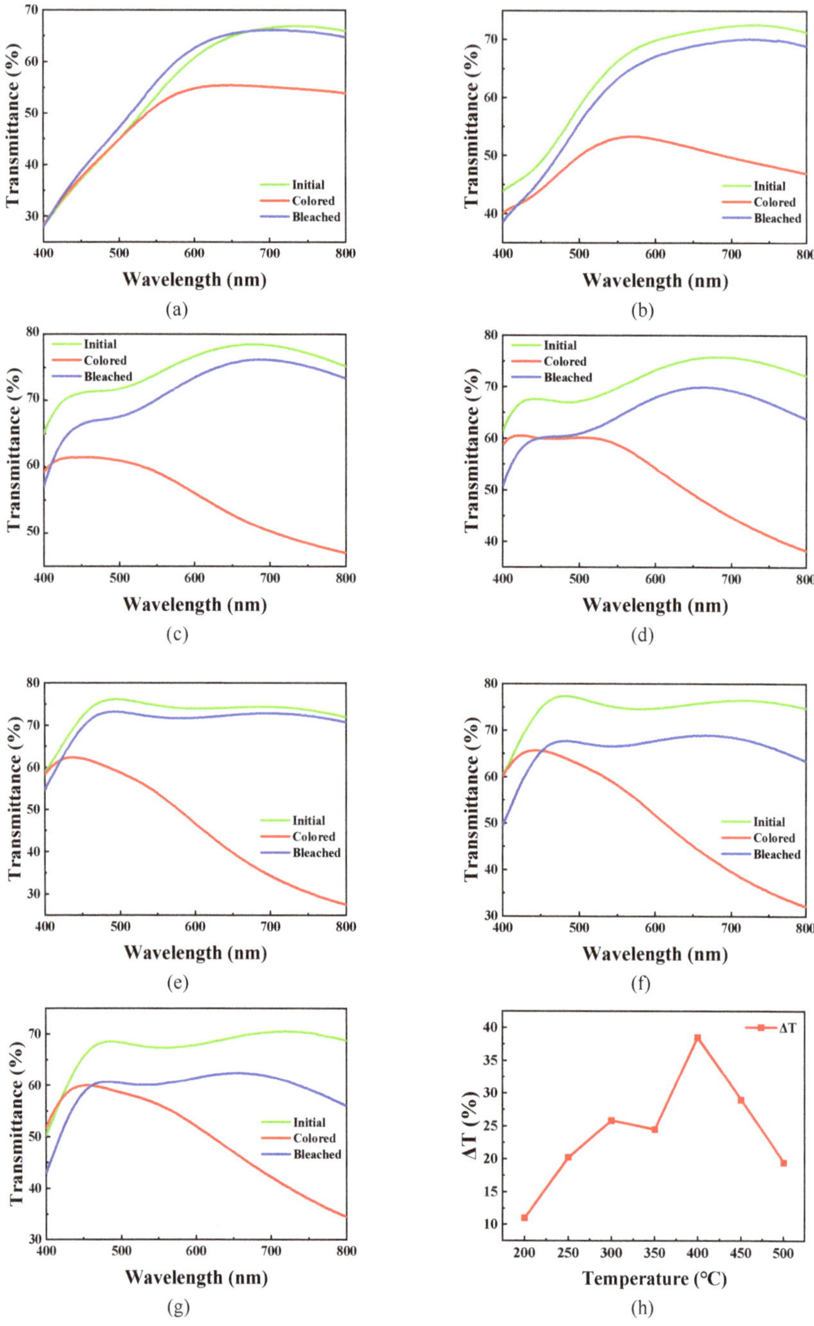

Figure 6. The optical transmittance spectra of the WO_3 films annealed at different temperatures at the initial state, colored state and bleached state: (**a**) 200 °C, (**b**) 250 °C, (**c**) 300 °C, (**d**) 350 °C, (**e**) 400 °C, (**f**) 450 °C, (**g**) 500 °C, respectively. (**h**) The optical transmittance modulation ability (ΔT at 700 nm wavelength) with ±3.0 V voltages.

The ΔT of the film at 400 °C is the best for the possible reasons as follows: Firstly, the crystalline WO_3 film possesses a higher carrier mobility [37,45], WO_3 is an n-type semiconductor [2] and electrons are the majority carrier, which is beneficial to electron injection in the electrochromic double injection of ions and electrons model [28,46]. Secondly, the crystallization degree of the film is not very high at 400 °C, and there are a lot of grain boundaries within the film in a mixed crystalline phase structure. Besides, different crystalline phases would produce a lot of crystallographic defects owing to the lattice constant mismatch. These grain boundaries and defects are essentially W–O-dangling bond, which are good ion binding sites, contributing to electrochromism. Hence, the ΔT of the film is the best at 400 °C.

However, with the further increase in annealing temperature, the crystal phase of the film is mainly monoclinic, and the diffraction peaks become sharper in Figure 5, which indicates that the crystallization degree of the film increases, suggesting that the grain boundaries in the film decrease and Li^+ binding sites decrease. Moreover, the densification of the film is not conducive to the injection and extraction of Lithium ions [44,47]. In consequence, the ΔT of thin film decreases gradually in the temperature stage of 400–500 °C.

The optical properties of the WO_3 films annealed at different temperatures are characterized in Figure 7. The optical band gap of these films can be calculated from the following formula [48,49]:

$$\alpha h\nu = A(h\nu - E_g)^n \tag{2}$$

In this formula, A is a content; α, the absorption coefficient, can be computed from the transmittance; the Planck constant (h) is 6.626×10^{-34} J s; ν, the photon frequency, can be converted from the wavelength; n is 2, because WO_3 is an indirect band gap semiconductor; E_g, the optical band gap, is the value of the fitting line horizontal intercept of the curve of $(\alpha h\nu)^{1/2}$ versus the photon energy $h\nu$.

As shown in Figure 7b, the optical band gap (E_g) of the WO_3 films are observed at the 3.48 eV, 3.45 eV, 3.47 eV and 3.37 eV ranges from 350 °C to 500 °C, respectively. The E_g of the WO_3 films prepared in this work decreased first and then increased, and afterward decreased again with the increase in annealing temperature. The average bandgap of the WO_3 films with a mixed crystalline structure, obtaining a minimum, is 3.37 eV at 500 °C [2].

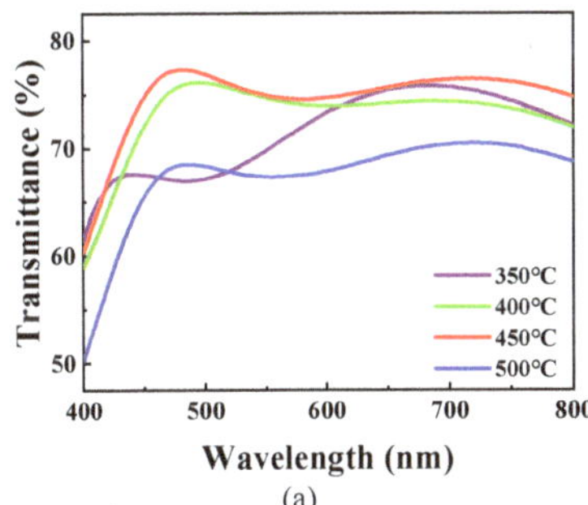

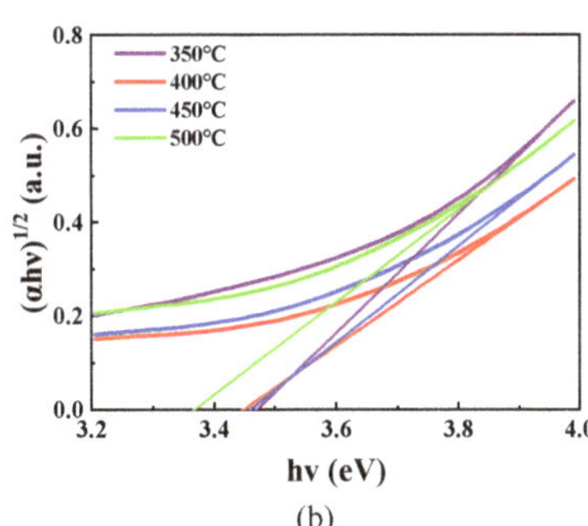

Figure 7. (**a**) The transmittance of WO_3 films prepared using the sample solution with 25% volume fraction glycerol annealed at different temperatures. The transmittance of air is specified as 100%. (**b**) The curve of $(\alpha h\nu)^{1/2}$ versus $h\nu$ for WO_3 films.

Figure 8a,b illustrate the change of transmittance at 680 nm wavelength of the WO_3 films prepared using the sample solution with 25% volume fraction glycerol and the change of transmittance at 680 nm wavelength under ±3.0 V voltages, respectively. Additionally, for sufficiently evaluating the electrochromic performance, coloration efficiency (CE) as an important parameter is defined by the following formula [50,51]:

$$CE = \Delta OD/\Delta Q = \log[T_b/T_c]/(Q/A) \tag{3}$$

$$Q = \int I\,dt \quad (4)$$

In the formula, ΔOD is the optical density of the change between two optical states at 680 nm wavelength; T_b and T_c were defined above; the injecting and extracting charge density (Q) were calculated by integrating the current of a cycle of the electrochromic test; A is the electrochromic total area. The coloration efficiency (CE) value of sample was calculated to be 46.3 $cm^2 \cdot C^{-1}$ for WO_3 thin films annealed at 400 °C in this work, as shown in Figure 8b. In general the CE value of sol-gel-prepared films ranges from 25 $cm^2 \cdot C^{-1}$ to 75 $cm^2 \cdot C^{-1}$, which shows that the CE value of sol-gel-free WO_3 films as reliable as normal sol-gel WO_3 films [37,52]. Further studies will be done about the electrical resistance and the electrochromic reliability.

It is worth comparing the performance in this work with other reported studies by other synthesis methods. The comparison of coloration efficiency and optical band gap between this work and other reported works would be demonstrated in Table 1. It is indicated that this work, using the sol-gel-free method, maintains a proper balance between optical band gap and coloration efficiency better than other works by synthesis methods in reference.

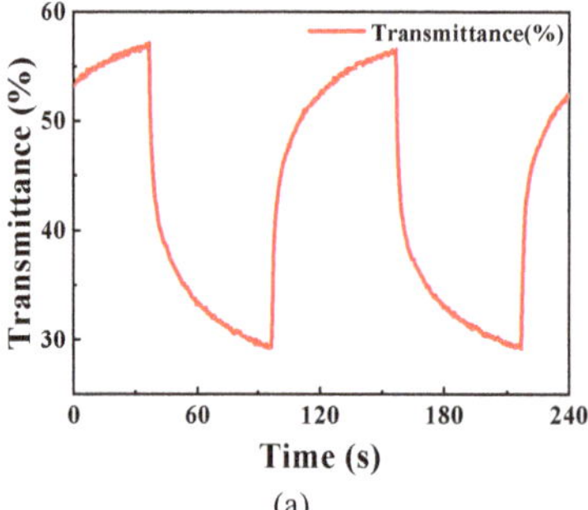

(a)

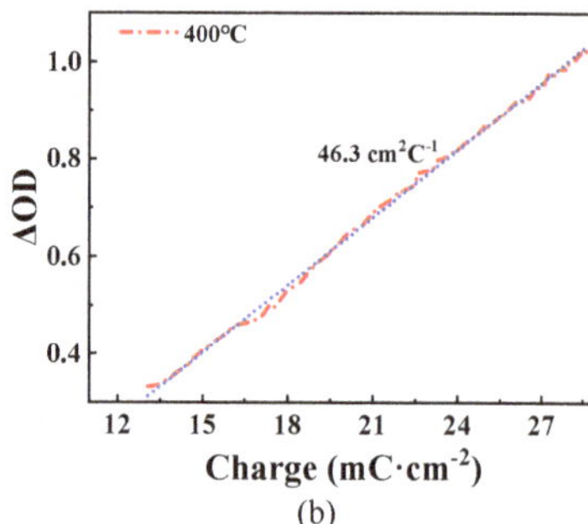

(b)

Figure 8. (**a**) Change curve of transmittance at 680 nm of WO_3 films prepared using the sample solution with 25% volume fraction glycerol. (**b**) Variation of the optical density (OD) versus charge density for WO_3 film. The applied voltage was range from −3.0 V to +3.0 V.

Table 1. The comparison of optical band gap and coloration efficiency between this work and the reported works by synthesis methods.

Result from	Synthesis Method	Crystallinity	Optical Band Gap(eV)	Coloration Efficiency (cm^2/C)
[5]	DC magnetron sputtering	Amorphous	3.51	40.5
[37]	Sol-gel	Amorphous	3.31	45.3
[37]	Sol-gel	Crystalline	3.10	27.4
[52]	Sol-gel	Crystalline	3.01	46.7
This work	Sol-gel free	Crystalline	3.37	46.3

4. Conclusions

The WO_3 thin films were successfully prepared on ITO glass substrate by spin coating technique using the glycerol-modified ammonium tungstate precursor solution. For better surface wetting, the sol-gel free ammonium tungstate solution with the 25% volume fraction of glycerol was selected to promote the growth of WO_3 thin films. It is concluded that the ammonium tungstate transformed to WO_3 range from 220 °C to 300 °C, and the crystalline structure of the WO_3 films was observed when films were annealed above 300 °C.

Additionally, as the annealing temperature increases, the component content of WO_3 within the film increased. The WO_3 films prepared in this work were smoother and more homogeneous than the sol-gel-prepared films. The optical band gap of the WO_3 films was variational from 3.48 eV to 3.37 eV with the increase in the annealing temperature above 350 °C. Furthermore, the coloration efficiency of the WO_3 films attained 46.3 $cm^2 \cdot C^{-1}$.

Author Contributions: Conceptualization, J.L. and H.N.; methodology, R.Y. and D.G.; validation, M.S., X.F. and J.C.; resources, J.P. and R.Y.; data curation, G.Z., T.Q. and K.G.; writing—Original draft preparation, J.L. and G.Z.; writing—Review and editing, J.P. and H.N.; All authors have read and agreed to the published version of the manuscript.

Funding: This work was supported by Key-Area Research and Development Program of Guangdong Province (No.2019B010934001), National Natural Science Foundation of China (Grant No.51771074, 61804029 and 61574061), the Major Integrated Projects of National Natural Science Foundation of China (Grant No.U1601651), the Basic and Applied Basic Research Major Program of Guangdong Province (No.2019B030302007), the Guangdong Natural Science Foundation (No.2018A030310353), Science and Technology Project of Guangzhou (No.201904010344), the Fundamental Research Funds for the Central Universities (No.2019MS012), the Open Project of Guangdong Province Key Lab of Display Material and Technology (No.2017B030314031), 2019 Guangdong University Student Science and Technology Innovation Special Fund ("Climbing Plan" Special Fund) (No.pdjh2019a0028 and pdjh2019b0041), National College Students' Innovation and Entrepreneurship Training Program (No.201910561005 and 201910561007), South China University of Technology 100 Step Ladder Climbing Plan Research Project (No.j2tw201902475, j2tw201902203 and j2tw202004095).

Conflicts of Interest: The authors declare no conflict of interest.

References

1. Thongpan, W.; Louloudakis, D.; Pooseekheaw, P.; Kumpika, T.; Kantarak, E.; Panthawan, A.; Tuantranont, A.; Thongsuwan, W.; Singjai, P. Electrochromic properties of tungsten oxide films prepared by sparking method using external electric field. *Thin Solid Films* **2019**, *682*, 135–141. [CrossRef]
2. Mukherjee, R.; Sahay, P.P. Effect of precursors on the microstructural, optical, electrical and electrochromic properties of WO_3 nanocrystalline thin films. *J. Mater. Sci. Mater. Electron.* **2015**, *26*, 6293–6305. [CrossRef]
3. Mardare, C.C.; Hassel, A.W. Review on the versatility of tungsten oxide coatings. *Phys. Status Solidi A* **2019**, *216*, 1900047. [CrossRef]
4. Bourdin, M.; Gaudon, M.; Weill, F.; Duttine, M.; Gayot, M.; Messaddeq, Y.; Cardinal, T. Nanoparticles (NPs) of WO_{3-x} compounds by polyol route with enhanced photochromic properties. *Nanomater.-Basel* **2019**, *9*, 1555. [CrossRef] [PubMed]
5. Yu, H.; Guo, J.; Wang, C.; Zhang, J.; Liu, J.; Dong, G.; Zhong, X.; Diao, X. Essential role of oxygen vacancy in electrochromic performance and stability for WO_{3-y} films induced by atmosphere annealing. *Electrochim. Acta* **2020**, *332*, 135504. [CrossRef]
6. Pan, J.; Wang, Y.; Zheng, R.; Wang, M.; Wan, Z.; Jia, C.; Weng, X.; Xie, J.; Deng, L. Directly grown high-performance WO_3 films by a novel one-step hydrothermal method with significantly improved stability for electrochromic applications. *J. Mater. Chem. A* **2019**, *7*, 13956–13967. [CrossRef]
7. Wang, B.; Man, W.; Yu, H.; Li, Y.; Zheng, F. Fabrication of Mo-Doped WO_3 Nanorod Arrays on FTO Substrate with Enhanced Electrochromic Properties. *Materials (Basel, Switzerland)* **2018**, *11*, 1627. [CrossRef]
8. Liang, Y.; Chang, C. Preparation of Orthorhombic WO_3 Thin Films and Their Crystal Quality-Dependent Dye Photodegradation Ability. *Coatings* **2019**, *9*, 90. [CrossRef]
9. Zheng, J.Y.; Haider, Z.; Van, T.K.; Pawar, A.U.; Kang, M.J.; Kim, C.W.; Kang, Y.S. Tuning of the crystal engineering and photoelectrochemical properties of crystalline tungsten oxide for optoelectronic device applications. *CrystEngComm* **2015**, *17*, 6070–6093. [CrossRef]
10. Zhang, G.; Lu, K.; Zhang, X.; Yuan, W.; Shi, M.; Ning, H.; Tao, R.; Liu, X.; Yao, R.; Peng, J. Effects of annealing temperature on optical band gap of sol-gel tungsten trioxide films. *Micromach.-Basel* **2018**, *9*, 377. [CrossRef]
11. Rozman, M.; Žener, B.; Matoh, L.; Godec, R.F.; Mourtzikou, A.; Stathatos, E.; Bren, U.; Lukšič, M. Flexible electrochromic tape using steel foil with WO_3 thin film. *Electrochim. Acta* **2020**, *330*, 135329. [CrossRef]
12. Yuan, J.; Wang, B.; Wang, H.; Chai, Y.; Jin, Y.; Qi, H.; Shao, J. Electrochromic behavior of WO_3 thin films prepared by GLAD. *Appl. Surf. Sci.* **2018**, *447*, 471–478. [CrossRef]
13. Wang, J.C.; Shi, W.; Sun, X.Q.; Wu, F.Y.; Li, Y.; Hou, Y. Enhanced Photo-Assisted Acetone Gas Sensor and Efficient Photocatalytic Degradation Using Fe-Doped Hexagonal and Monoclinic WO_3 Phase-Junction. *Nanomater.-Basel* **2020**, *10*, 398. [CrossRef] [PubMed]

14. Liang, Y.; Chang, C. Improvement of Ethanol Gas-Sensing Responses of ZnO-WO_3 Composite Nanorods through Annealing Induced Local Phase Transformation. *Nanomater.-Basel* **2019**, *9*, 669. [CrossRef]
15. Leidinger, M.; Huotari, J.; Sauerwald, T.; Lappalainen, J.; Schütze, A. Selective detection of naphthalene with nanostructured WO_3 gas sensors prepared by pulsed laser deposition. *J. Sens. Sens. Syst.* **2016**, *5*, 147–156. [CrossRef]
16. Zhou, Q.; Chen, Y.; Li, X.; Qi, T.; Peng, Z.; Liu, G. Preparation and electrochromism of pyrochlore-type tungsten oxide film. *Rare Metals* **2018**, *37*, 604–612. [CrossRef]
17. Kumar, A.; Singh, P.; Kulkarni, N.; Kaur, D. Structural and optical studies of nanocrystalline V_2O_5 thin films. *Thin Solid Films* **2008**, *516*, 912–918. [CrossRef]
18. Kim, H.; Lee, H.; Maeng, W.J. Applications of atomic layer deposition to nanofabrication and emerging nanodevices. *Thin Solid Films* **2009**, *517*, 2563–2580. [CrossRef]
19. De Andrade, J.R.; Cesarino, I.; Zhang, R.; Kanicki, J.; Pawlicka, A. Properties of electrodeposited WO_3 thin films. *Mol. Cryst. Liq. Cryst.* **2014**, *604*, 71–83. [CrossRef]
20. Pal, S.; Jacob, C. The influence of substrate temperature variation on tungsten oxide thin film growth in an HFCVD system. *Appl. Surf. Sci.* **2007**, *253*, 3317–3325. [CrossRef]
21. Blackman, C.S.; Parkin, I.P. Atmospheric pressure chemical vapor deposition of crystalline monoclinic WO_3 and WO_{3-x} thin films from reaction of WCl_6 with O-containing solvents and their photochromic and electrochromic properties. *Chem. Mater.* **2005**, *17*, 1583–1590. [CrossRef]
22. Kangkun, N.; Kiama, N.; Saito, N.; Ponchio, C. Optical properties and photoelectrocatalytic activities improvement of WO_3 thin film fabricated by fixed-potential deposition method. *Optik* **2019**, *198*, 163235. [CrossRef]
23. Zhang, G.; Zhang, X.; Ning, H.; Chen, H.; Wu, Q.; Jiang, M.; Li, C.; Guo, D.; Wang, Y.; Yao, R.; et al. Tungsten doped stannic oxide transparent conductive thin film using preoxotungstic acid dopant. *Superlattices Microstruct* **2019**, *130*, 277–284. [CrossRef]
24. Fardindoost, S.; Iraji Zad, A.; Rahimi, F.; Ghasempour, R. Pd doped WO_3 films prepared by sol-gel process for hydrogen sensing. *Int. J. Hydrog. Energy* **2010**, *35*, 854–860. [CrossRef]
25. Avellaneda, C.O.; Bueno, P.R.; Faria, R.C.; Bulhões, L.O.S. Electrochromic properties of lithium doped WO_3 films prepared by the sol-gel process. *Electrochim. Acta* **2001**, *46*, 1977–1981. [CrossRef]
26. Danks, A.E.; Hall, S.R.; Schnepp, Z. The evolution of 'sol-gel' chemistry as a technique for materials synthesis. *Mater. Horiz.* **2016**, *3*, 91–112. [CrossRef]
27. Xie, Z.; Liu, Q.; Zhang, Q.; Lu, B.; Zhai, J.; Diao, X. Fast-switching quasi-solid state electrochromic full device based on mesoporous WO_3 and NiO thin films. *Sol. Energy Mater. Sol. Cells* **2019**, *200*, 110017. [CrossRef]
28. Wen-Cheun Au, B.; Chan, K.; Knipp, D. Effect of film thickness on electrochromic performance of sol-gel deposited tungsten oxide (WO_3). *Opt Mater.* **2019**, *94*, 387–392. [CrossRef]
29. Sonavane, A.C.; Inamdar, A.I.; Shinde, P.S.; Deshmukh, H.P.; Patil, R.S.; Patil, P.S. Efficient electrochromic nickel oxide thin films by electrodeposition. *J. Alloys Compd.* **2010**, *489*, 667–673. [CrossRef]
30. Jin, L.H.; Bai, Y.; Li, C.S.; Wang, Y.; Feng, J.Q.; Lei, L.; Zhao, G.Y.; Zhang, P.X. Growth of tungsten oxide nanostructures by chemical solution deposition. *Appl. Surf. Sci.* **2018**, *440*, 725–729. [CrossRef]
31. Zhao, B.; Zhang, X.; Dong, G.; Wang, H.; Yan, H. Efficient electrochromic device based on sol-gel prepared WO_3 films. *Ionics* **2015**, *21*, 2879–2887. [CrossRef]
32. Vidmar, T.; Topič, M.; Dzik, P.; Opara Krašovec, U. Inkjet printing of sol-gel derived tungsten oxide inks. *Sol. Energy Mater. Sol. Cells* **2014**, *125*, 87–95. [CrossRef]
33. García-García, F.J.; Mosa, J.; González-Elipe, A.R.; Aparicio, M. Sodium ion storage performance of magnetron sputtered WO_3 thin films. *Electrochim. Acta* **2019**, *321*, 134669. [CrossRef]
34. Leitzke, D.W.; Cholant, C.M.; Landarin, D.M.; Lucio, C.S.; Krüger, L.U.; Gündel, A.; Flores, W.H.; Rodrigues, M.P.; Balboni, R.D.C.; Pawlicka, A.; et al. Electrochemical properties of WO_3 sol-gel thin films on indium tin oxide/poly(ethylene terephthalate) substrate. *Thin Solid Films* **2019**, *683*, 8–15. [CrossRef]
35. Djaoued, Y.; Ashrit, P.V.; Badilescu, S.; Bruning, R. Synthesis and characterization of macroporous tungsten oxide films for electrochromic application. *J. Sol-Gel Sci. Technol.* **2003**, *28*, 235–244. [CrossRef]
36. Caruso, T.; Castriota, M.; Policicchio, A.; Fasanella, A.; De Santo, M.P.; Ciuchi, F.; Desiderio, G.; La Rosa, S.; Rudolf, P.; Agostino, R.G.; et al. Thermally induced evolution of sol-gel grown WO_3 films on ITO/glass substrates. *Appl. Surf. Sci.* **2014**, *297*, 195–204. [CrossRef]

37. Ge, C.; Wang, M.; Hussain, S.; Xu, Z.; Liu, G.; Qiao, G. Electron transport and electrochromic properties of sol-gel WO_3 thin films: Effect of crystallinity. *Thin Solid Films* **2018**, *653*, 119–125. [CrossRef]
38. Mukherjee, R.; Sahay, P.P. Structural, morphological, optical and electrical properties of spray-deposited Sb-doped WO_3 nanocrystalline thin films prepared using ammonium tungstate precursor. *J. Mater. Sci. Mater. Electron.* **2015**, *26*, 2697–2708. [CrossRef]
39. Yang, H.; Yu, J.; Jeong, R.H.; Boo, J. Enhanced electrochromic properties of nanorod based WO_3 thin films with inverse opal structure. *Thin Solid Films* **2018**, *660*, 596–600. [CrossRef]
40. Xia, T.; Zhang, W.; Wang, Z.; Zhang, Y.; Song, X.; Murowchick, J.; Battaglia, V.; Liu, G.; Chen, X. Amorphous carbon-coated TiO_2 nanocrystals for improved lithium-ion battery and photocatalytic performance. *Nano Energy* **2014**, *6*, 109–118. [CrossRef]
41. Wang, Z.; Tian, W.; Liu, X.; Yang, R.; Li, X. Synthesis and electrochemical performances of amorphous carbon-coated Sn-Sb particles as anode material for lithium-ion batteries. *J. Solid State Chem.* **2007**, *180*, 3360–3365. [CrossRef]
42. Madhavi, V.; Kondaiah, P.; Hussain, O.M.; Uthanna, S. Structural, optical and electrochromic properties of RF magnetron sputtered WO_3 thin films. *Phys. B: Condens. Matter* **2014**, *454*, 141–147. [CrossRef]
43. Kim, H.; Choi, D.; Kim, K.; Chu, W.; Chun, D.; Lee, C.S. Effect of particle size and amorphous phase on the electrochromic properties of kinetically deposited WO_3 films. *Sol. Energy Mater. Sol. Cells* **2018**, *177*, 44–50. [CrossRef]
44. Koo, B.; Ahn, H. Fast-switching electrochromic properties of mesoporous WO_3 films with oxygen vacancy defects. *Nanoscale* **2017**, *9*, 17788–17793. [CrossRef]
45. Darmawi, S.; Burkhardt, S.; Leichtweiss, T.; Weber, D.A.; Wenzel, S.; Janek, J.; Elm, M.T.; Klar, P.J. Correlation of electrochromic properties and oxidation states in nanocrystalline tungsten trioxide. *Phys. Chem. Chem. Phys.* **2015**, *17*, 15903–15911. [CrossRef]
46. Kharade, R.R.; Mane, S.R.; Mane, R.M.; Patil, P.S.; Bhosale, P.N. Synthesis and characterization of chemically grown electrochromic tungsten oxide. *J. Sol-Gel Sci. Technol.* **2010**, *56*, 177–183. [CrossRef]
47. Muthu Karuppasamy, K.; Subrahmanyam, A. The electrochromic and photocatalytic properties of electron beam evaporated vanadium-doped tungsten oxide thin films. *Sol. Energy Mater. Sol. Cells* **2008**, *92*, 1322–1326. [CrossRef]
48. Hoseinzadeh, S.; Ghasemiasl, R.; Bahari, A.; Ramezani, A.H. Effect of post-annealing on the electrochromic properties of layer-by-layer arrangement FTO-WO_3-Ag-WO_3-Ag. *J. Electron. Mater.* **2018**, *47*, 3552–3559. [CrossRef]
49. Najafi-Ashtiani, H.; Bahari, A. Optical, structural and electrochromic behavior studies on nanocomposite thin film of aniline, o-toluidine and WO_3. *Opt. Mater.* **2016**, *58*, 210–218. [CrossRef]
50. Najafi-Ashtiani, H.; Bahari, A.; Gholipour, S. Investigation of coloration efficiency for tungsten oxide-silver nanocomposite thin films with different surface morphologies. *J. Mater. Sci. Mater. Electron.* **2018**, *29*, 5820–5829. [CrossRef]
51. Hoseinzadeh, S.; Ghasemiasl, R.; Bahari, A.; Ramezani, A.H. The injection of Ag nanoparticles on surface of WO_3 thin film: Enhanced electrochromic coloration efficiency and switching response. *J. Mater. Sci. Mater. Electron.* **2017**, *28*, 14855–14863. [CrossRef]
52. Zhi, M.; Huang, W.; Shi, Q.; Wang, M.; Wang, Q. Sol-gel fabrication of WO_3/RGO nanocomposite film with enhanced electrochromic performance. *RSC Adv.* **2016**, *6*, 67488–67494. [CrossRef]

Article

Silver Nanorings Fabricated by Glycerol-Based Cosolvent Polyol Method

Zhihang Li [1], Dong Guo [2], Peng Xiao [3], Junlong Chen [1], Honglong Ning [1,*], Yiping Wang [4], Xu Zhang [1], Xiao Fu [1], Rihui Yao [1,5,*] and Junbiao Peng [1]

1 Institute of Polymer Optoelectronic Materials and Devices, State Key Laboratory of Luminescent Materials and Devices, South China University of Technology, Guangzhou 510640, China; mslzhscut@mail.scut.edu.cn (Z.L.); msjlchen@gmail.com (J.C.); yyxlxz@126.com (X.Z.); 201630343721@mail.scut.edu.cn (X.F.); psjbpeng@scut.edu.cn (J.P.)

2 School of Medical Instrument & Food Engineering, University of Shanghai for Science and Technology, No.516 Jungong Road, Shanghai 200093, China; guodong99@tsinghua.org.cn

3 School of Physics and Optoelectronic Engineering, Foshan University, Foshan 528000, China; xiaopeng@fosu.edu.cn

4 State Key Laboratory of Mechanics and Control of Mechanical Structures, Nanjing University of Aeronautics and Astronautics, Nanjing 210016, China; yipingwang@nuaa.edu.cn

5 Guangdong Province Key Lab of Display Material and Technology, Sun Yat-sen University, Guangzhou 510275, China

* Correspondence: ninghl@scut.edu.cn (H.N.); yaorihui@scut.edu.cn (R.Y.); Tel.: +86-20-8711-4525 (H.N.)

Received: 17 January 2020; Accepted: 22 February 2020; Published: 25 February 2020

Abstract: The urgent demand for transparent flexible electrodes applied in wide bandgap devices has promoted the development of new materials. Silver nanoring (AgNR), known as a special structure of silver nanowire (AgNW), exhibits attractive potential in the field of wearable electronics. In this work, an environmentally friendly glycerol-based cosolvent polyol method was investigated. The Taguchi design was utilized to ascertain the factors that affect the yield and ring diameter of AgNRs. Structural characterization showed that AgNR seeds grew at a certain angle during the early nucleation period. The results indicated that the yield and ring diameter of AgNRs were significantly affected by the ratio of cosolvent. Besides, the ring diameter of AgNRs was also tightly related to the concentration of polyvinylpyrrolidone (PVP). The difference of reducibility between glycerol, water, and ethylene glycol leads to the selective growth of (111) plane and is probably the main reason AgNRs are formed. As a result, AgNRs with a ring diameter range from 7.17 to 42.94 μm were synthesized, and the quantity was increased significantly under the optimal level of factors.

Keywords: wide bandgap semiconductors; flexible devices; silver nanoring; silver nanowire; polyol method; cosolvent

---

1. Introduction

With the development of flexible wide bandgap (WBG) semiconductor devices, such as flexible thin-film transistors (TFT) [1,2], wearable display devices [3,4], photovoltaic devices [5], and energy storage devices [6], the requirement for flexible electrodes, especially ones that are transparent and printable, has become urgent. As a traditional transparent conductive thin film (TCF), indium tin oxide (ITO) is brittle and requires high temperature preparation, which limits its application in flexible TCFs [7,8]. To meet such demand, several materials like graphene [9], carbon nanotubes [10], metal grid [11,12], and metal nanowires [13,14] have been studied over the years and achieved some substantial results.

Among these materials, silver nanowire (AgNW), known as the most suitable material to prepare TCFs, is considered to be a potential material because of its excellent electrical, optical, and mechanical properties as well as simple solution preparation [15–19]. Recently, a new form of silver nanostructure called silver nanoring (AgNR) was prepared as the by-product of AgNW. It showed high flexible performance with little effect on light transmission and also exhibited some attractive properties to be applied in optical nanoantennae, plasmonic devices, and optical manipulation [20,21]. Azani et al. simulated random networks deposited with conductive rings or wires and declared that AgNR can reach better optoelectrical property than AgNW at the same sheet resistance [22]. AgNR retains intrinsic chemical stability and low electrical resistivity like other silver materials and has strong localized surface plasmon resonance (LSPR) [23] to be used in energy conversion or signal enhancement. When AgNR and AgNW are deposited in flexible substrates, the two-dimensional structure of AgNR can increase the probability of overlapping to form a conductive network, and the maximum number of the contact point is twice as many as that of AgNW, which gives AgNR greater resistance against disconnect. At the same time, the ring shape gives the conductive network stronger tensile resistance [24]. Owing to these advantages, AgNR can be fabricated into printable flexible transparent electrodes to form WBG semiconductor devices with superior performance. However, like other nanorings composed of Cu [25], GaN [26], AlN [27], and ZnO [28], the preparation of AgNR still faces the difficulties of low yield and less purity, which seriously affect its application. Ethylene glycol (EG), the most commonly used solvent in the polyol method to prepare AgNWs, has subacute and chronic toxicity to humans. Glycerol, another polyol with higher hydroxyl content ratio, is far less harmful to humans and the environment and is commonly used in food and medicine. In this study, AgNRs were synthesized through an environmentally friendly glycerol-based cosolvent polyol method. The factors that influence the yield and average ring diameter were analyzed by the Taguchi design, and the growth mechanism was investigated by combining the results of structural characterization.

2. Materials and Methods

The AgNRs were synthesized with a glycerol-based cosolvent polyol method. Polyvinylpyrrolidone (PVP, Shanghai Maclean Biochemical Co., Ltd., Shanghai, China) of different molecular weights (Mw) (K60, Mw ~360,000, K90 Mw ~1,300,000) was added in glycerol (Shanghai Richjoint Co., Ltd., Shanghai, China). The mixture was then heated to 160 °C with vigorous stirring. After 30 min, 26.88 mM NaCl (Shanghai Richjoint Co., Ltd.) or tetramethylammonium chloride (TMA-C, Shanghai Maclean Biochemical Co., Ltd., Shanghai, China) and 9 mM NaBr (Shanghai Richjoint Co., Ltd.) or tetrapropylammonium bromide (TPA-B, Shanghai Maclean Biochemical Co., Ltd.) dissolved in glycerol were added to the mixture and stirred for another 30 min. Next, 137.95 mM $AgNO_3$ (Sinopharm Chemical Reagent Co., Ltd., Shanghai, China) dissolved in glycerol mixed with EG (Shanghai Richjoint Co., Ltd.) or deionized (DI) water was added dropwise to the mixture for 10 min. Then, the mixture was further kept at 160 °C for 1 h without stirring until the reaction finished. Under pressure conditions, the reaction was conducted in a hydrothermal kettle, and the reaction time was increased to 7 h. After cooling, the solution was purified with acetone and ethanol and suspended in ethanol. AgNR dispersion in ethanol was spin-coated on silicon wafers and then annealed at 130 °C for 15 min for characterization. SEM images were taken by Hitachi Regulus8100 field-emission SEM, and the quantity and ring diameter of silver nanorings were measured with Olympus 3D confocal laser scanning microscope (CLSM) OLS5000 (Olympus Corporation, Tokyo, Japan).

3. Results and Discussion

To fabricate AgNRs, a glycerol-based polyol method with a reaction time of 45 min was used. As shown in Figure 1a,b, AgNRs could hardly be observed except for several AgNRs before closure (Figure 1a) and some silver nanoarcs (Figure 1b). By completing the rest of the silver nanoarcs in Figure 1b, we could see that nearly all of them fitted well with the standard ring shape. This indicated that such arcs could grow into AgNRs if the reaction time was longer. The AgNRs seemingly grew along the track of the rings instead of becoming ring-shaped because of internal force after the two ends of AgNWs met. This suggests that AgNRs grow differently from AgNWs at the early stage, which implies that decahedral seeds [29] probably grow at a certain angle during the early nucleation period. To improve the yield of AgNRs, the reaction time was increased to 1 h. This resulted in AgNRs being synthesized with perfect geometry (Figure 1c), exhibiting the possibility of fabricating AgNRs with a higher yield.

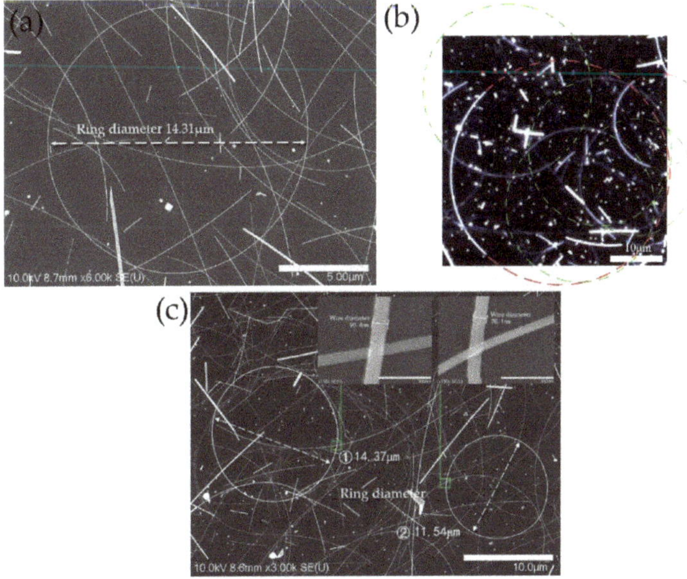

Figure 1. (**a**) SEM image of silver nanorings (AgNRs) before closure; (**b**) confocal laser scanning microscope (CLSM) image of silver nanoarcs; (**c**) SEM images of AgNRs after increasing reaction time to 1 h.

To increase the number of decahedral seeds that grow along a certain angle and finally raise the yield of AgNRs, a cosolvent method was introduced to provide different nucleation effect and reduction potential [30]. EG, the most commonly used solvent in the polyol method [31], and water, the solvent for preparing AgNWs by the hydrothermal method [32], were chosen as other solvents with less unpredictable influence on the reaction. To further select the main factors that affect the quantity and average ring diameter of the prepared AgNRs, an $L_8(2^7)$ orthogonal array was designed, and the corresponding results are presented in Table 1. The quantity of AgNRs was obtained by counting 50 images of 250 μm × 250 μm that were randomly selected by CLSM. The optimal level to be chosen and the rank of factors analyzed by the Taguchi design are listed in Table 2. The delta value in Table 2 represents the degree of influence on the result, and the rank values are the sequences of delta values. Both the concentration of PVP and the ratio of cosolvent strongly affected the quantity and average ring diameter, while the Mw of PVP and the type of chloride salt had less effect on the reaction. To increase the yield of AgNRs, the concentration of PVP, the ratio of cosolvent, and the type of cosolvent were chosen as the main factors for further experiment, while other factors were kept at the optimal level for greater quantity, as shown in Table 2. It is worth noting that the quantity listed in Table 1 is much less than that of our abovementioned work with a glycerol-based polyol method using NaCl, NaBr, and 82.8 mM PVP-K90 without any cosolvent and applying pressure. Among these factors, the addition of another solvent played an important role, as shown in Table 2, and a smaller ratio seemed to benefit the generation of AgNRs to some extent. Moreover, with excess addition of EG or DI water, the viscosity of the solution decreased sharply, which would suppress AgNWs from curving into the ring shape. Therefore, the ratio of cosolvent was reduced to ≈10% of the original.

Table 1. $L_8(2^7)$ orthogonal array and the corresponding result.

Sample	Mw of PVP	Concentration of PVP /mM	Chloride Salt	Bromide Salt	Type of Cosolvent	Ratio of Cosolvent/%	Pressure	Quantity of AgNRs	Average Ring Diameter/μm
1	K90	41.4	Na^+	Na^+	EG	5	with	28	14.77
2	K90	41.4	Na^+	TPA-	DI	10	without	5	10.38
3	K90	82.8	TMA-	Na^+	EG	10	without	76	14.65
4	K90	82.8	TMA-	TPA-	DI	5	with	75	14.80
5	K60	41.4	TMA-	Na^+	DI	5	without	24	13.25
6	K60	41.4	TMA-	TPA-	EG	10	with	10	12.40
7	K60	82.8	Na^+	Na^+	DI	10	with	14	15.00
8	K60	82.8	Na^+	TPA-	EG	5	without	108	14.90

Table 2. The optimal level and the rank of factors (larger the better).

Results	Factors	Mw of PVP	Concentration of PVP/mM	Chloride Salt	Bromide Salt	Type of Cosolvent	Ratio of Cosolvent/%	Pressure
Quantity of AgNRs	Level	K90	82.8	TMA-	TPA-	EG	5	without
	Delta	7.00	51.50	7.50	14.00	26.00	32.50	21.50
	Rank	7	1	6	5	3	2	4
Average ring diameter	Level	K60	82.8	TMA-	TPA-	DI	5	with
	Delta	0.24	2.14	0.01	1.30	0.82	1.32	0.95
	Rank	6	1	7	3	5	2	4

Another $L_{16}(2^1 4^2)$ orthogonal array was designed to study the optimal level of concentration of PVP, the ratio of cosolvent, and the type of cosolvent (Table 3). The analysis results for the quantity of AgNRs with the Taguchi design are shown in Figure 2 and Table 4. The mean quantity presented in Figure 2 was achieved by the following regression equation:

$$\begin{aligned}
&\text{Quantity}\\
&= 63.25 - 6.0\ \text{Concentration of PVP_1}\\
&+ 17.8\ \text{Concentration of PVP_2}\\
&+ 3.8\ \text{Concentration of PVP_3}\\
&- 8.0\ \text{Concentration of PVP_4}\\
&- 36.3\ \text{Ratio of cosolvent/\%_1}\\
&- 19.7\ \text{Ratio of cosolvent/\%_2}\\
&+ 56.0\ \text{Ratio of cosolvent/\%_3}\\
&+ 0.0\ \text{Ratio of cosolvent/\%_4}\\
&+ 1.88\ \text{Type of cosolvent_1}\\
&+ 1.88\ \text{Type of cosolvent_2}
\end{aligned} \tag{1}$$

The number behind the factor refers to its level, and it was the value employed to substitute into the equation. According to Figure 2, the best level of above 3 factors was achieved by adding 1% DI water into glycerol and 82.8 mM PVP. Only the p-value of the ratio of cosolvent was less than 0.05, which meant the ratio of cosolvent had a significant effect on the quantity. The mean quantity rose slightly from 27.00 to 43.50 as 0.5% cosolvent was added and then rose significantly to 119.25 when the ratio was 1%; it finally rapidly declined to 63.25. Despite the significant effect of ratio, the mean quantity hardly changed with a delta value of merely 3.75 when EG was replaced with DI water. Similarly, with the increase in concentration of PVP, the mean quantity rose slightly and reached a peak of 81.00 at 82.8 mM, then eventually decreased to 55.25.

Table 3. $L_{16}(2^1 4^2)$ orthogonal array and the corresponding result.

Sample	Concentration of PVP/mM	Ratio of Cosolvent/%	Type of Cosolvent	Quantity	Average Ring Diameter/μm
9	41.4	0	EG	32	12.80
10	41.4	0.5	EG	18	13.47
11	41.4	1	DI	140	15.74
12	41.4	1.5	DI	39	13.57
13	82.8	0	EG	42	15.18
14	82.8	0.5	EG	65	17.15
15	82.8	1	DI	162	16.56
16	82.8	1.5	DI	55	14.43
17	124.2	0	DI	27	15.45
18	124.2	0.5	DI	41	16.31
19	124.2	1	EG	97	16.72
20	124.2	1.5	EG	73	15.28
21	165.6	0	DI	7	15.65
22	165.6	0.5	DI	50	17.48
23	165.6	1	EG	78	16.49
24	165.6	1.5	EG	86	14.77

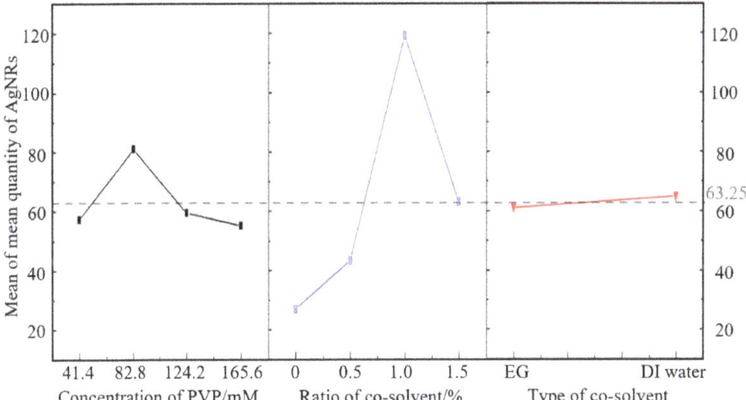

Figure 2. Plot of the main effects on the mean quantity of AgNRs.

Table 4. Analysis of variance with the general linear model for the quantity of AgNRs.

Source	DF	Adj SS	Adj MS	Delta	*F*-Value	*p*-Value
Concentration of PVP	3	1716.5	572.17	25.75	0.79	0.531
Ratio of cosolvent	3	19360.5	6453.5	92.25	8.96	0.006
Type of cosolvent	1	56.3	56.25	3.75	0.08	0.787
Error	8	5761.7	720.22	–	–	–
Total	15	26895.0	–	–	–	–

Correspondingly, the mean ring diameter was analyzed by the following regression equation:

$$\begin{aligned} &\text{Ring diameter} \\ &= 15.441 \\ &- 1.546 \text{ Concentration of PVP\_1} \\ &+ 0.389 \text{ Concentration of PVP\_2} \\ &- 0.499 \text{ Concentration of PVP\_3} \\ &- 0.657 \text{ Concentration of PVP\_4} \\ &- 0.671 \text{ Ratio of cosolvent/\%\_1} \\ &- 0.662 \text{ Ratio of cosolvent/\%\_2} \\ &+ 0.937 \text{ Ratio of cosolvent/\%\_3} \\ &+ 0.928 \text{ Ratio of cosolvent/\%\_4} \\ &- 0.208 \text{ Type of cosolvent\_1} \\ &+ 0.208 \text{ Type of cosolvent\_2} \end{aligned} \quad (2)$$

As can be seen from the results in Figure 3 and Table 5, the optimal level was 165.6 mM PVP and the addition of 1% DI water into glycerol. Like the quantity, the type of cosolvent had little to do with the mean ring diameter. However, the p-value of both the ratio of cosolvent and concentration of PVP was less than 0.05, showing a significant effect. However, the mean ring diameter rose more gently as the ratio of cosolvent increased. It rose greatly when the concentration of PVP increased to 82.8 mM, then rose mildly as more PVP was introduced. Putting the two figures together, the peak of the ratio of cosolvent appeared at 1%, and no matter how much EG or DI water was added, both of them produced a noticeable effect without much difference.

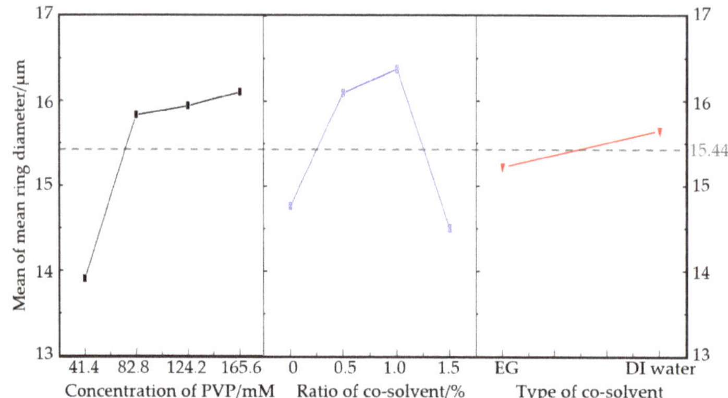

Figure 3. Plot of the main effects for mean ring diameter.

Table 5. Analysis of variance with the general linear model for ring diameter.

Source	DF	Adj SS	Adj MS	Delta	F-Value	p-Value
Concentration of PVP	3	12.8857	4.2952	2.20	9.05	0.006
Ratio of cosolvent	3	10.5079	3.5026	1.86	7.38	0.011
Type of cosolvent	1	0.6931	0.6931	0.42	1.46	0.261
Error	8	3.7951	0.4744	–	–	–
Total	15	27.8817	–	–	–	–

The EG or DI water added to glycerol to form the cosolvent had some features distinct from glycerol. To explain the mechanism of the effect of the cosolvent on the synthesis of AgNRs, a schematic is illustrated in Figure 4. On the one hand, EG and water were less reductive than glycerol. At the early nucleation period of AgNRs, due to the concentration fluctuation, a small amount of EG or water adsorbed on several (111) planes of decahedral seeds diluted the concentration of glycerol in the micro area, which provided slower growth rate than those reduced by glycerol on these planes. As a result, these seeds grew with a certain angle. Meanwhile, EG or water on these planes was localized on account of the supplement of the reactant. Therefore, the growth direction of AgNRs was maintained, and the two ends finally met and combined to become rings. When excess EG or water was added, the difference between the planes became smaller because the concentration of each area was similar, resulting in the growth of AgNWs instead of AgNRs [33]. On the other hand, the viscosity of EG and water is far less than that of glycerol. Although glycerol has stronger reducibility, its high viscosity slowed down the migration velocity, resulting in an incomplete reaction. By adding an appropriate ratio of EG or water, the viscosity could fall to a more suitable level [34]. What is more, they could serve as the medium to promote the proton hopping on the surface of the reactant, facilitating the charge transfer process [35]. The cosolvent system also provided better solubility for $AgNO_3$ than glycerol. All these features increased the ion mobility in the solution and then facilitated the process of reaction. However, when the ratio was too high, the condition beneficial for the growth of AgNRs provided by glycerol was disturbed, causing a reduction of AgNRs.

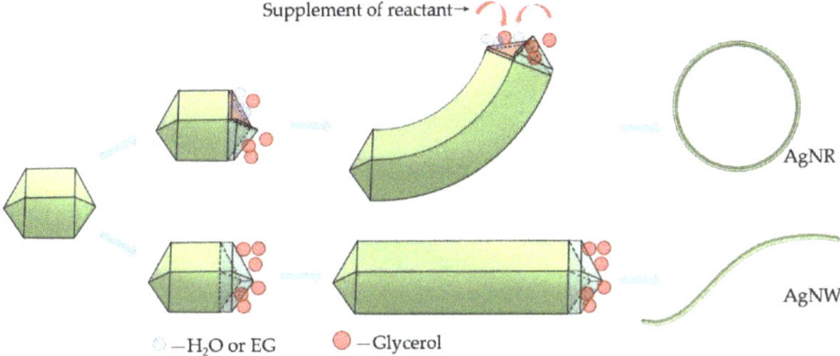

Figure 4. Schematic of the growth of AgNR with glycerol-based cosolvent polyol method.

In the case of added PVP, which acts as the capping agent [36], when the concentration increased to 82.8 mM, the effect of restricting the wire diameter by covering the surface of AgNRs became significant. More early seeds were controlled to grow into AgNRs or AgNWs, resulting in higher precursor utilization. In other words, AgNRs with larger ring diameter and higher yield tended to be generated. With excess PVP, such capping effect was almost saturated [37], showing that the ring diameter only slightly increased. Besides, as a material that easily absorbs moisture, excess PVP would absorb the water or EG molecule and restrict it from diffusion, then nullify the cosolvent effect mentioned above, leading to a low yield of AgNRs. With the optimal level of factors for a greater quantity of AgNRs (same as Sample 15), AgNRs were synthesized with ring diameter ranging from 7.17 to 42.94 μm and wire diameter of about 76 nm, as shown in Figure 5, and the quantity increased significantly under such conditions. Figure 6 compares the UV−vis adsorption spectra of AgNRs and AgNWs; the AgNWs here were synthesized using EG as solvent. As can be seen, the adsorption spectra of both materials had two surface plasmon resonance (SPR) peaks at ~361 nm and ~373 nm, but the spectra of AgNWs had another peak around ~390 nm with high intensity, which corresponded to transverse plasmon resonance [38,39]. Such a strong absorption peak at ~390 nm could influence the optical properties of the film; hence, AgNRs thin films have the potential to substitute AgNWs as TCF materials. This work provides a new, environmentally friendly method to fabricate AgNRs with a solvent consisting of glycerol and water instead of EG [40]. This makes the synthesis of AgNRs a different process from the former method, which is a by-product of AgNWs [20,21].

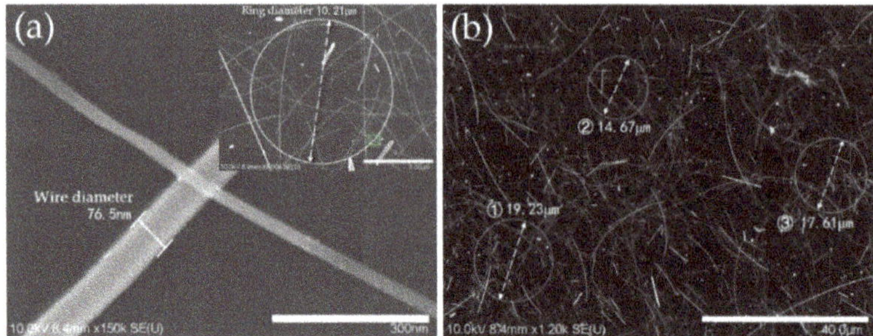

Figure 5. SEM images of AgNRs synthesized with the optimal level for greater quantity of AgNRs at high (**a**) and low (**b**) magnification.

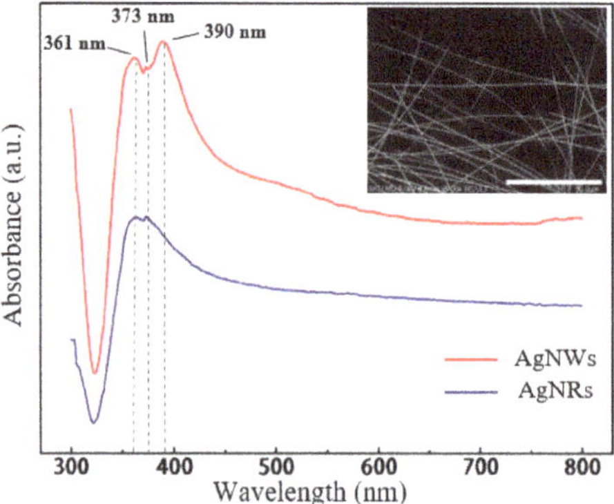

Figure 6. UV−vis adsorption spectra of AgNRs and silver nanowires (AgNWs). The inset is the SEM image of AgNWs.

4. Conclusions

In conclusion, we have reported a modified, environmentally friendly polyol method to synthesize AgNRs with higher yield using a cosolvent consisting of glycerol and DI water, with other factors determined by employing the Taguchi design. Structural characterization revealed that AgNRs grew along the track of rings from the early stage of the reaction. To achieve high yield of AgNRs, the addition of DI water or EG was significant, which acted as the medium to increase ion mobility and another reductant that is less reductive to control the growth direction. Otherwise, the ratio of it should be precisely controlled. As a result, AgNRs with ring diameter ranging from 7.17 to 42.94 μm and wire diameter of about 76 nm were synthesized with higher yield. The research indicates the potential to synthesize pure AgNRs and to further apply them in flexible WBG semiconductor devices.

Author Contributions: Conceptualization, Z.L. and H.N.; methodology, R.Y. and D.G.; validation, Z.L., X.F., and J.C.; resources, J.P. and R.Y.; data curation, P.X., Y.W., and X.Z.; writing—original draft preparation, Z.L. and D.G.; writing—review and editing, J.P. and H.N. All authors have read and agreed to the published version of the manuscript.

Funding: This work was supported by Key Area Research and Development Program of Guangdong Province (No.2019B010934001), National Natural Science Foundation of China (Grant No.51771074, 61804029 and 61574061), Major Integrated Projects of National Natural Science Foundation of China (Grant No.U1601651), Basic and Applied Basic Research Major Program of Guangdong Province (No.2019B030302007), the Guangdong Natural Science Foundation (No.2018A030310353), Science and Technology Project of Guangzhou (No.201904010344), Fundamental Research Funds for the Central Universities (No.2019MS012), Open Project of Guangdong Province Key Lab of Display Material and Technology (No.2017B030314031), and Open Fund of Foshan University.

Conflicts of Interest: The authors declare no conflict of interest.

References

1. Maita, F.; Maiolo, L.; Pecora, A.; Minotti, A.; Fortunato, G.; Smecca, E.; Alberti, A. Low-temperature flexible piezoelectric AlN capacitor integrated on ultra-flexible poly-Si TFT for advanced tactile sensing. *Sensors* **2014**, *2014*, 1730–1733.
2. John, R.A.; Ko, J.; Kulkarni, M.R.; Tiwari, N.; Chien, N.A.; Ing, N.G.; Leong, W.L.; Mathews, N. Flexible Ionic-Electronic Hybrid Oxide Synaptic TFTs with Programmable Dynamic Plasticity for Brain—Inspired Neuromorphic Computing. *Small* **2017**, *13*, 1701193. [CrossRef] [PubMed]
3. Chung, K.; Yoo, H.; Hyun, J.K.; Oh, H.; Tchoe, Y.; Lee, K.; Baek, H.; Baek, M.; Yi, G.C. Flexible GaN light—Emitting diodes using GaN microdisks epitaxial laterally overgrown on graphene dots. *Adv. Mater.* **2016**, *28*, 7688–7694. [CrossRef] [PubMed]
4. Liu, Z.; Chong, W.C.; Wong, K.M.; Lau, K.M. GaN-based LED micro-displays for wearable applications. *Microelectron. Eng.* **2015**, *148*, 98–103. [CrossRef]
5. Wang, Y.; Zheng, D.; Li, L.; Zhang, Y. Enhanced Efficiency of Flexible GaN/Perovskite Solar Cells Based on the Piezo-Phototronic Effect. *ACS Appl. Energy Mater.* **2018**, *1*, 3063–3069. [CrossRef]
6. Chen, Y.; Zhang, X.; Xie, Z. Flexible nitrogen doped SiC nanoarray for ultrafast capacitive energy storage. *ACS Nano* **2015**, *9*, 8054–8063. [CrossRef]
7. Zhang, P.; Wyman, I.; Hu, J.; Lin, S.; Zhong, Z.; Tu, Y.; Huang, Z.; Wei, Y. Silver nanowires: Synthesis technologies, growth mechanism and multifunctional applications. *Mater. Sci. Eng. B* **2017**, *223*, 1–23. [CrossRef]
8. Kumar, A.; Zhou, C. The race to replace tin-doped indium oxide: Which material will win? *ACS Nano* **2010**, *4*, 11–14. [CrossRef]
9. He, M.; Jung, J.; Qiu, F.; Lin, Z. Graphene-based transparent flexible electrodes for polymer solar cells. *J. Mater. Chem.* **2012**, *22*, 24254–24264. [CrossRef]
10. Ge, J.; Cheng, G.; Chen, L. Transparent and flexible electrodes and supercapacitors using polyaniline/single-walled carbon nanotube composite thin films. *Nanoscale* **2011**, *3*, 3084–3088. [CrossRef]
11. Schneider, J.; Rohner, P.; Thureja, D.; Schmid, M.; Galliker, P.; Poulikakos, D. Electrohydrodynamic nanodrip printing of high aspect ratio metal grid transparent electrodes. *Adv. Funct. Mater.* **2016**, *26*, 833–840. [CrossRef]
12. Park, J.H.; Lee, D.Y.; Kim, Y.H.; Kim, J.K.; Lee, J.H.; Park, J.H.; Lee, T.W.; Cho, J.H. Flexible and transparent metallic grid electrodes prepared by evaporative assembly. *ACS Appl. Mater. Interfaces* **2014**, *6*, 12380–12387. [CrossRef] [PubMed]
13. Ran, Y.; He, W.; Wang, K.; Ji, S.; Ye, C. A one-step route to Ag nanowires with a diameter below 40 nm and an aspect ratio above 1000. *Chem. Commun.* **2014**, *50*, 14877–14880. [CrossRef] [PubMed]
14. Chen, Z.; Ye, S.; Stewart, I.E.; Wiley, B.J. Copper nanowire networks with transparent oxide shells that prevent oxidation without reducing transmittance. *ACS nano* **2014**, *8*, 9673–9679. [CrossRef] [PubMed]
15. Liu, Y.; Zhang, J.; Gao, H.; Wang, Y.; Liu, Q.; Huang, S.; Guo, C.F.; Ren, Z. Capillary-force-induced cold welding in silver-nanowire-based flexible transparent electrodes. *Nano Lett.* **2017**, *17*, 1090–1096. [CrossRef]
16. Yao, R.; Li, X.; Li, Z.; Shi, M.; Zhou, S.; Yuan, W.; Ning, H.; Peng, J.; Fang, Z. Fabrication and Properties of Silver Nanowire Flexible Transparent. In Proceedings of the 19th International Conference on Electronic Packaging Technology (ICEPT), Shanghai, China, 8–11 August 2018; pp. 454–456.
17. Zhang, Y.; Guo, J.; Xu, D.; Sun, Y.; Yan, F. One-pot synthesis and purification of ultralong silver nanowires for flexible transparent conductive electrodes. *ACS Appl. Mater. Interfaces* **2017**, *9*, 25465–25473. [CrossRef]
18. Li, Z.; Ning, H.; Li, X.; Tao, R.; Liu, X.; Cai, W.; Chen, J.; Wang, L.; Yao, R.; Peng, J. Synthesis of Silver Nanowires by Mixing Different Nucleating Control Agents. *Mater.-Rep.* **2019**, *33*, 303–306.
19. Ge, Y.; Duan, X.; Zhang, M.; Mei, L.; Hu, J.; Hu, W.; Duan, X. Direct room temperature welding and chemical protection of silver nanowire thin films for high performance transparent conductors. *J. Am. Chem. Soc.* **2017**, *140*, 193–199. [CrossRef]
20. Zhou, L.; Fu, X.F.; Yu, L.; Zhang, X.; Yu, X.F.; Hao, Z.H. Crystal structure and optical properties of silver nanorings. *Appl. Phys. Lett.* **2009**, *94*, 153102. [CrossRef]
21. Gong, H.M.; Zhou, L.; Su, X.R.; Xiao, S.; Liu, S.D.; Wang, Q.Q. Illuminating dark plasmons of silver nanoantenna rings to enhance exciton–plasmon interactions. *Adv. Funct. Mater.* **2009**, *19*, 298–303. [CrossRef]

22. Azani, M.R.; Hassanpour, A.; Tarasevich, Y.Y.; Vodolazskaya, I.V.; Eserkepov, A.V. Transparent electrodes with nanorings: A computational point of view. *J. Appl. Phys.* **2019**, *125*, 234903. [CrossRef]
23. Zhang, Q.; Shan, X.Y.; Zhou, L.; Zhan, T.R.; Wang, C.X.; Li, M.; Jia, J.F.; Zi, J.; Wang, Q.Q.; Xue, Q.K. Scattering focusing and localized surface plasmons in a single Ag nanoring. *Appl. Phys. Lett.* **2010**, *97*, 261107. [CrossRef]
24. Zhang, Y.; Guo, J.; Xu, D.; Sun, Y.; Yan, F. Synthesis of ultrathin semicircle-shaped copper nanowires in ethanol solution for low haze flexible transparent conductors. *Nano Res.* **2018**, *11*, 3899–3910. [CrossRef]
25. Zhan, Y.; Lu, Y.; Peng, C.; Lou, J. Solvothermal synthesis and mechanical characterization of single crystalline copper nanorings. *J. Cryst. Growth* **2011**, *325*, 76–80. [CrossRef]
26. Li, Z.J.; Chen, X.L.; Li, H.J.; Tu, Q.Y.; Yang, Z.; Xu, Y.P.; Hu, B.Q. Synthesis and Raman scattering of GaN nanorings, nanoribbons and nanowires. *Appl. Phys. A* **2001**, *72*, 629–632. [CrossRef]
27. Duan, J.; Yang, S.; Liu, H.; Gong, J.; Huang, H.; Zhao, X.; Tang, J.; Zhang, R.; Du, Y. AlN nanorings. *J. Cryst. Growth* **2005**, *283*, 291–296. [CrossRef]
28. Hughes, W.L.; Wang, Z.L. Controlled synthesis and manipulation of ZnO nanorings and nanobows. *Appl. Phys. Lett.* **2005**, *86*, 043106. [CrossRef]
29. Da Silva, R.R.; Yang, M.; Choi, S.I.; Chi, M.; Luo, M.; Zhang, C.; Li, Z.Y.; Camargo, P.H.C.; Ribeiro, S.J.L.; Xia, Y. Facile synthesis of sub-20 nm silver nanowires through a bromide-mediated polyol method. *ACS Nano* **2016**, *10*, 7892–7900. [CrossRef]
30. Yin, Z.; Song, S.K.; You, D.J.; Ko, Y.; Cho, S.; Yoo, J.; Park, S.Y.; Piao, Y.; Chang, S.T.; Kim, Y.S. Novel synthesis, coating, and networking of curved copper nanowires for flexible transparent conductive electrodes. *Small* **2015**, *11*, 4576–4583. [CrossRef]
31. Schuette, W.M.; Buhro, W.E. Silver chloride as a heterogeneous nucleant for the growth of silver nanowires. *Acs Nano* **2013**, *7*, 3844–3853. [CrossRef]
32. Bari, B.; Lee, J.; Jang, T.; Won, P.; Ko, S.H.; Alamgir, K.; Arshad, M.; Guo, L.J. Simple hydrothermal synthesis of very-long and thin silver nanowires and their application in high quality transparent electrodes. *J. Mater. Chem. A* **2016**, *4*, 11365–11371. [CrossRef]
33. Yin, Z.; Song, S.K.; Cho, S.; You, D.J.; Yoo, J.; Chang, S.T.; Kim, Y.S. Curved copper nanowires-based robust flexible transparent electrodes via all-solution approach. *Nano Res.* **2017**, *10*, 3077–3091. [CrossRef]
34. Abbasi, N.M.; Wang, L.; Yu, H.; Akram, M.; Khalid, H.; Yongshen, C.; Deng, Z. Glycerol and Water Mediated Synthesis of Silver Nanowires in the Presence of Cobalt Chloride as Growth Promoting Additive. *J. Inorg. Organomet. Polym. Mater.* **2016**, *26*, 680–690. [CrossRef]
35. Yang, C.; Tang, Y.; Su, Z.; Zhang, Z.; Fang, C. Preparation of silver nanowires via a rapid, scalable and green pathway. *J. Mater. Sci. Technol.* **2015**, *31*, 16–22. [CrossRef]
36. Hwang, J.; Shim, Y.; Yoon, S.M.; Lee, S.H.; Park, S.H. Influence of polyvinylpyrrolidone (PVP) capping layer on silver nanowire networks: Theoretical and experimental studies. *RSC Adv.* **2016**, *6*, 30972–30977. [CrossRef]
37. Song, Y.J.; Wang, M.; Zhang, X.Y.; Wu, J.Y.; Zhagn, T. Investigation on the role of the molecular weight of polyvinyl pyrrolidone in the shape control of high-yield silver nanospheres and nanowires. *Nanoscale Res. Lett.* **2014**, *9*, 17. [CrossRef]
38. Azani, M.R.; Hassanpour, A. Synthesis of Silver Nanowires with Controllable Diameter and Simple Tool to Evaluate their Diameter, Concentration and Yield. *Chem. Select.* **2019**, *4*, 2716–2720. [CrossRef]
39. Niu, Z.; Cui, F.; Kuttner, E.; Xie, C.; Chen, H.; Sun, Y.; Dehestani, A.; Schierle-Arndt, K.; Yang, P. Synthesis of silver nanowires with reduced diameters using benzoin-derived radicals to make transparent conductors with high transparency and low haze. *Nano Lett.* **2018**, *18*, 5329–5334. [CrossRef]
40. Azani, M.R.; Hassanpour, A. Silver nanorings: New generation of transparent conductive films. *Chem. Eur. J.* **2018**, *24*, 19195–19199. [CrossRef]

 *micromachines*

Article

Gallium Nitride (GaN) High-Electron-Mobility Transistors with Thick Copper Metallization Featuring a Power Density of 8.2 W/mm for Ka-Band Applications

Y. C. Lin [1], S. H. Chen [2], P. H. Lee [2], K.H. Lai [2], T. J. Huang [3], Edward Y. Chang [3] and Heng-Tung Hsu [3,*]

1 Department of Materials Science and Engineering, National Chiao Tung University, Hsinchu 300, Taiwan; nctulin@yahoo.com.tw
2 Department of Electronics Engineering, National Chiao Tung University, Hsinchu 300, Taiwan; tommy4920@yahoo.com.tw (S.H.C.); tom2657140@gmail.com (P.H.L.); khlai.mse03@g2.nctu.edu.tw (K.H.L.)
3 International College of Semiconductor Technology, National Chiao Tung University, Hsinchu 300, Taiwan; joehuang@nctu.edu.tw (T.J.H.); edc@nctu.edu.tw (E.Y.C.)
* Correspondence: hthsu@nctu.edu.tw

Received: 19 January 2020; Accepted: 19 February 2020; Published: 21 February 2020

Abstract: Copper-metallized gallium nitride (GaN) high-electron-mobility transistors (HEMTs) using a Ti/Pt/Ti diffusion barrier layer are fabricated and characterized for Ka-band applications. With a thick copper metallization layer of 6.8 μm adopted, the device exhibited a high output power density of 8.2 W/mm and a power-added efficiency (PAE) of 26% at 38 GHz. Such superior performance is mainly attributed to the substantial reduction of the source and drain resistance of the device. In addition to improvement in the Radio Frequency (RF) performance, the successful integration of the thick copper metallization in the device technology further reduces the manufacturing cost, making it extremely promising for future fifth-generation mobile communication system applications at millimeter-wave frequencies.

Keywords: high-electron-mobility transistors; copper metallization; millimeter wave

---

1. Introduction

Gallium nitride (GaN) high-electron-mobility transistors (HEMTs) have become one of the most popular devices for high-frequency and high-power applications in recent years. Compared to traditional silicon devices, GaN material has several remarkable properties, such as better electron mobility at high electric field, wider energy bandgap (3.4 eV), higher breakdown electric field and higher saturation electron drift velocity [1–3]. Such excellent material properties have made AlGaN/GaN devices the streamline technology for high-frequency and high-power applications for next-generation wireless communication systems at millimeter-wave frequencies [4–6].

For the allocation of sufficient bandwidth to meet the stringent demand of ultrahigh data rates, operating at millimeter-wave frequencies has been a common practice for next-generation wireless communication networks. One of the main challenging issues is the unavoidable higher level of signal attenuation in free space as well as the losses induced in the transmission media. In that sense, device performance is strongly affected by the skin effect at high operating frequencies since the parasitic resistance tends to increase due to the limited cross-sectional area for current flow. Such parasitic resistance could possibly be minimized through thick metal deposition for interconnects at the device level.

Gold is usually selected as the interconnect material for III–V devices. However, the price of the material makes production cost inevitably high, making commercialization difficult. To address this issue, Au-free process technology was developed [7], which demonstrated CMOS-compatible AlGaN/GaN Metal-Insulator-Semiconductor HEMT (MIS-HEMT) device configuration for power electronics applications. In [8], an Au-free process was also reported in a thick metal deposition process using aluminum- and copper-based material as the interconnects. Detailed process steps overcoming the main fabrication challenges were included, and power devices with enhancement-mode and depletion-mode performances were demonstrated. In our approach, a thick copper metallization process is adopted as an alternative in the GaN device because copper has lower resistivity and higher thermal conductivity with lower cost than gold. Therefore, copper is considered a good candidate to replace gold for high-frequency device interconnection. Nevertheless, copper material suffers from the interdiffusion effect. A high-quality diffusion barrier of copper metallization is then required. Some reports showed that TaN, TiN, WNx and Pt can be used as diffusion barriers for copper metallization [9–12]. Among them, Pt material has the lowest resistivity. Therefore, a Pt diffusion barrier is adopted due to its extremely low resistivity, low electrical degradation features and better temperature stability in this work.

The objective of this study focuses on the investigation of the effect of thick copper metallization on device performance at millimeter-wave frequencies. In the following sections, device performance based on small-signal and large-signal characterization will be compared. In order to quantize the effect of the thick copper metallization, we have also extracted the corresponding parameters of the small-signal equivalent circuit for comparison purposes.

2. Device Fabrication

From the top to the bottom, the epitaxial layer structure of our device consists of an AlGaN barrier layer, an AlN spacer layer, the GaN channel layer, a thick GaN buffer layer and the SiC substrate. The device process can be divided into four major parts including the ohmic contact, mesa isolation, gate formation and thick copper metallization. First, the fabrication process started with ohmic contact forming. The ohmic region was defined by the mask aligner using the photoresist; then, deposition of the Ti/Al/Ni/Au multilayer was conducted by e-gun evaporation, followed by the lift-off process. The multilayer metal was then annealed at 850 °C for 30 s in an N_2 ambient environment by a rapid thermal annealing system (RTA). The device mesa isolation was then performed, which defined the active region by lithography; then, an inductively coupled plasma (ICP) machine was used with Cl_2 in an Ar ambient to etch the AlGaN and GaN layer for around 180 nm. For device gate formation, two-step e-beam lithography with spatial offset techniques were applied to achieve the Γ-gate structure and small gate length. A 100 nm SiNx passivation layer was deposited through plasma-enhanced chemical vapor deposition (PECVD). Then, the ditch for the gate stem was fabricated by e-beam lithography and SiNx etching by ICP. The second e-beam lithography pattern shifted 100 nm away from the previous location, which formed an overlap region. The size of the overlap region eventually determined the device gate length, which was around 90 nm in this study. To further enhance the gate controllability and improve the device transconductance, gate recess was performed. The gate metal deposition was then formed by Ni/Au metal stacks, followed by a lift-off process. Finally, a 100 nm SiNx layer was deposited with the nitride via the fabricated device pad region.

The thick copper metallization process started with triple photoresist coating using AZ5214E to reach a minimum thickness of 9 μm for a thick copper lift-off process. Then, exposure and development were carried out to define the pattern. Finally, the thick metal stacks with Ti (30 nm)/Pt (40 nm)/Ti (10 nm)/Cu (6800 nm) structure and thickness were deposited by e-gun evaporation. Ti layers were used to enhance the adhesion ability between Pt and ohmic as well as the adhesion of Cu–Pt interface in this study [13–15].

Figure 1 shows the overall epitaxial configuration and the device structure. The scanning electron microscope (SEM) image of the gate was also included in the figure. The source-to-drain distance of the device was 2 μm with the gate positioned at the center. The schematic of thick copper metallization technology and the SEM image of thick copper metallization cross-section are shown in Figure 2.

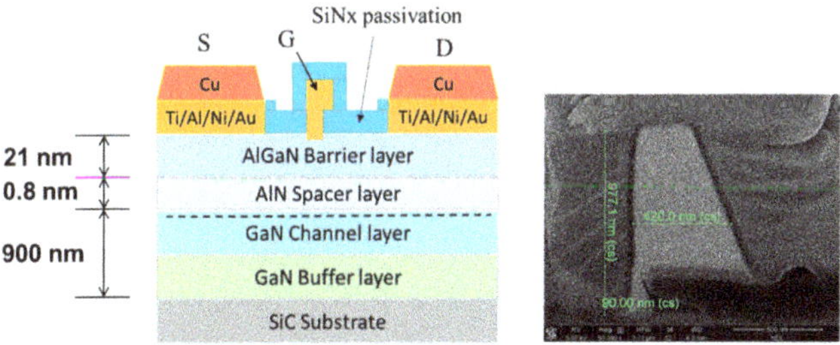

Figure 1. AlGaN/GaN high-electron-mobility transistors (HEMTs) epitaxial configuration and device structure with scanning electron microscope (SEM) image of the gate.

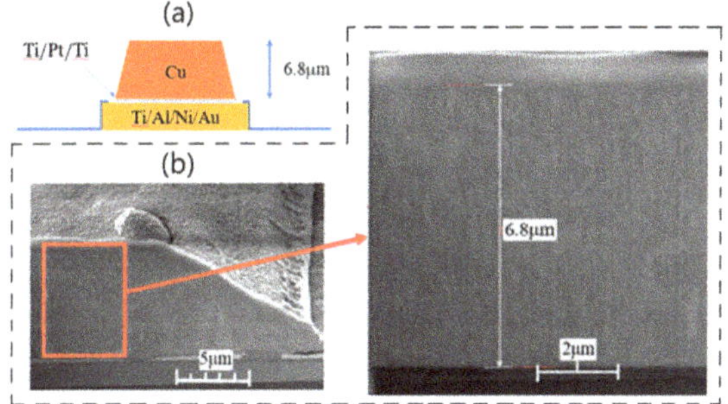

Figure 2. (**a**) Schematic of thick copper metallization structure. (**b**) SEM image of cross-section of thick copper metallization for GaN HEMT.

3. Results and Discussions

The two-port network analysis method with a small signal model was used to analyze the relationship between the drain–source current (I_{DS}) and the transconductance (G_m) versus source resistance and drain resistance. The DC and RF measurement results of devices with and without thick copper metallization were then compared. By utilizing load-pull measurement methodology, output power and power-added efficiency (PAE) characteristics could be obtained [16]. The impact of gate width on the device performance are then able to be discussed.

3.1. Two-Port Network Analysis

With a two-port network, the small signal model of the AlGaN/GaN HEMT device can be depicted as in Figure 3 [17,18]. Utilizing the y-parameter analysis, the drain–source current (I_{DS}) and the transconductance (G_m) of the device can be derived as:

$$I_{DS} = \frac{y_{21}v'_i + y_{22}v'_o}{1 + y_{21}R'_S + y_{22}R'_S + y_{22}R'_D} \tag{1}$$

$$G_m = \frac{dI_{DS}}{dv'_i} = \frac{y_{21}}{1 + y_{21}R'_S + y_{22}R'_S + y_{22}R'_D} \tag{2}$$

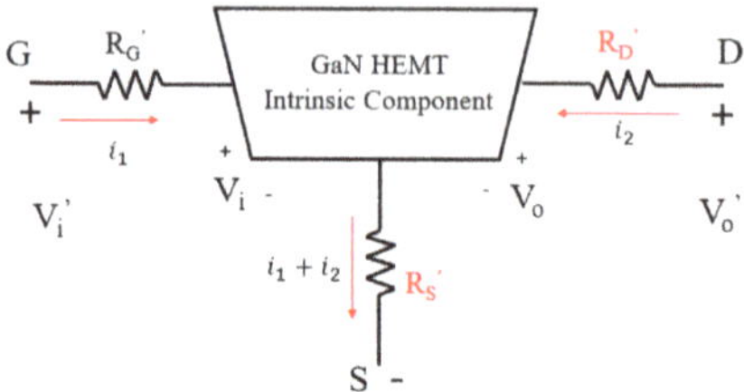

Figure 3. Small signal model of GaN HEMT.

It is apparent from Equations (1) and (2) that the decrease of R_S' and R_D' results in the increase of I_{DS} and G_m levels.

3.2. DC Characteristics

The contact resistance for devices with and without thick copper metallization was measured through the transmission line method (TLM). Figure 4 shows the measurement results. The least squares regression method was adopted to find the best fit for the sets of measured data points. As expected, linear behavior was obtained, and the intersecting points with the vertical axis were extracted as the contact resistance. It was observed that the contact resistances of the samples with (green line) and without (red line) thick copper metallization were 2.5×10^{-6} Ω·cm^2 and 1.7×10^{-6} Ω·cm^2, respectively, leading to a difference in metal resistance of 0.96 Ω between the cases with and without thick copper metallization.

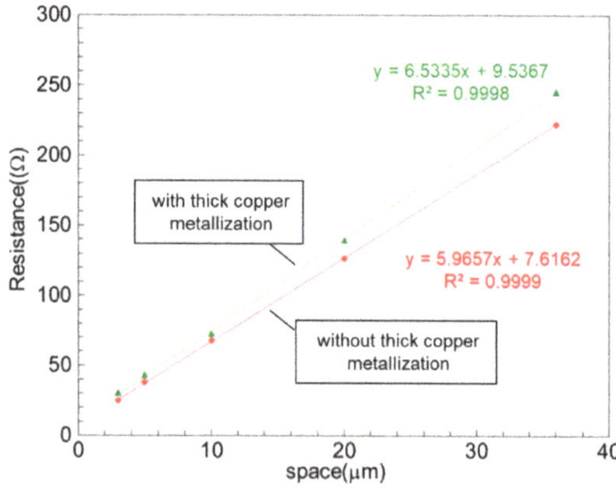

Figure 4. Transmission line method (TLM) measurement results before and after thick copper metallization.

To evaluate the effect of the thick copper metallization on device performances, a test device with a total gate periphery of 40 μm—which was composed of two fingers, with each finger of 20 μm in length—was fabricated. The DC characteristics, including current–voltage (I_{DS}–V_{GS}) relationship and the transfer curve (G_m–V_{GS}) of the device with and without thick Cu metallization, are plotted in Figures 5 and 6, respectively. As observed, the device without thick copper metallization exhibited an I_{DS} of 1010 mA/mm and a maximum G_m of 350 mS/mm at V_{DS} = 10 V. For the device with thick copper metallization at V_{DS} = 10 V, the measured I_{DS} was 1110 mA/mm and the maximum G_m was 380 mS/mm. Such improvement in the DC characteristics was mainly attributed to the reduction in the source and drain parasitic resistance contributed by the thick copper metallization. Figure 7 shows the comparison of DC I–V curves for the device with and without thick copper metallization. The on-resistance (RON) was extracted to be 1.53 Ω·mm for the device with thick copper metallization and 1.67 Ω·mm for the device without thick copper metallization.

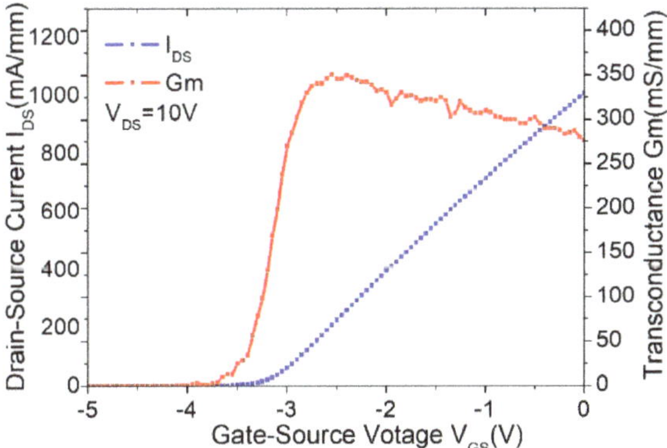

Figure 5. DC characteristic of the 2 × 20 μm device without thick copper metallization.

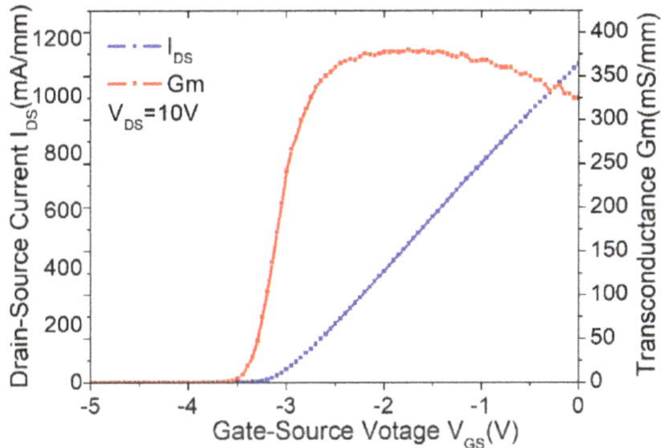

Figure 6. DC characteristic of the 2 × 20 μm device with thick copper metallization.

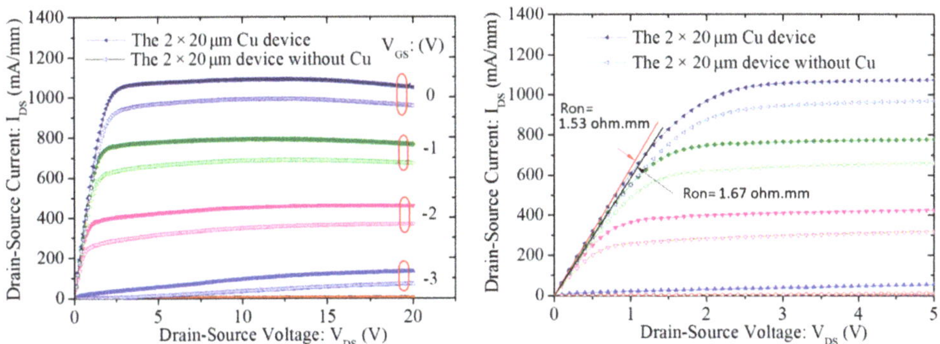

Figure 7. The comparison of DC I–V curves for the device with and without thick copper metallization.

3.3. RF Characteristics

Figure 8 shows the comparison of the measured small-signal performance for the cases with and without thick copper metallization. The measurement was performed using a vector signal analyzer in an on-wafer probing system up to 67 GHz. With the DC bias set at the maximum transconductance, the unit-current-gain cutoff frequency (fT) and the maximum oscillation frequency (fmax) were also extracted for the extrinsic device without de-embedding. As observed, the fT (fmax) of the device with thick copper metallization was 42 GHz (115 GHz) compared to that of 32 GHz (100 GHz) for the device without thick copper metallization.

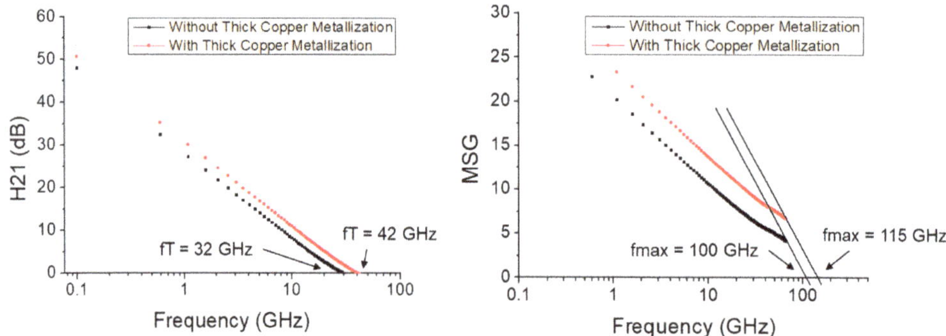

Figure 8. Measured small-signal performance up to 67 GHz with the extracted fT and fmax values for the devices with and without thick copper metallization.

To further quantize the effect of the thick copper metallization, we performed the extraction of the parameters of the small-signal equivalent circuit for the devices following the same procedures outlined in [19]. All the corresponding parasitic components extrinsic to the active region of the device were extracted using both cold forward and cold pinchoff bias conditions as defined. Figure 9 shows the S-parameters measured and predicted using the small-signal equivalent circuit model. Good agreement between the measurement and prediction was obtained up to 67 GHz. The corresponding parameter values were also included for comparison. As observed, devices with thick copper metallization generally exhibited lower parasitic resistance values, which contributed to the higher fmax measured. Additionally, lower gate capacitances were extracted from the measurement for the device with thick copper metallization.

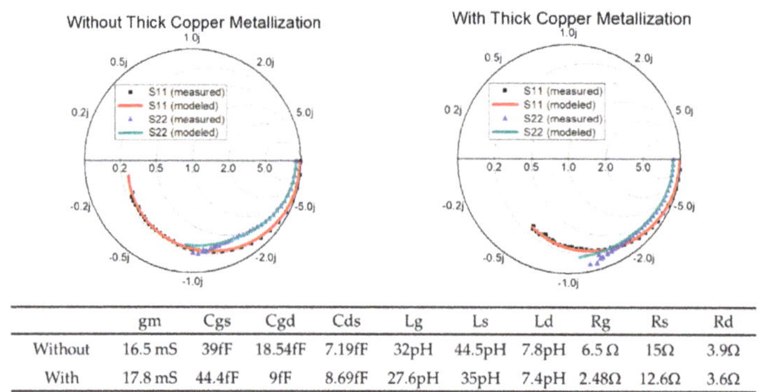

	gm	Cgs	Cgd	Cds	Lg	Ls	Ld	Rg	Rs	Rd
Without	16.5 mS	39fF	18.54fF	7.19fF	32pH	44.5pH	7.8pH	6.5 Ω	15Ω	3.9Ω
With	17.8 mS	44.4fF	9fF	8.69fF	27.6pH	35pH	7.4pH	2.48Ω	12.6Ω	3.6Ω

Figure 9. Measured and predicted S-parameters for the 2 × 20 μm device with (**right**) and without (**left**) thick copper metallization. The small-signal circuit model was extracted using the procedure defined in [19].

On-wafer load-pull characterization (continuous mode) was also performed to investigate the power performance at 38 GHz using an automatic tuning system; the measurement results for the device without and with thick copper metallization are shown in Figures 10 and 11, respectively. The output power and power-added efficiency (PAE) were compared at 3-dB gain compression with respect to the small-signal gain. With the drain bias set at 20 V, the gate bias for the device with thick copper metallization was set at −2 V and that for the one without thick copper metallization was −1.7 V. The corresponding quiescent drain current was 21 mA for the device with thick copper metallization and 19 mA for the one without thick copper metallization, with both being close to Class A operation. The measured power density and PAE for the device with thick copper metallization were 5.9 W/mm and 28.7%, compared to those of 4.5 W/mm and 24.8%, respectively, for the one without thick copper metallization.

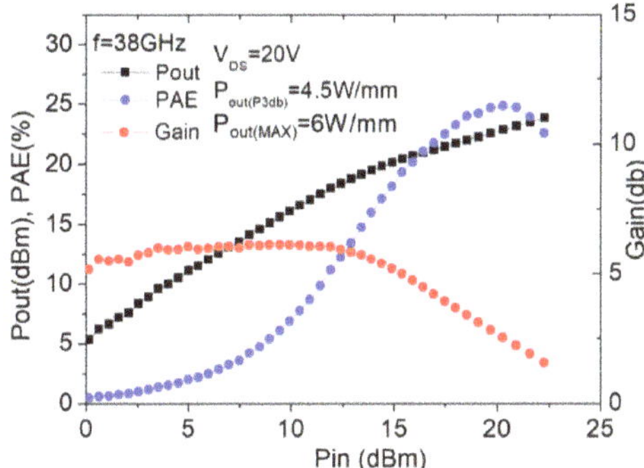

Figure 10. Large-signal performance of the 2 × 20 μm device without copper metallization.

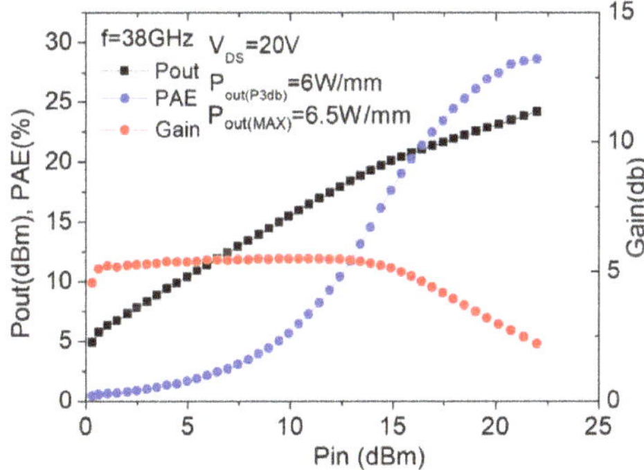

Figure 11. Large-signal performance of the 2 × 20 μm device with copper metallization.

As mentioned, for operation at millimeter-wave frequencies, the skin effect would force the current to flow on the surface of the interconnects, leading to the limitation of the effective area for current distribution, which in turn gives rise to the effective resistances at RF frequencies. Such effect could be even worse at higher frequencies since the skin depth is inversely proportional to the square root of the operating frequency. This is the major reason that the conductor loss is always dominant for planar circuits. Apparently, utilizing thick copper metallization in the device fabrication process provides a straightforward solution to such problem. From the measurement results of the 2 × 20 μm test device, it is obvious that performance improvements in the DC characteristics, the small-signal gain and the large-signal power/PAE are achieved.

3.4. Experimental Study of the Effect of Gate Width on the Device Performance

Based on the previous conclusions, we have fabricated and characterized the devices with different gate peripheries, namely, 2 × 25 μm and 2 × 15 μm, with thick copper metallization. The corresponding results of the large-signal performance characterized using on-wafer load-pull system at 38 GHz are shown in Figures 12 and 13 for the 2 × 25 μm and 2 × 15 μm devices, respectively. As shown, the device with larger gate periphery (2 × 25 μm) exhibited a power density of 7.7 W/mm at the maximum output power and the peak PAE was measured to be 36% (at the corresponding power level of 6.2 W/mm). As for the device with total gate width of 2 × 15 μm, we obtained a slightly higher power density of 8.2 W/mm at the maximum output power and the peak PAE was measured to be 26% (at the corresponding power level of 7.0 W/mm). The higher PAE achieved for the device with larger gate width of 2 × 25 μm was mainly due to the higher gain resulting from the reduction of the parasitic resistance associated with the device. Table 1 lists the performance comparisons of our devices with previously published works at the Ka band. As observed, the device exhibited the power density performance comparable to the state-of-the-art device technologies.

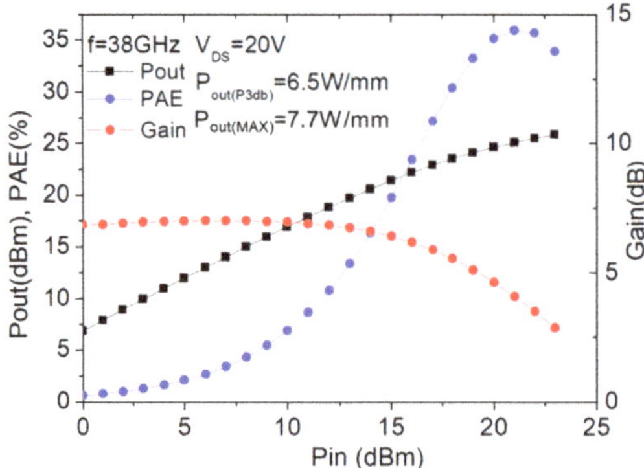

Figure 12. Large-signal performance of the device with gate width of 2 × 25 μm at 38 GHz.

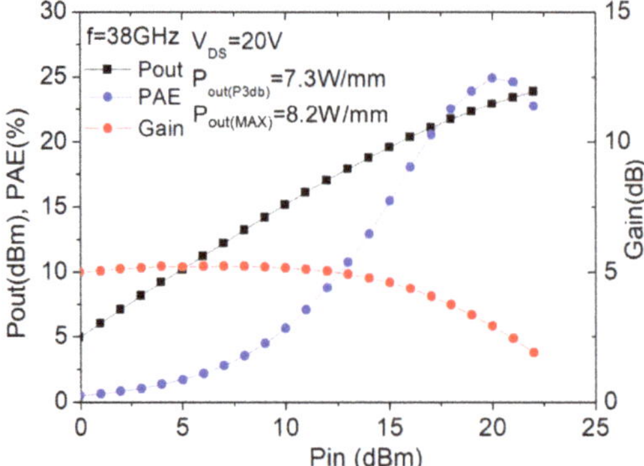

Figure 13. Large-signal performance of the device with gate width of 2 × 15 μm at 38 GHz.

Table 1. Power performance comparison of the Cu metallization with other reports.

References	Freq. (GHz)	V_{DS} (V)	L_g (nm)	Device Size (μm)	Pout @PAEmax (W/mm)	PAE (%)	$P_{out, max}$ (W/mm)
This Work—Device 1 with Cu metallization	38	20	90	2 × 25	6.2	36.0	7.7
This Work—Device 1 without Cu metallization	38	20	90	2 × 25	5.5	32	6.5
This Work—Device 2 with Cu metallization	38	20	90	2 × 15	7.0	26.0	8.2
This Work—Device 2 without Cu metallization	38	20	90	2 × 15	6.2	23.2	7.3
[20]	30	30	60	2 × 50	2.9	21.3	–
[21]	30	25	100	2 × 50	–	46.8	6.0
[22]	30	20	150	2 × 50	5.0	39.0	6.0
[23]	35	25	200	2 × 50	5.1	42.8	–
[24]	40	25	75	2 × 50	2.7	12.5	–
[25]	40	15	100	2 × 25	2.2	18.0	2.5
[26]	40	20	200	2 × 75	1.8	18.5	2.1
[27]	40	15	60	2 × 30	–	20.1	3.3
[28]	40	15	225	2 × 50	2.0	13.0	–
[29]	40	30	160	2 × 75	–	33	10.5

4. Conclusions

In this study, the copper metallization technique for GaN HEMT devices operating at millimeter-wave frequencies has been realized. Thick copper metallization contributes to the reduction of R_S' and R_D' and alleviates skin effect under high-frequency operation, which in turn improves the DC and RF characteristics of the devices. Experimental verifications revealed that the device with 2 × 15 μm gate width exhibited a record-high maximum output power density of 8.2 W/mm, which is the highest among the state-of-the-art published results at the Ka band. Such superior results have proven the feasibility of integrating thick copper metallization in the device process and make it promising for next-generation wireless communication system applications.

Author Contributions: Conceptualization and methodology, Y.C.L., H.-T.H., and E.Y.C.; data curation, S.H.C., P.H.L., K.H.L., and T.J.H.; software and validation, T.J.H.; writing—original draft preparation, Y.C.L.; writing—review and editing, H.-T.H., and E.Y.C.; supervision, project administration and funding acquisition, E.Y.C., and H.-T.H. All authors have read and agreed to the published version of the manuscript.

Funding: This research was funded by the Ministry of Science and Technology, Taiwan, under Grants 107-2221-E-009-093-MY2 and 108-2911-I-009-502. This work was financially supported by the "Center for the Semiconductor Technology Research" from The Featured Areas Research Center Program within the framework of the Higher Education Sprout Project by the Ministry of Education (MOE) in Taiwan. Also supported in part by the Ministry of Science and Technology, Taiwan, under Grant MOST-108-3017-F-009-003.

Conflicts of Interest: The authors declare no conflict of interest.

References

1. Kikkawa, T.; Makiyama, K.; Ohki, T.; Kanamura, M.; Imanishi, K.; Hara, N.; Joshin, K. High performance and high reliability AlGaN/GaN HEMTs. *Phys. Status Solidi A* **2009**, *206*, 1135–1144. [CrossRef]
2. Mishra, U.; Parikh, P.; Wu, Y.-F. AlGaN/GaN HEMTs-an overview of device operation and applications. *Proc. IEEE* **2002**, *90*, 1022–1031. [CrossRef]
3. Vetury, R.; Zhang, N.; Keller, S.; Mishra, U. The impact of surface states on the DC and RF characteristics of AlGaN/GaN HFETs. *IEEE Trans. Electron. Devices* **2001**, *48*, 560–566. [CrossRef]
4. Meneghini, M.; Meneghesso, G.; Zanoni, E. *Power GaN Devices Materials, Applications and Reliability*, 1st ed.; Springer: Berlin, Germany, 2017.

5. Sheppard, S.; Doverspike, K.; Pribble, W.; Allen, S.; Palmour, J.; Kehias, L.; Jenkins, T. High-power microwave GaN/AlGaN HEMTs on semi-insulating silicon carbide substrates. *IEEE Electron. Device Lett.* **1999**, *20*, 161–163. [CrossRef]
6. Shen, L.; Heikman, S.; Moran, B.; Coffie, R.; Zhang, N.-Q.; Buttari, D.; Smorchkova, I.; Keller, S.; DenBaars, S.; Mishra, U. AlGaN/AlN/GaN high-power microwave HEMT. *IEEE Electron. Device Lett.* **2001**, *22*, 457–459. [CrossRef]
7. De Jaeger, B.; Van Hove, M.; Wellekens, D.; Kang, X.; Liang, H.; Mannaert, G.; Geens, K.; Decoutere, S. Au-free CMOS-compatible AlGaN/GaN HEMT Processing on 200 mm Si Substrates. In Proceedings of the 24th International Symposium on Power Semiconductor Devices and ICs, Bruges, Belgium, 3–7 June 2012; pp. 49–52.
8. Marcon, D.; De Jaeger, B.; Halder, S.; Vranckx, N.; Mannaert, G.; Van Hove, M.; Decoutere, S. Manufacturing Challenges of GaN-on-Si HEMTs in a 200 mm CMOS Fab. *IEEE Trans. Semicond. Manuf.* **2013**, *26*, 361–367. [CrossRef]
9. Wong, S.S.; Lee, H.; Ryu, C.; Kwon, K.-W. Barriers for Copper Interconnections. *MRS Proc.* **1998**, *514*, 75. [CrossRef]
10. Kizil, H.; Kim, G.; Steinbrüchel, C.; Zhao, B. TiN and TaN diffusion barriers in copper interconnect technology: Towards a consistent testing methodology. *J. Electron. Mater.* **2001**, *30*, 345–348. [CrossRef]
11. Takeyama, M. Properties of TaNx films as diffusion barriers in the thermally stable Cu/Si contact systems. *J. Vac. Sci. Technol. B Microelectron. Nanometer Struct.* **1996**, *14*, 674. [CrossRef]
12. Vijayendran, A.; Danek, M. Copper barrier properties of ultrathin PECVD W/sub x/N. In Proceedings of the IEEE 1999 International Interconnect Technology Conference (Cat. No.99EX247), San Francisco, CA, USA, 26 May 1999; pp. 122–124.
13. Lin, Y.-C.; Kuo, T.-Y.; Chuang, Y.-L.; Wu, C.-H.; Chang, C.-H.; Huang, K.-N.; Ha, M.T.H. Ti/Pt/Ti/Cu-Metallized Interconnects for GaN High-Electron-Mobility Transistors on Si Substrate. *Appl. Phys. Express* **2012**, *5*, 066503. [CrossRef]
14. Lyu, Y.-T.; Jaw, K.-L.; Lee, C.-T.; Tsai, C.-D.; Lin, Y.-J.; Cherng, Y.-T. Ohmic performance comparison for Ti/Ni/Au and Ti/Pt/Au on InAs/graded InGaAs/GaAs layers. *Mater. Chem. Phys.* **2000**, *63*, 122–126. [CrossRef]
15. Chang, S.-W.; Ha, M.T.H.; Biswas, D.; Lee, C.-S.; Chen, K.-S.; Tseng, C.-W.; Hsieh, T.-L.; Wu, W.-C. Gold-Free Fully Cu-Metallized InGaP/GaAs Heterojunction Bipolar Transistor. *Jpn. J. Appl. Phys.* **2005**, *44*, 8–11. [CrossRef]
16. Ghannouchi, F.M.; Hashmi, M. *Load-Pull Techniques with Applications to Power Amplifier Design*; Springer: Berlin, Germany, 2013.
17. Alim, M.A.; Rezazadeh, A.; Gaquiere, C. Thermal characterization of DC and small-signal parameters of 150 nm and 250 nm gate-length AlGaN/GaN HEMTs grown on a SiC substrate. *Semicond. Sci. Technol.* **2015**, *30*, 125005. [CrossRef]
18. Liu, Z. High Frequency Small-Signal Modelling of GaN High Electron Mobility Transistors for RF Applications. Master's Thesis, University of Victoria, Victoria, BC, Canada, 9 December 2016.
19. Anwar, J.; Gunter, K. A New Small-Signal Modeling Approach Applied to GaN Devices. *IEEE Trans. Microw. Theory Tech.* **2005**, *53*, 3440–3448.
20. Higashiwaki, M.; Pei, Y.; Chu, R.; Mishra, U.K. Effects of Barrier Thinning on Small-Signal and 30-GHz Power Characteristics of AlGaN/GaN Heterostructure Field-Effect Transistors. *IEEE Trans. Electron Devices* **2011**, *58*, 1681–1686. [CrossRef]
21. Mi, M.; Ma, X.; Yang, L.; Lu, Y.; Hou, B.; Zhu, J.; Zhang, M.; Zhang, H.-S.; Zhu, Q.; Yang, L.-A.; et al. Millimeter-Wave Power AlGaN/GaN HEMT Using Surface Plasma Treatment of Access Region. *IEEE Trans. Electron Devices* **2017**, *64*, 4875–4881. [CrossRef]
22. Aubry, R.; Jacquet, J.C.; Oualli, M.; Patard, O.; Piotrowicz, S.; Chartier, E.; Michel, N.; Trinh-Xuan, L.; Lancereau, D.; Potier, C.; et al. ICP-CVD SiN Passivation for High-Power RF InAlGaN/GaN/SiC HEMT. *IEEE Electron Device Lett.* **2016**, *37*, 629–632. [CrossRef]
23. Zhang, Y.; Wei, K.; Huang, S.; Wang, X.; Zheng, Y.; Liu, G.; Chen, X.; Li, Y.; Liu, X. High-Temperature-Recessed Millimeter-Wave AlGaN/GaN HEMTs with 42.8% Power-Added-Efficiency at 35 GHz. *IEEE Electron Device Lett.* **2018**, *39*, 727–730. [CrossRef]

24. Altuntas, P.; Lecourt, F.; Cutivet, A.; Defrance, N.; Okada, E.; Lesecq, M.; Rennesson, S.; Agboton, A.; Cordier, Y.; Hoel, V.; et al. Power Performance at 40 GHz of AlGaN/GaN High-Electron Mobility Transistors Grown by Molecular Beam Epitaxy on Si(111) Substrate. *IEEE Electron Device Lett.* **2015**, *36*, 303–305. [CrossRef]
25. Medjdoub, F.; Zegaoui, M.; Grimbert, B.; Ducatteau, D.; Rolland, N.; Rolland, P.A. First Demonstration of High-Power GaN-on-Silicon Transistors at 40 GHz. *IEEE Electron Device Lett.* **2012**, *33*, 1168–1170. [CrossRef]
26. Marti, D.; Tirelli, S.; Alt, A.R.; Roberts, J.; Bolognesi, C.R. 150-GHz Cutoff Frequencies and 2-W/mm Output Power at 40 GHz in a Millimeter-Wave AlGaN/GaN HEMT Technology on Silicon. *IEEE Electron Device Lett.* **2012**, *33*, 1372–1374. [CrossRef]
27. Soltani, A.; Gerbedoen, J.-C.; Cordier, Y.; Ducatteau, D.; Rousseau, M.; Chmielowska, M.; Ramdani, M.; De Jaeger, J.-C. Power Performance of AlGaN/GaN High-Electron-Mobility Transistors on (110) Silicon Substrate at 40 GHz. *IEEE Electron Device Lett.* **2013**, *34*, 490–492. [CrossRef]
28. Lecourt, F.; Agboton, A.; Ketteniss, N.; Behmenburg, H.; Defrance, N.; Hoel, V.; Kalisch, H.; Vescan, A.; Heuken, M.; De Jaeger, J.-C. Power Performance at 40 GHz on Quaternary Barrier InAlGaN/GaN HEMT. *IEEE Electron Device Lett.* **2013**, *34*, 978–980. [CrossRef]
29. Palacios, T.; Chakraborty, A.; Rajan, S.; Poblenz, C.; Keller, S.; DenBaars, S.P.; Speck, J.; Mishra, U. High-power AlGaN/GaN HEMTs for Ka-band applications. *IEEE Electron Device Lett.* **2005**, *26*, 781–783. [CrossRef]

micromachines

Article

Investigation of Recessed Gate AlGaN/GaN MIS-HEMTs with Double AlGaN Barrier Designs toward an Enhancement-Mode Characteristic

Tian-Li Wu *, Shun-Wei Tang and Hong-Jia Jiang

International College of Semiconductor Technology, National Chiao Tung University, Hsinchu 30010, Taiwan; tangandy.icst07g@nctu.edu.tw (S.-W.T.); dk0754013.icst07g@nctu.edu.tw (H.-J.J.)
* Correspondence: tlwu@nctu.edu.tw; Tel.: +886-3-571212

Received: 31 December 2019; Accepted: 2 February 2020; Published: 3 February 2020

Abstract: In this work, recessed gate AlGaN/GaN metal-insulator-semiconductor high-electron-mobility transistors (MIS-HEMTs) with double AlGaN barrier designs are fabricated and investigated. Two different recessed depths are designed, leading to a 5 nm and a 3 nm remaining bottom AlGaN barrier under the gate region, and two different Al% (15% and 20%) in the bottom AlGaN barriers are designed. First of all, a double hump trans-conductance (g_m)–gate voltage (V_G) characteristic is observed in a recessed gate AlGaN/GaN MIS-HEMT with a 5 nm remaining bottom $Al_{0.2}Ga_{0.8}N$ barrier under the gate region. Secondly, a physical model is proposed to explain this double channel characteristic by means of a formation of a top channel below the gate dielectric under a positive V_G. Finally, the impacts of Al% content (15% and 20%) in the bottom AlGaN barrier and 5 nm/3 nm remaining bottom AlGaN barriers under the gate region are studied in detail, indicating that lowering Al% content in the bottom can increase the threshold voltage (V_{TH}) toward an enhancement-mode characteristic.

Keywords: GaN; metal-insulator-semiconductor high-electron-mobility transistor (MIS-HEMT); recessed gate; double barrier

---

1. Introduction

AlGaN/GaN high-electron-mobility transistors (HEMTs) are promising for power switching applications due to the wide band gap, large breakdown electric field, and the inherent high electron mobility due to two-dimensional electron gas (2DEG) [1,2]. Normally, conventional AlGaN/GaN Schottky HEMTs suffer high gate leakage, resulting in an unfavorable power loss during an off-state condition and a low gate overdrive during an on-state condition. In order to tackle this issue, Metal-insulator-semiconductors high electrons mobility transistors (MIS-HEMTs) have gained attentions recently [2–4]. Inserting a dielectric in the interface between AlGaN and gate metal significantly reduces the gate leakage current, allowing a high gate overdrive to have a fast switch from off-state to on-state operation. Due to a piezoelectric and polarization effects, a two dimensional electron gas (2DEG) is naturally formed in the interface between GaN and AlGaN, leading to depletion mode (V_{TH}<0) characteristics. However, an enhancement mode characteristic is more favorable in practical applications due to a lower power consumption, less failure issues, and a flexible integration. So far, there are several approaches to realize an enhancement mode operation, such as a recessed gate structure [5–7], p-GaN/p-AlGaN gate [2,8], Fluoride-based plasma treatment [9], the metal–oxide–semiconductor field-effect transistor structure [10], cascode-based topology in connecting high voltage D-mode HEMTs with a low voltage Si MOSFETs or E-mode HEMTs [11,12], etc. HEMTs with a p-GaN/p-AlGaN gate generally suffer the challenges to effectively dope the Mg into top GaN or AlGaN layer and to remove the top p-GaN/p-AlGaN layer in the access region. The thermal stability of the charges induced by Fluoride-based plasma remains a challenge. The MOS-type GaN FETs suffer the low electron mobility due to the disappearance of the 2DEG. The cascode-based topology with dual GaN-based transistors increase the active area and complicates the layout designs. Therefore, the recessed gate-based HEMT is one of the most popular architectures to obtaining an enhancement-mode characteristic because the 2DEG can be reduced under the gate by using a simple etching process, i.e., Reactive-ion etching (RIE)-based etching or atomic layer etching (ALE). Recently, the recessed gate AlGaN/GaN-based devices show a promising performance toward an enhancement-mode characteristic [5–7].

Typically, AlGaN/GaN-based HEMTs are fabricated with the single AlGaN barrier layer. A multi barrier device was demonstrated first in GaAs-based devices in 1985 [13]. Since 1999, a depletion mode double AlGaN barrier design was first demonstrated by Gaska et al. [14]. Afterwards, depletion-mode double AlGaN barrier HEMTs have been explored in details in [15–18], showing a high current drive due to a second transconductance (gm) and lower access resistance. Furthermore, the recent demonstration using AlGaN/AlN/GaN/AlN epitaxy stack to achieve the double channel has attracted a lot of attentions [19]. Although the demonstration of the GaN-based HEMTs with a double barrier exhibits the promising characteristics, the impacts of the Al% in the top and bottom AlGaN barriers are still unclear. Furthermore, the investigation of the combination of the recessed-based approach in the HEMTs with double AlGaN barrier designs is lacking as well, which can further provide the insightful analysis to understand the device physics and structure designs toward an enhancement-mode characteristic.

In this work, recessed gate MIS-HEMTs with double AlGaN barrier designs with different recessed depths (3 nm or 5 nm remaining bottom AlGaN barrier under the gate region) and different Al% content in the bottom AlGaN barrier (15% and 20%) are fabricated and investigated. First of all, we observed a double hump g_m–V_G characteristic in a recessed gate AlGaN/GaN MIS-HEMT with a 5nm remaining bottom $Al_{0.2}Ga_{0.8}N$ barrier under the gate region. Then, a physical model considering the formation of the top channel under a positive V_G is proposed to explain this double hump in the g_m–V_G characteristic. Furthermore, the impacts from the Al% (20% and 15%) in bottom AlGaN barrier and recessed depth (3 nm and 5 nm remaining bottom AlGaN barrier under the gate region) are discussed to understand the device characteristics and the V_{TH} can be increased by designing the device with a lower Al% in the bottom AlGaN barrier.

2. Device Fabrications

Figure 1 shows the schematic of the epitaxy structure in this study and Figure 2 shows an example of the transmission electron microscopy (TEM) image in an epitaxy structure with a double AlGaN barrier. This structure was grown by metal–organic chemical vapor deposition (MOCVD) on a silicon (111) substrate and consists of an AlN nucleation layer, a GaN channel, a bottom AlGaN barrier with two different Al contents (20% and 15%), a top AlGaN barrier with 30% Al content, and a 1 nm GaN cap. The Al% (20% and 30%) in AlGaN barrier is calibrated with the XPS by using a single AlGaN barrier hetero-structure. The calibrated growing conditions in MOCVD are used for the double AlGaN barrier hetero-structure. Figure 3 shows the simulated band diagram with the double AlGaN barriers, clearly indicating the existence of the electrons in the interface between top AlGaN/bottom AlGaN and bottom AlGaN/GaN. The recessed gate structure was formed by reactive ion etching (RIE). In order to control the etching depth, low etching rate of 5Å/sec is achieved by the mixed BCl_3 (10 sccm)/Cl_2 (15 sccm) gas. A gate recessed process is performed and a 15-nm Plasma-enhanced chemical vapor deposition (PECVD) Si_3N_4 is deposited as a surface passivation layer in the access region and a gate dielectric. TiN is used as the gate metal. The Ti/Al-based Au-free Ohmic contacts were formed by etching the Si_3N_4 layers and etching the AlGaN barrier. This was followed by annealing at 600 °C for 1 min in N_2, resulting in 1 ohm.mm of R_c (contact resistance). Figure 4 shows the schematic of recessed gate MIS-HEMTs with a double AlGaN barrier. The important varied parameters are summarized in Table 1. The devices with Lg = 1 um, Lgs = 2 um, and Lgd = 6 um are fabricated for electrical characterizations.

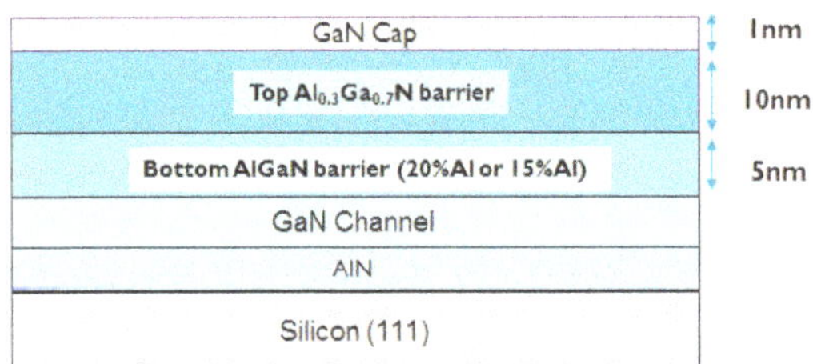

Figure 1. Schematic of the epitaxy structure with a double AlGaN barrier used in this study.

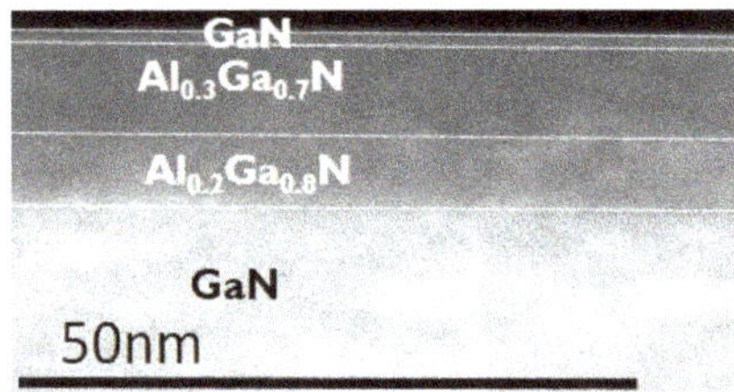

Figure 2. Transmission electron microscopy (TEM) images of the epitaxy structure with a double AlGaN barrier.

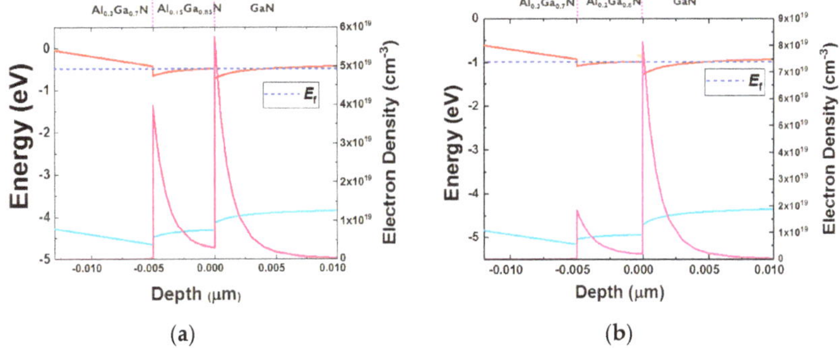

Figure 3. Simulated band diagrams with an $Al_{0.3}Ga_{0.7}N/Al_{0.15}Ga_{0.85}N$ barrier (**a**) and an $Al_{0.3}Ga_{0.7}N/Al_{0.2}Ga_{0.8}N$ barrier (**b**).

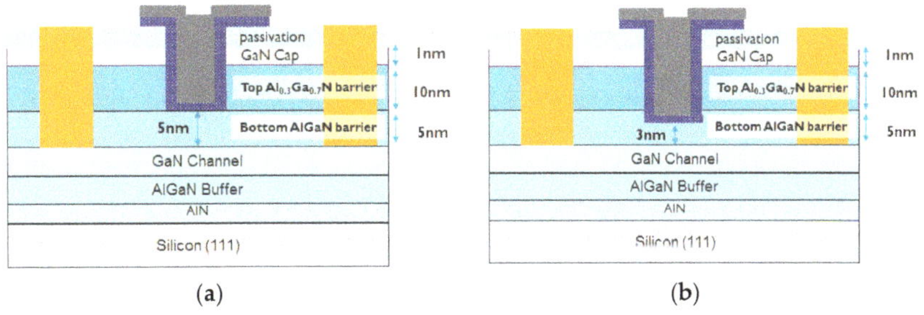

Figure 4. Schematic of device structures of recessed gate AlGaN/GaN MIS-HEMTs with a 5 nm remaining bottom AlGaN barrier under the gate region (**a**) and a 3 nm remaining bottom AlGaN barrier under the gate region (**b**).

Table 1. Summary of the Varied Parameters.

Parameters	Top AlGaN Barrier	Bottom AlGaN Barrier
Thickness	10 nm	5 nm
Al%	30%	20% 15%
Recessed depth		
Remaining bottom AlGaN barrier under the gate region	–	5 nm 3 nm

3. Results

In the case of devices with a 5 nm remaining bottom AlGaN barrier under the gate area, a gate recess process is performed to etch until it reaches the surface of the bottom AlGaN barrier (Figure 4a). The I_D–V_G, I_G-V_G, and g_m–V_G characteristics are shown in Figure 5. Note that all I_D–V_G characteristics in this work are measured from a lower V_G till a higher V_G. A double hump of g_m–V_G characteristic is observed in Figure 5c, which is similar to the literature [15–18]. The double hump of g_m–V_G characteristics in these references [15–18] arises from a shrinking of the depletion region below the gate, due to the depletion-mode characteristics. In our case, the recessed gate AlGaN/GaN MIS-HEMTs has a double barrier design. A double channel model that considers the electron transfer from the bottom channel to the top channel is proposed to explain the double hump g_m-V_G characteristics. First of all, once the V_G is larger than V_{TH}, the bottom channel is gradually turned on (Figures 6a and 7a), resulting in a first g_m peak as shown in Figure 5c. It is worth noting that at this stage the top channel from source to drain is initially disconnected below the gate dielectric due to a recessed gate process. However, when the V_G is above 5 V, the top channel could be connected again, as shown in Figure 6b. In this scenario, the electrons can be transferred from the lateral 2DEG channel in the access region and/or interface below the bottom AlGaN barrier to the interface between the dielectric and the bottom AlGaN barrier [20,21] (Figure 7b). Then, the electrons can be accumulated under the gate dielectric [20,22]. This leads to the formation of the second channel under the gate, further connecting the source and drain to form the top channel leading to a second g_m peak (Figure 5c).

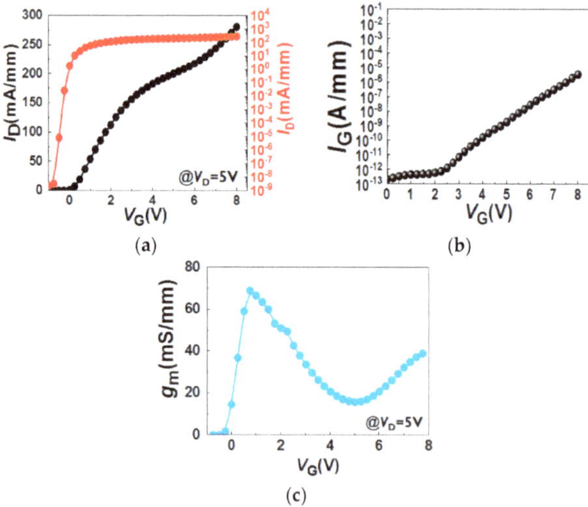

Figure 5. I_D–V_G (**a**), I_G–V_G (**b**) and g_m–V_G (**c**) characteristics in a recessed gate MIS-HEMTs with a 5nm remaining bottom $Al_{0.2}Ga_{0.8}N$ barrier under the gate. By designing a 5 nm bottom AlGaN barrier with 20% Al content, V_{TH} ~0 V is realized (V_{TH} is defined at V_G of I_D = 0.1 mA/mm).

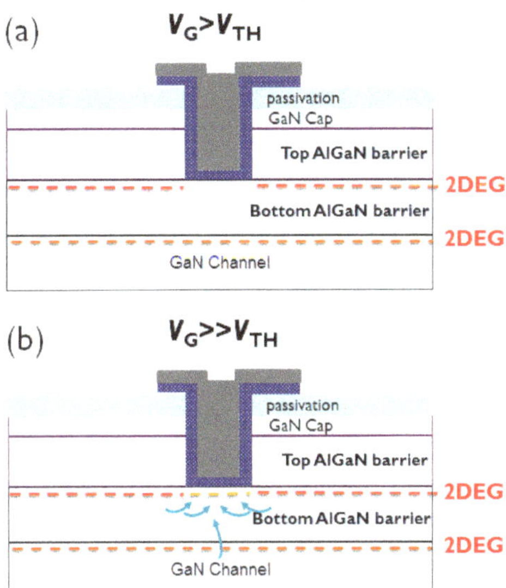

Figure 6. The schematic of proposed double channel model (**a**) when $V_G > V_{TH}$ and (**b**) $V_G >>> V_{TH}$.

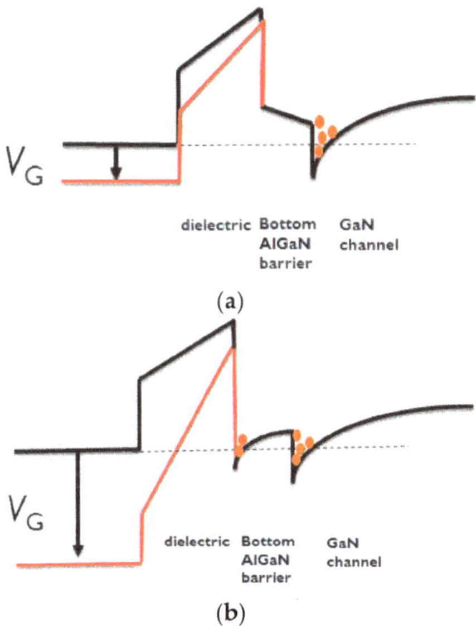

Figure 7. Schematic of the band diagram under (**a**) $V_G > V_{TH}$ and (**b**) $V_G >> V_{TH}$.

Figure 8 shows an example of Capacitance-Voltage (CV) measurement in the device with 3 nm remaining bottom $Al_{0.15}Ga_{0.85}N$ barrier. The capacitance is increase when the gate voltage is larger than 0 V, indicating the formation of the first channel (Figures 6a and 7a). Once the gate voltage is applied larger enough, the capacitance is increased again, which is mainly due to the formation of the channel between dielectric and bottom AlGaN barrier (Figures 6b and 7b), consistent with the reported literature [19].

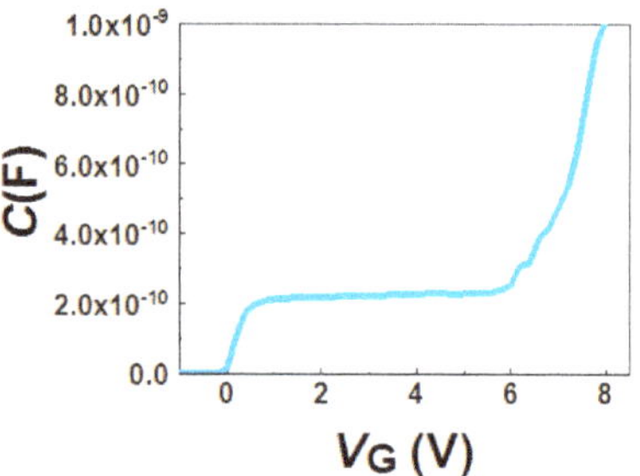

Figure 8. Capacitance-Voltage (CV) measurement in the device with 3 nm remaining bottom $Al_{0.15}Ga_{0.85}N$ barrier.

Figure 9 shows the I_D–V_G characteristics in the devices with different Al% content ($Al_{0.2}Ga_{0.8}N$ and $Al_{0.15}Ga_{0.85}N$) in the bottom barriers and two different recessed depths (5 nm and 3 nm bottom AlGaN thickness under the gate area). The I_D decreases but the V_{TH} increases with a thinner remaining bottom AlGaN barrier (Figure 9b). Furthermore, the subthreshold slope (SS) is increased once the Al% in the bottom AlGaN barrier is decreased, which is mainly due to the low electron density in the channel between bottom AlGaN/GaN. By designing with 3nm remaining bottom $Al_{0.15}Ga_{0.85}N$ barrier under the gate region, V_{TH} ~3.25 V is achieved (V_{TH} is defined at V_G of I_D = 0.1 mA/mm).

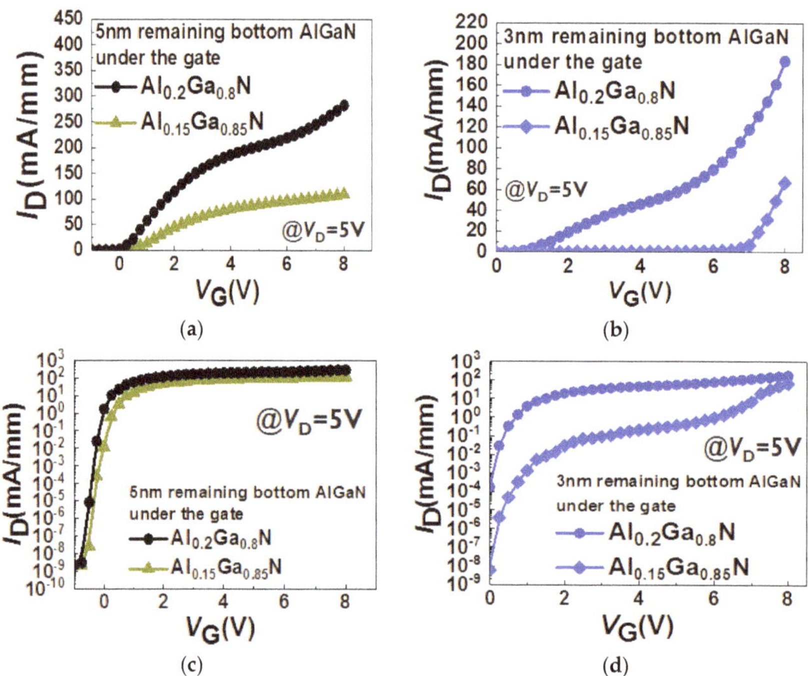

Figure 9. I_D-V_G in a linear scale (**a**,**b**) and a logarithmic scale (**c**,**d**) in the devices with 5 nm or 3 nm remaining bottom AlGaN layer under the gate region with different Al% in the bottom AlGaN barrier.

Figure 10 shows the g_m–V_G characteristics in the devices with different Al% content ($Al_{0.2}Ga_{0.8}N$ and $Al_{0.15}Ga_{0.85}N$) in the bottom barriers and two different recessed depths (5 nm and 3 nm bottom AlGaN thickness under the gate area). In the case of the devices with a 5 nm remaining bottom AlGaN barrier, lowering the Al% in the bottom AlGaN barrier decreases the g_m peaks (Figure 10a).

Furthermore, double hump g_m–V_G characteristics can still be observed in the device with a 3 nm remaining bottom $Al_{0.2}Ga_{0.8}N$ barrier under the gate region. First, the first g_m peak decreases with a thinner remaining bottom AlGaN barrier under the gate dielectric, suggesting that the remaining bottom AlGaN barrier under the gate region limits the current contribution from the bottom channel. Second, the g_m increases after 4 V in the device with a 3 nm remaining bottom $Al_{0.2}Ga_{0.8}N$ barrier under the gate region (Figure 10b). Whereas, the g_m increases after 5 V in the devices with a 5 nm remaining bottom $Al_{0.2}Ga_{0.8}N$ barrier under the gate region (Figure 10a). This is in agreement with the model proposed above: Due to a thinner AlGaN barrier, a lower gate voltage can allow the electrons to transfer from the bottom channel to the area below the gate. Third, the second g_m is higher than the first g_m in the devices with a 3 nm remaining bottom AlGaN barrier under the gate region (Figure 10b). However, the first g_m is higher than the second g_m in the devices with a 5 nm remaining bottom AlGaN barrier under the gate region (Figure 10a). These observations suggest that the main current contribution in the devices with a 5 nm remaining bottom AlGaN barrier under the gate region is the bottom channel. However, in the devices with a 3 nm remaining bottom AlGaN barrier under the gate region, the main current contribution is derived from the top channel.

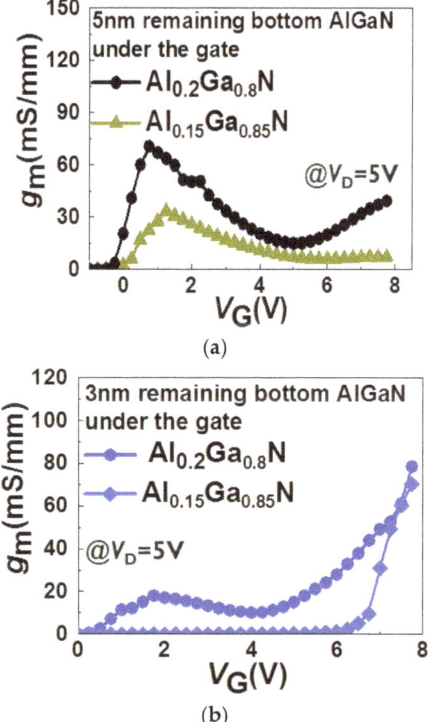

Figure 10. g_m–V_G characteristics in the devices with a (**a**) 5 nm or (**b**) 3 nm remaining bottom AlGaN layer under the gate region with different Al% in the bottom AlGaN barrier.

Table 2 summarizes the comparisons of this work with other recent reports in double channel HEMTs, indicating our work shows the promising characteristics in terms of Ion/Ioff ratio and V_{TH} toward an enhancement mode characteristic.

Table 2. Reference for the double channel GaN HEMTs.

Reference	Barrier designs	I_{on}/I_{off}	SS (mV/dec)	V_{TH} (V) @0.1mA/mm	1st g_m (mS/mm)	2nd g_m (mS/mm)
Our work	$Al_{0.3}Ga_{0.7}N/Al_{0.2}Ga_{0.8}N/GaN$ 5nm remaining bottom $Al_{0.2}Ga_{0.8}N$	1.2×10^{11}	80.7	~0	68.9	39
	$Al_{0.3}Ga_{0.7}N/Al_{0.15}Ga_{0.85}N/GaN$ 5nm remaining bottom $Al_{0.15}Ga_{0.85}N$	4.8×10^{10}	87.9	0.25	34.8	7.6
	$Al_{0.3}Ga_{0.7}N/Al_{0.2}Ga_{0.8}N/GaN$ 3nm remaining bottom $Al_{0.2}Ga_{0.8}N$	-	153.1	0.5	18.8	90
	$Al_{0.3}Ga_{0.7}N/Al_{0.15}Ga_{0.85}N/GaN$ 3nm remaining bottom $Al_{0.15}Ga_{0.85}N$	5.5×10^{10}	229.3	3.25	-	71.2
Kamath et al. [17]	$Al_{0.3}Ga_{0.7}N/GaN/Al_{0.15}Ga_{0.75}N/GaN$	-	-	~−7.8	34	89
Wei et al. [19]	AlGaN/AlN/GaN/AlN with 1.5 nm over-etch upper GaN layer	~10^9	72	0.5	170	103
Lee et al. [23]	AlGaN/GaN/AlGaN/GaN Fin Structure	-	-	0.2	133	-
Deen et al. [24]	AlN/GaN/AlN/GaN	-	-	~−4	190	-

4. Conclusions

In summary, recessed gate AlGaN/GaN MIS-HEMTs with double AlGaN barrier designs are investigated and discussed. A double hump of the g_m–V_G characteristic can be observed in the recessed gate AlGaN/GaN MIS-HEMTs with double AlGaN barrier designs. A physical model is proposed to explain the double channel characteristics, which is mainly due to the formation of the top channel under a high V_G bias. Once the gate voltage is applied at a high enough level, the top channel is formed, leading to an increase in drain current due to the current contribution from the top channel. Furthermore, by lowering the Al% in the bottom AlGaN barrier, the devices show a more positive V_{TH} with the same recessed depth, indicating that a double AlGaN barrier design in recessed gate MIS-HEMTs can be an alternative strategy to achieve an enhancement mode characteristic.

Author Contributions: Data curation, T.-L.W., S.-W.T., and H.-J.J.; writing—original draft preparation, T.-L.W.; writing—review and editing, T.-L.W., S.-W.T., and H.-J.J. All authors have read and agreed to the published version of the manuscript.

Funding: This work was financially supported by the "Center for the Semiconductor Technology Research" from the Featured Areas Research Center Program within the framework of the Higher Education Sprout Project by the Ministry of Education (MOE) in Taiwan. It is also supported in part by the Ministry of Science and Technology, Taiwan, under Grant MOST-108-3017-F-009-003, MOST-106-2218-E-009-025-MY3 and by the Young Scholar Fellowship Program under Grant MOST 108-2636-E-009-006.

Conflicts of Interest: The authors declare no conflict of interest.

References

1. Amano, H.; Baines, Y.; Beam, E.; Borga, M.; Bouchet, T.; Chalker, P.R.; Charles, M.; Chen, K.J.; Chowdhury, N.; Chu, R.; et al. The 2018 GaN power electronics roadmap. *J. Phys. D Appl. Phys.* **2018**, *51*, 1–48. [CrossRef]
2. Chen, K.J.; Häberlen, O.; Lidow, A.; Tsai, C.L.; Ueda, T.; Uemoto, Y.; Wu, Y. GaN-on-Si power technology: Devices and applications. *IEEE Trans. Electron Devices* **2017**, *64*, 779–795. [CrossRef]
3. Van Hove, M.; Boulay, S.; Bahl, S.R.; Stoffels, S.; Kang, X.; Wellekens, D.; Geens, K.; Delabie, A.; Decoutere, S. CMOS Process-Compatible High-Power Low-Leakage AlGaN/GaN MISHEMT on Silicon. *IEEE Electron Device Lett.* **2012**, *33*, 667–669. [CrossRef]
4. Moens, P.; Liu, C.; Banerjee, A.; Vanmeerbeek, P.; Coppens, P.; Ziad, H.; Constant, A.; Li, Z.; de Vleeschouwer, H.; RoigGuitart, J.; et al. An Industrial Process for 650 V Rated GaN-on-Si Power Devices Using in-Situ SiN as a Gate Dielectric. In Proceedings of the International Symposium on Power Semiconductor Devices and ICs (ISPSD), Waikoloa, HI, USA, 15–19 June 2014; pp. 374–377. [CrossRef]

5. Hua, M.; Zhang, Z.; Wei, J.; Lei, J.; Tang, G.; Fu, K.; Cai, Y.; Zhang, B.; Chen, K.J. Integration of LPCVD-SiNx Gate Dielectric with Recessed-Gate E-Mode GaN MIS-FETs: Toward High Performance, High Stability and Long TDDB Lifetime. In Proceedings of the IEEE International Electron Devices Meeting (IEDM), San Francisco, CA, USA, 3–7 December 2016; pp. 10.4.1–10.4.4. [CrossRef]
6. Wu, T.-L.; Marcon, D.; de Jaeger, B.; van Hove, M.; Bakeroot, B.; Lin, D.; Stoffels, S.; Kang, X.; Roelofs, R.; Groeseneken, G.; et al. The Impact of the Gate Dielectric Quality in Developing Au-Free D-Mode and E-Mode Recessed Gate AlGaN/GaN Transistors on a 200 mm Si Substrate. In Proceedings of the International Symposium on Power Smicond Devices and ICs (ISPSD), Hong Kong, China, 10–14 May 2015; pp. 225–228. [CrossRef]
7. De Jaeger, B.; van Hove, M.; Wellekens, D.; Kang, X.; Liang, H.; Mannaert, G.; Geens, K.; Decoutere, S. Au-Free CMOS-Compatible AlGaN/GaN HEMT Processing on 200 mm Si Substrates. In Proceedings of the 2012 24th International Symposium on Power Semiconductor Devices and ICs, Bruges, Belgium, 3–7 June 2012; pp. 49–52. [CrossRef]
8. Wu, T.-L.; Marcon, D.; You, S.; Posthuma, N.; Bakeroot, B.; Stoffels, S.; van Hove, M.; Groeseneken, G.; Decoutere, S. Forward bias gate breakdown mechanism in enhancement-mode p-GaN gate AlGaN/GaN high-electron mobility transistors. *IEEE Electron Device Lett.* **2015**, *36*, 1001–1003. [CrossRef]
9. Cai, Y.; Zhou, Y.; Chen, K.J.; Lau, K.M. High-performance enhancement-mode AlGaN/GaN HEMTs using fluoride-based plasma treatment. *IEEE Electron Device Lett.* **2005**, *26*, 435–437. [CrossRef]
10. Huang, W.; Khan, T.; Chow, T.P. Enhancement-mode n-channel GaN MOSFETs on p and n-GaN/sapphire substrates. *IEEE Electron Device Lett.* **2006**, *27*, 796–798. [CrossRef]
11. Lu, B.; Saadat, O.I.; Palacios, T. High-performance integrated dual-gate AlGaN/GaN enhancement-mode transistor. *IEEE Electron Device Lett.* **2010**, *31*, 990–992. [CrossRef]
12. Wang, R.; Wu, Y.; Chen, K.J. Gain improvement of enhancement-mode AlGaN/GaN high-electron-mobility transistors using dual-gate architecture. *Jpn. J. Appl. Phys.* **2013**, *47*, 2820. [CrossRef]
13. Sheng, N.H.; Lee, C.P.; Chen, R.T.; Miller, D.L.; Lee, S.J. Multiple-Channel GaAs/AlGaAs High Electron Mobility Transistors. *IEEE Electron Device Lett.* **1985**, *6*, 307–310. [CrossRef]
14. Gaska, R.; Shur, M.S.; Fjeldly, T.A.; Bykhovski, A.D. Two-channel AlGaN/GaN heterostructure field effect transistor for high power applications. *J. Appl. Phys.* **1999**, *85*, 3009–3011. [CrossRef]
15. Chu, R.; Zhou, Y.; Liu, J.; Wang, D.; Chen, K.J.; Lau, K.M. AlGaN-GaN Double-Channel HEMTs. *IEEE Trans. Electron Devices* **2005**, *52*, 438–446. [CrossRef]
16. Palacios, T.; Chini, A.; Buttari, D.; Heikman, S.; Chakraborty, A.; Keller, S.; DenBaars, S.P.; Mishra, U.K. Use of Double-Channel Heterostructures to Improve the Access Resistance and Linearity in GaN-Based HEMTs. *IEEE Trans. Electron Devices* **2006**, *53*, 562–565. [CrossRef]
17. Kamath, A.; Patil, T.; Adari, R.; Bhattacharya, I.; Ganguly, S.; Aldhaheri, R.W.; Hussain, M.A.; Saha, D. Double-Channel AlGaN/GaN High Electron Mobility Transistor with Back Barriers. *IEEE Electron Device Lett.* **2012**, 33. [CrossRef]
18. Xue, J.; Zhang, J.; Zhang, K.; Zhao, Y.; Zhang, L.; Ma, X.; Li, X.; Meng, F.; Hao, Y. Fabrication and characterization of InAlN/GaN-based double-channel high electron mobility transistors for electronic applications. *J. Appl. Phys.* **2012**, *111*, 114513. [CrossRef]
19. Wei, J.; Liu, S.; Li, B.; Tang, X.; Lu, Y.; Liu, C.; Hua, M.; Zhang, Z.; Tang, G.; Chen, K.J. Enhancement-Mode GaN Double-Channel MOS-HEMT with Low On-Resistance and Robust Gate Recess. In Proceedings of the IEEE International Electron Devices Meeting (IEDM), Washington, DC, USA, 7–9 December 2015; pp. 9.4.1–9.4.4. [CrossRef]
20. Mizue, C.; Hori, Y.; Miczek, M.; Hashizume, T. Capacitance-Voltage Characteristics of Al_2O_3/AlGaN/GaN Structures and State Density Distribution at Al_2O_3/AlGaN Interface. *Jpn. J. Appl. Phys.* **2011**, *50*, 021001. [CrossRef]
21. Wu, T.-L.; Marcon, D.; Ronchi, N.; Bakeroot, B.; You, S.; Stoffels, S.; van Hove, M.; Bisi, D.; Meneghini, M.; Groeseneken, G.; et al. Analysis of Slow De-trapping Phenomena after a Positive Gate Bias on AlGaN/GaN MIS-HEMTs with in-situ Si_3N_4/Al_2O_3 Bilayer Gate Dielectrics. *Solid-State Electron.* **2015**, *103*, 127–130. [CrossRef]

22. Lagger, P.; Steinschifter, P.; Reiner, M.; Denifl, G.; Naumann, A.; Wilde, L.; Sundqvist, J.; Pogany, D.; Ostermaier, C. Role of the dielectric for the charging dynamics of the dielectric/barrier interface in AlGaN/GaN based metal-insulator-semiconductor structures under forward gate bias stress. *Appl. Phys. Lett.* **2014**, *105*, 033512. [CrossRef]
23. Lee, J.-H.; Kim, J.-G.; Ju, J.-M.; Ahn, W.-H.; Kang, S.-H.; Lee, J.-H. AlInGaN/GaN double-channel FinFET with high on-current and negligible current collapse. *Solid-State Electron.* **2020**, *164*, 107678. [CrossRef]
24. Deen, D.A.; Miller, R.A.; Osinsky, A.V.; Downey, B.P.; Storm, D.F.; Meyer, D.J.; Katzer, D.S.; Nepal, N. Polarization-mediated Debye-screening of surface potential fluctuations in dual-channel AlN/GaN high electron mobility transistors. *Appl. Phys. Lett.* **2016**, *120*, 235704. [CrossRef]

Article

Design and Implementation of a GaN-Based Three-Phase Active Power Filter

Chao-Tsung Ma * and Zhen-Huang Gu

Department of Electrical Engineering, CEECS, National United University, Miaoli 36063, Taiwan; M0621002@smail.nuu.edu.tw

* Correspondence: ctma@nuu.edu.tw; Tel.: +886-37-382482; Fax: +886-37-382488

Received: 25 December 2019; Accepted: 22 January 2020; Published: 24 January 2020

Abstract: Renewable energy (RE)-based power generation systems and modern manufacturing facilities utilize a wide variety of power converters based on high-frequency power electronic devices and complex switching technologies. This has resulted in a noticeable degradation in the power quality (PQ) of power systems. To solve the aforementioned problem, advanced active power filters (APFs) with improved system performance and properly designed switching devices and control algorithms can provide a promising solution because an APF can compensate for voltage sag, harmonic currents, current imbalance, and active and reactive powers individually or simultaneously. This paper demonstrates, for the first time, the detailed design procedure and performance of a digitally controlled 2 kVA three-phase shunt APF system using gallium nitride (GaN) high electron mobility transistors (HEMTs). The designed digital control scheme consists of three type II controllers with a digital signal processor (DSP) as the control core. Using the proposed APF and control algorithms, fast and accurate compensation for harmonics, imbalance, and reactive power is achieved in both simulation and hardware tests, demonstrating the feasibility and effectiveness of the proposed system. Moreover, GaN HEMTs allow the system to achieve up to 97.2% efficiency.

Keywords: gallium nitride (GaN); power switching device; active power filter (APF); power quality (PQ)

---

1. Introduction

In recent years, the rapid increase in the penetration level of renewable energy (RE)-based distributed power generation (DPG) has resulted in noticeable degradation in the voltage stability and power quality (PQ) of existing power systems. The intrinsic features of DPG systems include (1) intermittent power flow caused by the utilization of maximum power point tracking (MPPT) control functions and (2) harmonic current injections caused by various power converters using switching type control techniques. Moreover, modern technologies such as automatic manufacturing systems (AMS), Internet of Things (IoT), and Industry 4.0 require a large number of power converters based on high-switching frequency power electronic devices. It can be imagined that the combination of a variety of different harmonics in load currents and unbalanced active and reactive powers can negatively affect the PQ of power networks. In practice, there are a number of compensating schemes and devices commonly used for PQ improvement applications, such as capacitor banks, passive filters, series active power filters, shunt active power filters and their combinations. Of the reported methods, the active power filter (APF) is a very widely adopted solution because, with appropriate control strategy, it is capable of providing simultaneous compensation for voltage sag, harmonic current, imbalance, and active and reactive powers. Moreover, with an external energy source and/or energy storage devices, it can also be used as an uninterruptable power supply (UPS). The main unit of an APF system is a power converter that performs the desired PQ control functions though proper switching control of power electronic devices. An APF is usually connected to the point of common coupling (PCC) in parallel or series to perform its designed functions; however, shunt APFs are more commonly used because they provide more flexible compensation functions though current-injecting control and require less auxiliary equipment. When a series APF and a shunt APF are connected with a common direct current (DC) link, they can be used simultaneously, known as a unified power quality conditioner (UPQC) [1–4].

In open literature, there are a lot of papers investigating the application of APF to the mitigation of PQ issues, such as the improvement of PCC voltage stability [5,6], compensation of harmonic currents [5–7] and unbalance load currents [7,8], injection of active power [8,9], regulation of reactive power [6,9], etc. In order to develop high-performance APFs, it is crucial to improve the switching performance of power converters. In other words, it is very desirable to achieve higher efficiency, shorter response time, and higher power density of power converters used in APFs. To realize the aforementioned objective, wide-bandgap (WBG)-based power switching devices offer a promising solution. Gallium nitride (GaN) is a widely discussed WBG semiconductor material that benefits power-switching device technology with higher voltage, higher switching frequency, higher power, and better high-temperature capability in power switches compared with conventional silicon (Si)-based technologies. It has been expected that GaN high electron-mobility transistors (HEMTs) can greatly enhance the performance of power converters with less than 1-kV power switching requirement [10–12]. However, in open literature, the published papers on GaN HEMTs mostly address the manufacturing, device characteristics, driving, and switching performances [13]. Only two papers regarding GaN-based single-phase APFs were found in the Institute of Electrical and Electronics Engineers (IEEE)/Institution of Engineering and Technology (IET) Electronic Library (IEL) and ScienceDirect OnSite (SDOS) databases [14,15]. In [14], a GaN-based single-phase APF with a modified sigma-delta modulator technique was proposed and verified with simulation studies. In [15], a 5kW single-phase hybrid APF with a new control system was proposed to improve the system performance. There are currently few papers addressing the design issues and reporting performances of GaN-based three-phase inverters in practical application cases. As a result, this paper presents the key design procedure and demonstrates for the first time the performance of a 2-kVA GaN-based three-phase shunt APF system.

Following the introduction in this section, the next section will briefly describe the features of GaN HEMTs and its driving requirements. The third section addresses the mathematical modeling and control strategy of the proposed GaN-based three-phase APF based on synchronous reference frame (SRF) theory. In the fourth section, the proposed APF and control system are simulated using a comprehensive PQ control scenario. Hardware implementation and a test of a 2 kVA prototype are carried out in the fifth section. The sixth section provides some discussions on key technical issues related to the proposed GaN based three-phase APF. Finally, this paper is concluded in the last section.

2. Gallium Nitride (GaN) High Electron-Mobility Transistor (HEMT) and Its Driving Requirements

GaN HEMTs are believed to be the most promising solution for low- to medium-power applications because of their advantages such as higher breakdown voltage, lower on-resistance, and higher switching speed compared with conventional Si-based switching devices; these advantages can increase system efficiency and power density significantly and thus lead to new opportunities for achieving power converters with improved performance. Commercially available GaN HEMTs now achieve up to 650 V/50 A and 900 V/15 A [13].

There are normally on and normally off GaN HEMTs. Normally on GaN HEMTs, also known as depletion mode (D mode) GaN HEMTs, are not popular because normally off switching devices are a common requirement for power converter applications. On the other hand, normally off GaN HEMTs such as enhancement mode (E mode) GaN HEMTs and cascode GaN HEMTs can be turned on with positive V_{GS} and turned off with zero or negative V_{GS}. Generally, the turn-on threshold and highest allowed driving voltages of a GaN HEMT are much smaller than those of conventional Si-based switches. As a result, the careful design of driving circuit is necessary in order to avoid fault turn-on and high overshoot. Common suggestions include providing separate turn-on and turn-off driving paths, achieving minimized overlapping between driving and power loops, using Miller clamp and negative voltage sources to ensure reliable turn-offs, etc. [13].

3. Mathematical Modeling and Control Algorithms of GaN-Based Active Power Filter (APF)

3.1. GaN-Based Three-Phase Active Power Filter

The main power electronic circuit in a three-phase shunt APF system is a three-phase inverter. The main control functions in an APF are to adjust the DC link voltage of the three-phase inverter to the rated value and to compensate reactive power, unbalanced current and harmonic components of the load as required. The circuit architecture of the proposed three-phase inverter is shown in Figure 1. The DC link voltage control adopts dual-loop control schemes, where the inner loop controls inductor currents, and the outer loop controls DC link voltage. By controlling the inductor currents, the goals of regulating DC link voltage and compensating harmonics, unbalanced and reactive components of load currents can be achieved.

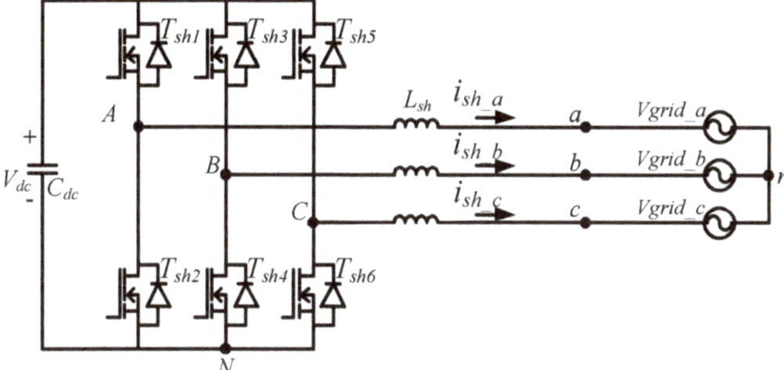

Figure 1. Shunt three-phase inverter.

The circuit specifications of the proposed three-phase inverter developed in this paper are the following: the three-phase line to line voltage of the grid = 110 V_{rms}, grid frequency = 60 Hz, rated power = 2 kVA, DC link voltage = 200 V, switching frequency = 50–100 kHz, DC voltage sensing factor = 0.012, AC current sensing factor = 0.05, AC voltage sensing factor = 0.0062, and DC voltage variation limit = 1%.

3.2. Design of Direct Current (DC) Capacitor and Filter Inductors

The main function of the DC link capacitor is to stabilize DC link voltage. If the DC link capacitance is too large, the dynamic response of the DC link voltage will be slow, and the cost of the APF hardware system will be increased; if the DC capacitor is too small, it will be difficult to suppress the disturbance caused by external power flow. In order to design the appropriate size of the capacitor, we first define instantaneous power of the DC link:

$$P_{dc} = V_{dc}I_{dc} = (\overline{V}_{dc} + \widetilde{V}_{dc})(\overline{I}_{dc} + \widetilde{I}_{dc}), \tag{1}$$

where V_{dc} and I_{dc} represent DC voltage and current, respectively, which can both be separated into their respective DC components ($\overline{V}_{dc}$ and $\overline{I}_{dc}$) and AC components ($\widetilde{V}_{dc}$ and $\widetilde{I}_{dc}$). In order to simplify the analysis, we make three assumptions: the conversion efficiency of the three-phase inverter is 100%,

$\widetilde{V}_{dc}$ is considered zero, and $\bar{I}_{dc}$ is considered zero because $\widetilde{I}_{dc}$ is generally far larger than $\bar{I}_{dc}$. As a result, we obtain the following:

$$P_{dc} \cong \overline{V}_{dc}\widetilde{I}_{dc}(t) = \overline{V}_{dc}C_{dc}\frac{d\widetilde{V}_{dc}(t)}{dt}, \tag{2}$$

where C_{dc} represents DC link capacitance. Then, we obtain the voltage variation of the DC link capacitor:

$$\widetilde{V}_{dc}(t) = \frac{1}{C_{dc}\overline{V}_{dc}}\int_0^t P_{dc}(t)dt, \tag{3}$$

where $\int_0^t P_{dc}(t)dt$ represents the capacity of the three-phase inverter. Then, we obtain DC link capacitance:

$$\Delta V_{dc} = \frac{S}{f_{sw}C_{dc}\overline{V}_{dc}} \Rightarrow C_{dc} = \frac{S}{f_{sw}\Delta V_{dc}\overline{V}_{dc}}, \tag{4}$$

where f_{sw} represents the switching frequency. It should be noted that if an electrolytic capacitor were used for this APF design case, a higher capacitor specification will be required.

The function of the filter inductors is to filter out current ripples caused by the switching of the shunt three-phase inverter. Large inductances suppress the ripples of inductor currents but reduce the response speed of current controllers. On the other hand, although small inductances improve the response speed of the current controller, they cause large current ripples. Therefore, the inductances can be adjusted according to the actual situation. In order to design the filter inductances, we first need the following inductor voltage equation:

$$v(t) = L_{sh}\frac{di_{sh}(t)}{dt}, \tag{5}$$

where L_{sh} represents inductance value, and i_{sh} represents inductor current. According to the relationship between voltage and current on an inductor, (5) can be expressed as follows.

$$\Delta I_{sh} = \frac{\frac{D}{2} \times T_{sw} \times (V_{dc} - V_{grid})}{L_{sh}}, \tag{6}$$

where ΔI_{sh} represents shunt inductor current ripple, D represents duty cycle, T_{sw} represents switching period, and V_{grid} represents grid voltage. The duty cycle can be expressed the following:

$$D(\omega t) = m_a \sin(\omega t), \tag{7}$$

where m_a represents modulation factor and equals modulation signal divided by triangular wave amplitude (v_{con}/v_{tri}). Then, we get output AC voltage:

$$V_{sh}(\omega t) = V_{dc}m_a \sin(\omega t). \tag{8}$$

Substituting (7) and (8) into (6) yields the following:

$$\Delta I_{sh} = \frac{V_{dc} \times T_{sw}}{2L_{sh}} m_a \sin(\omega t)[1 - m_a \sin(\omega t)]. \tag{9}$$

Then, we differentiate (9) and let the result be zero in order to obtain the maximum value of inductor current ripple:

$$\frac{d\Delta I_{sh}(\omega t)}{d\omega t} = \frac{V_{dc}T_{sw}}{2L_{sh}} m_a[\cos(\omega t) - 2m_a \sin(\omega t)\cos(\omega t)] = 0. \tag{10}$$

As a result,

$$\sin(\omega t) = \frac{1}{2m_a}. \tag{11}$$

Lastly, substituting (11) into (9) yields the following equation:

$$L_{sh} = \frac{V_{dc}}{8 f_{sw} \Delta I_{sh}}. \tag{12}$$

According to the circuit specifications of the three-phase inverter and commonly assumed inductor current ripple, 10% of output current, it is calculated that the required inductance should be at least larger than 500 μH.

3.3. Mathematical Modeling and Controller's Design for GaN-Based Shunt APF

3.3.1. Mathematical Modeling

The mathematical model of the shunt-connected three-phase inverter can be derived according to Figure 1. First, the following equations are obtained with Kirchhoff's voltage law:

$$L_{sh}\frac{di_{sh\_a}}{dt} = v_{AN} - v_{grid\_a} - v_{nN}, \tag{13}$$

$$L_{sh}\frac{di_{sh\_b}}{dt} = v_{BN} - v_{grid\_b} - v_{nN}, \tag{14}$$

$$L_{sh}\frac{di_{sh\_c}}{dt} = v_{CN} - v_{grid\_c} - v_{nN}, \tag{15}$$

where i_{sh_a}, i_{sh_b}, and i_{sh_c} represent three-phase inductor currents, v_{AN}, v_{BN}, and v_{CN} represent switching point voltages, V_{grid_a}, V_{grid_b}, and V_{grid_c} represent three-phase grid voltages, and v_{nN} represents the voltage between the grid ground and the inverter ground. Also, the three-phase three-wire system satisfies the following condition:

$$i_{sh\_a} + i_{sh\_b} + i_{sh\_c} = 0. \tag{16}$$

As a result, v_{nN} can be expressed as the following:

$$v_{nN} = \frac{(v_{AN} + v_{BN} + v_{CN}) - (v_{grid\_a} + v_{grid\_b} + v_{grid\_c})}{3}. \tag{17}$$

Substituting Equation (17) into Equations (13)–(15) yields the following:

$$\begin{bmatrix} L_{sh}\frac{di_{sh\_a}}{dt} \\ L_{sh}\frac{di_{sh\_b}}{dt} \\ L_{sh}\frac{di_{sh\_c}}{dt} \end{bmatrix} = \frac{2}{3}\begin{bmatrix} 1 & \frac{-1}{2} & \frac{-1}{2} \\ \frac{-1}{2} & 1 & \frac{-1}{2} \\ \frac{-1}{2} & \frac{-1}{2} & 1 \end{bmatrix}\left(\begin{bmatrix} v_{AN} \\ v_{BN} \\ v_{CN} \end{bmatrix} - \begin{bmatrix} v_{grid\_a} \\ v_{grid\_b} \\ v_{grid\_c} \end{bmatrix}\right). \tag{18}$$

In this study, pulse width modulation (PWM) is used in the control, where the three-phase modulation signals v_{cona}, v_{conb}, and v_{conc} are compared with v_{tri} respectively to trigger the switches of all three switching legs. The output voltages of the switching legs can be expressed as follows:

$$v_{aN} = (\frac{1}{2} + \frac{v_{cona}}{2v_{tri}})V_{dc}; \tag{19}$$

$$v_{bN} = (\frac{1}{2} + \frac{v_{conb}}{2v_{tri}})V_{dc}; \tag{20}$$

$$v_{cN} = (\frac{1}{2} + \frac{v_{conc}}{2v_{tri}})V_{dc}. \tag{21}$$

Substituting Equations (19)–(21) into Equation (18) and letting $V_{dc}/2V_{tri} = K_{pwm}$ yield the following:

$$\begin{bmatrix} L_{sh}\frac{di_{sh\_a}}{dt} \\ L_{sh}\frac{di_{sh\_b}}{dt} \\ L_{sh}\frac{di_{sh\_c}}{dt} \end{bmatrix} = \frac{2}{3}\begin{bmatrix} 1 & \frac{-1}{2} & \frac{-1}{2} \\ \frac{-1}{2} & 1 & \frac{-1}{2} \\ \frac{-1}{2} & \frac{-1}{2} & 1 \end{bmatrix}\left(K_{pwm}\begin{bmatrix} v_{cona} \\ v_{conb} \\ v_{conc} \end{bmatrix} - \begin{bmatrix} v_{grid\_a} \\ v_{grid\_b} \\ v_{grid\_c} \end{bmatrix}\right). \tag{22}$$

Using SRF theory, Equation (22) can be converted into the following:

$$\begin{bmatrix} L_{sh}\frac{dI_{sh\_d}}{dt} \\ L_{sh}\frac{dI_{sh\_q}}{dt} \\ L_{sh}\frac{dI_{sh\_0}}{dt} \end{bmatrix} = K_{pwm}\begin{bmatrix} 1 & 0 & 0 \\ 0 & 1 & 0 \\ 0 & 0 & 1 \end{bmatrix}\begin{bmatrix} v_{cond} \\ v_{conq} \\ v_{con0} \end{bmatrix} - \begin{bmatrix} 1 & 0 & 0 \\ 0 & 1 & 0 \\ 0 & 0 & 1 \end{bmatrix}\begin{bmatrix} V_{grid\_d} \\ V_{grid\_q} \\ V_{grid\_0} \end{bmatrix} - \begin{bmatrix} 0 & \omega L_{sh} & 0 \\ -\omega L_{sh} & 0 & 0 \\ 0 & 0 & 0 \end{bmatrix}\begin{bmatrix} I_{sh\_d} \\ I_{sh\_q} \\ I_{sh\_0} \end{bmatrix}. \tag{23}$$

3.3.2. Design of Current Controllers

According to Equation (23), we can obtain block diagrams of direct-quadrature axis (d-q axis) current loops with type-II controllers as shown in Figures 2 and 3, where k_s and k_v represent AC current and voltage-sensing factors, respectively. Under ideal feed-forward conditions, the transfer function of current loop (d-axis or q-axis) is as follows:

$$H_i(s) = \frac{k_s K_{pwm}}{sL_{sh}}. \tag{24}$$

The transfer function of the adopted type II controller, which consists of a proportional-integral (PI) controller and a low pass filter (LPF), is as follows:

$$G_i(s) = \frac{k(s+z)}{s(s+p)}. \tag{25}$$

The loop gain can be expressed as follows:

$$L_i(s) = G_i(s)H_i(s) = \frac{k(s+z)}{s(s+p)}\frac{k_s K_{pwm}}{sL_{sh}}. \tag{26}$$

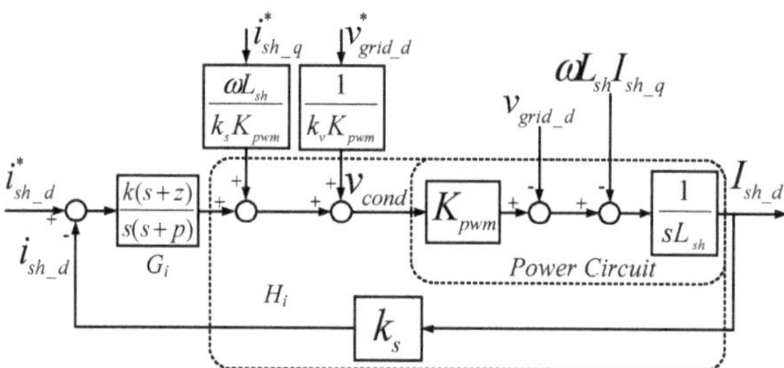

Figure 2. Block diagram of d-axis current controller.

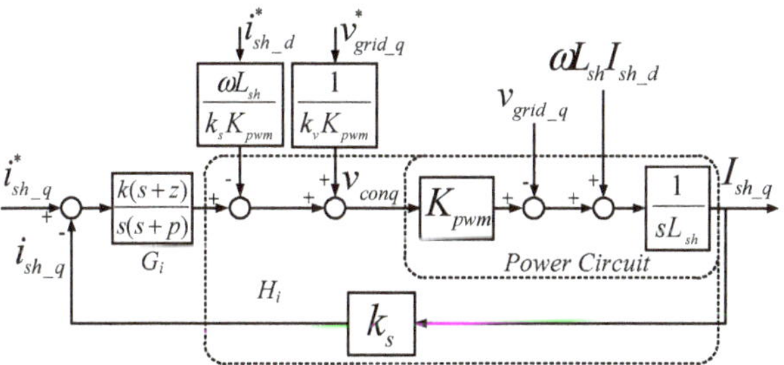

Figure 3. Block diagram of q-axis current controller.

In this application case, the crossover frequency of a Type II controller is designed within the range of 1/4 to 1/10 of the switching frequency. This paper chooses the controller crossover frequency to be 1/10 of the switching frequency, the zero is designed at 1/4 of the crossover frequency, and the cut-off frequency of the LPF is designed to be 15 kHz:

$$\omega_i = 0.1 \times 50k \times 2\pi = 31416 rad/s. \tag{27}$$

$$z = \omega_i/4 = 7854 rad/s. \tag{28}$$

$$p = 2\pi \times 15k = 94248 rad/s. \tag{29}$$

It follows that the gain of the plant at crossover frequency (Gain_{Hi}) is as follows:

$$\begin{array}{c} \text{Gain}_{\text{Hi}} = \frac{k_s K_{pwm}}{sL_{sh}} = \frac{2000}{j\omega} = 0 - j0.0637 \\ \Rightarrow |\text{Gain}_{\text{Hi}}| = 0.0637 \end{array} \tag{30}$$

The gain of the controller at crossover frequency (Gain_{Gi1}) is as follows:

$$\begin{array}{c} \text{Gain}_{\text{Gi1}} = \frac{(s+z)}{s(s+p)} = \frac{j\omega_i + 7854}{j\omega_i(j\omega_i + 94248)} = 8.7535 \times 10^{-6} - j5.5704 \times 10^{-6} \\ \Rightarrow |\text{Gain}_{\text{Gi1}}| = 1.0376 \times 10^{-5} \end{array} \tag{31}$$

Then, the required gain for compensation at crossover frequency can be calculated:

$$k = \frac{1}{|\text{Gain}_{\text{Hi}}| \times |\text{Gain}_{\text{Gi1}}|} = 1.5139 \times 10^6. \tag{32}$$

Finally, the transfer function is obtained as follows:

$$G_i(s) = \frac{1.5139 \times 10^6 (s + 7854)}{s(s + 94248)}. \tag{33}$$

The designed k_P and k_I are 16.0633 and 2.5232956, respectively. Figure 4 shows the Bode plot of the controller and plant, where the designed phase margin is 58 degrees.

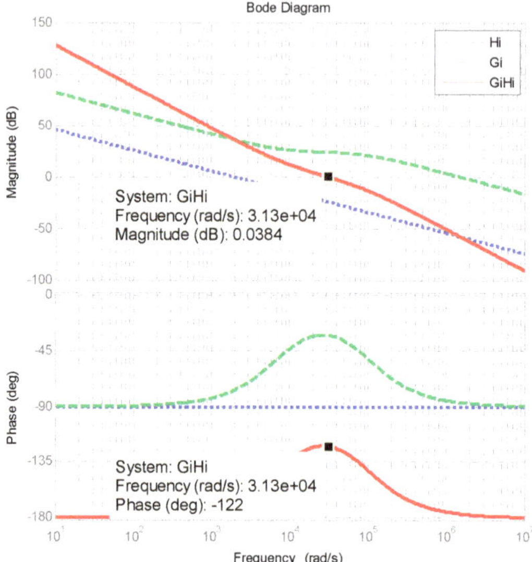

Figure 4. Bode plot of shunt active power filter (APF) inductor current control loop.

3.3.3. Design of DC Link Voltage Controller

The DC link voltage control loop regulates the real power balancing between the alternating current (AC) and DC terminals of the three-phase inverter. By ignoring steady-state operating point, we can obtain equivalent small signal model of the voltage loop as shown in Figure 5.

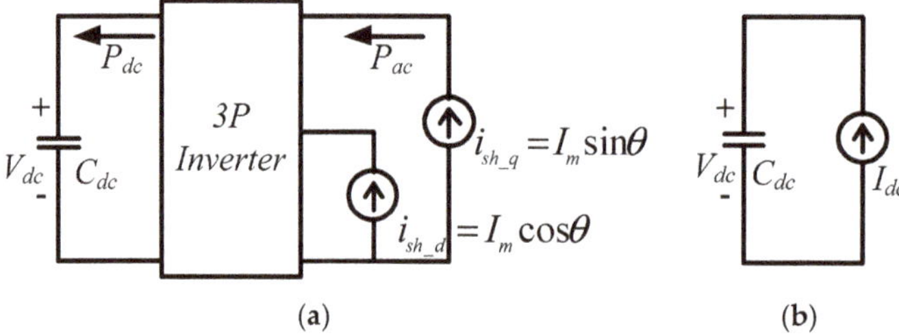

Figure 5. Equivalent circuits of the voltage control loop: (**a**) equivalent circuit under synchronous reference frame; (**b**) equivalent circuit on direct current (DC) side.

The instantaneous AC power at the AC side can be defined as follows:

$$P_{ac} = V_m \sin\theta * I_m \sin\theta + V_m \cos\theta * I_m \cos\theta, \tag{34}$$

where V_m and I_m represent the maximum voltage and current under dq axes, respectively. According to trigonometric functions, Equation (34) can be simplified as follows:

$$P_{ac} = V_m I_m. \tag{35}$$

Mapping the AC side signals onto the DC side and assuming that the inverter is lossless, we obtain the following:

$$P_{ac} = P_{dc}; \tag{36}$$

$$V_m I_m = V_{dc} I_{dc}. \tag{37}$$

Then, we can obtain the relationship between the DC side current and the AC side current:

$$I_{dc} = \frac{V_m}{V_{dc}} I_m = k_{dc} I_m; \tag{38}$$

$$C_{dc} \frac{dV_{dc}}{dt} = I_{dc} \Rightarrow V_{dc} = I_{dc} \frac{1}{sC_{dc}}, \tag{39}$$

where k_{dc} represents the conversion factor from AC side to DC side. According to Equations (38) and (39), we can obtain the transfer function of DC side voltage:

$$\frac{V_{dc}}{I_m} = \frac{k_{dc}}{sC_{dc}}, k_{dc} = \frac{V_m}{V_{dc}}. \tag{40}$$

According to the above derivations, we can obtain the block diagram of a DC link voltage control loop with a type-II controller as shown in Figure 6, where k_{vd} and k_s represent the sensing factors of DC voltage and AC current, respectively. Therefore, the transfer function of the DC voltage loop is as follows:

$$H_{dc}(s) = \frac{k_{vd} k_{dc}}{k_s C_{dc} s}. \tag{41}$$

The transfer function of the Type II controller is defined as follows:

$$G_v(s) = \frac{k(s+z)}{s(s+p)}. \tag{42}$$

It follows that the loop gain can be expressed as follows:

$$L_v(s) = G_v(s) H_{dc}(s) = \frac{k(s+z)}{s(s+p)} \frac{k_{vd} k_{dc}}{k_s C_{dc} s}. \tag{43}$$

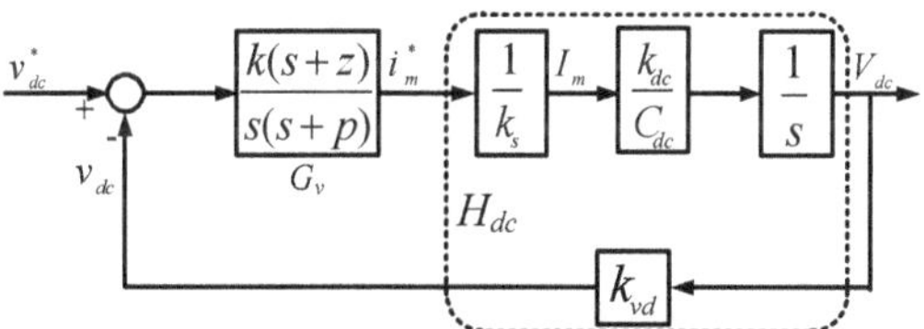

Figure 6. Block diagram of DC voltage loop type II controller.

The main purpose of the Type II controller is to use an LPF to reduce possible interference affecting the DC link when the APF system compensates for PQ problems such as imbalance and harmonics in three-phase load currents. The crossover frequency is set at 1/500 of that of the current loop, the cut-off frequency of the LPF is set at 49 Hz, and the zero is designed at 1/5 of the crossover frequency of the DC loop:

$$\omega_v = \omega_i \times 0.002 = 62.832 rad/s; \tag{44}$$

$$p = 2\pi \times 49 = 307.8768 rad/s; \tag{45}$$

$$z = \omega_v / 5 = 12.5664 rad/s. \tag{46}$$

The gain of the plant at crossover frequency (Gain_{Hdc}) can be calculated as follows:

$$\begin{aligned} \text{Gain}_{\text{Hdc}} &= \tfrac{k_{vd} k_{dc}}{k_s C_{dc} s} = \tfrac{118.9}{j\omega_v} \\ &= 0 - j1.8917 \Rightarrow |\text{Gain}_{\text{Hdc}}| = 1.8917 \end{aligned} \tag{47}$$

The gain of the controller at the crossover frequency (Gain_{Gv1}) is as follows:

$$\begin{aligned}\text{Gain}_{\text{Gv1}} &= \frac{(s+z)}{s(s+p)} = \frac{j\omega_v+12.57}{j\omega_v(j\omega_v+307.88)} \\ &= 0.003 - j0.0013 \\ &\Rightarrow |\text{Gain}_{\text{Gv1}}| = 0.0032\end{aligned} \tag{48}$$

Then, the required gain compensation at the designed crossover frequency can be calculated by:

$$k = \frac{1}{|\text{Gain}_{\text{Hdc}}| \times |\text{Gain}_{\text{Gv1}}|} = 165.1953. \tag{49}$$

Finally, the transfer function of the voltage controller is obtained:

$$G_v(s) = \frac{165.1953(s+12.5664)}{s(s+307.8768)}. \tag{50}$$

The designed k_P and k_I are 0.5286 and 0.0001329397, respectively. Figure 7 shows the Bode plot of the controller and plant, where phase margin is 67 degrees.

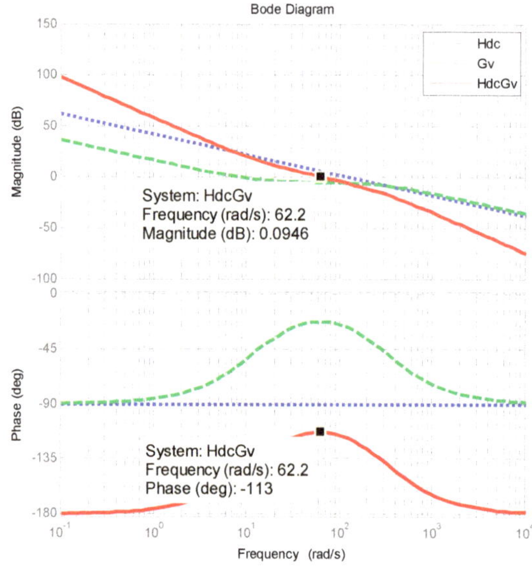

Figure 7. Bode plot of shunt APF inductor DC link voltage control loop.

3.3.4. Load Current Compensation Signals of APF

Using the SRF conversion technique, distorted and unbalanced three-phase load currents can be expressed as follows:

$$\begin{bmatrix} i_{Ld} \\ i_{Lq} \end{bmatrix} = \begin{bmatrix} \bar{i}_{Ld} \\ \bar{i}_{Lq} \end{bmatrix} + \begin{bmatrix} \tilde{i}_{Ld} \\ \tilde{i}_{Lq} \end{bmatrix}, \tag{51}$$

where i_{Ld} and i_{Lq} represent dq-axis load currents, $\bar{i}_{Ld}$ and $\bar{i}_{Lq}$ represent dq-axis load currents with the fundamental frequency, and $\tilde{i}_{Ld}$ and $\tilde{i}_{Lq}$ represent the dq-axis components that require compensation. In order to obtain the compensation signals of the active current (q axis) $i_{Lq}{}^*$, i_{Lq} is firstly filtered with an LPF and then subtracted from q-axis current feedback signal (i_{Lq}), while the compensation signals of the reactive current (d axis) $i_{Ld}{}^*$ equals the whole d-axis current feedback signal (i_{Ld}), as shown in Figure 8.

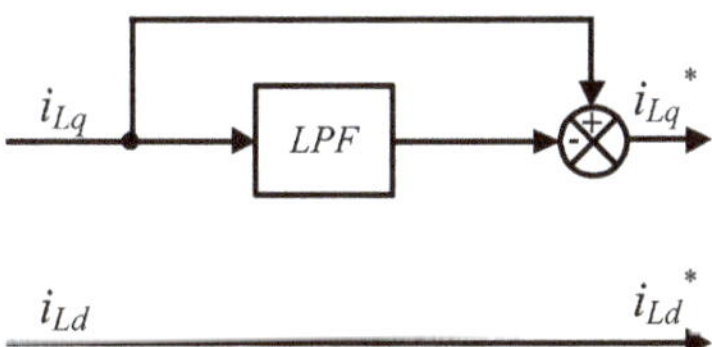

Figure 8. The direct-quadrature axis currents compensation signals of the APF.

3.4. Complete System of GaN-Based Shunt APF

According to Figure 8, DC link voltage controller, and inductor current controllers, we can obtain the circuit configuration of the proposed GaN based three-phase APF system with the block diagram of complete control architecture, as shown in Figure 9.

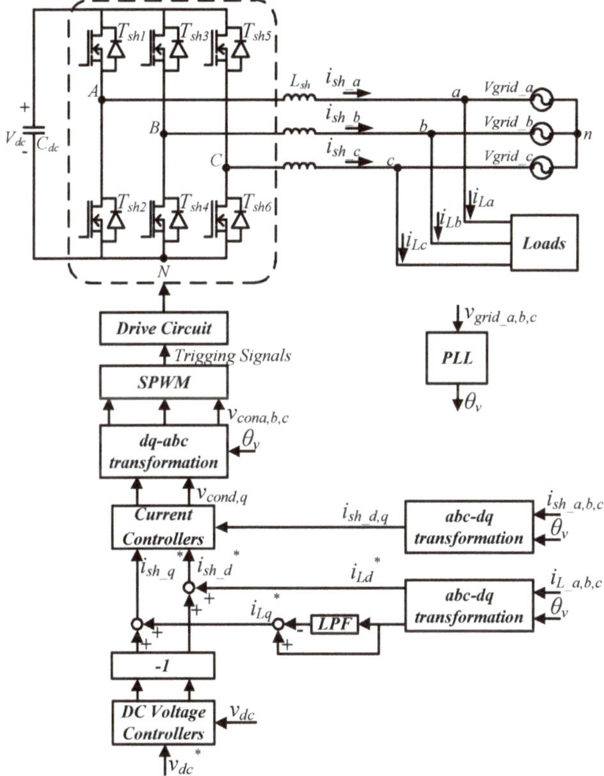

Figure 9. The circuit configuration of gallium nitride (GaN)-based shunt APF system and the block diagram of the control scheme.

4. Simulation Study and Results

With the design presented in the previous section, the proposed Gan-based shunt-type APF is tested for an integrated compensation of multiple power quality problems, including current harmonics, load current imbalance, and reactive currents. Powersim (PSIM) software is used to perform the simulation case of the abovementioned comparison tasks. The PSIM simulation model is shown in Figure 10.

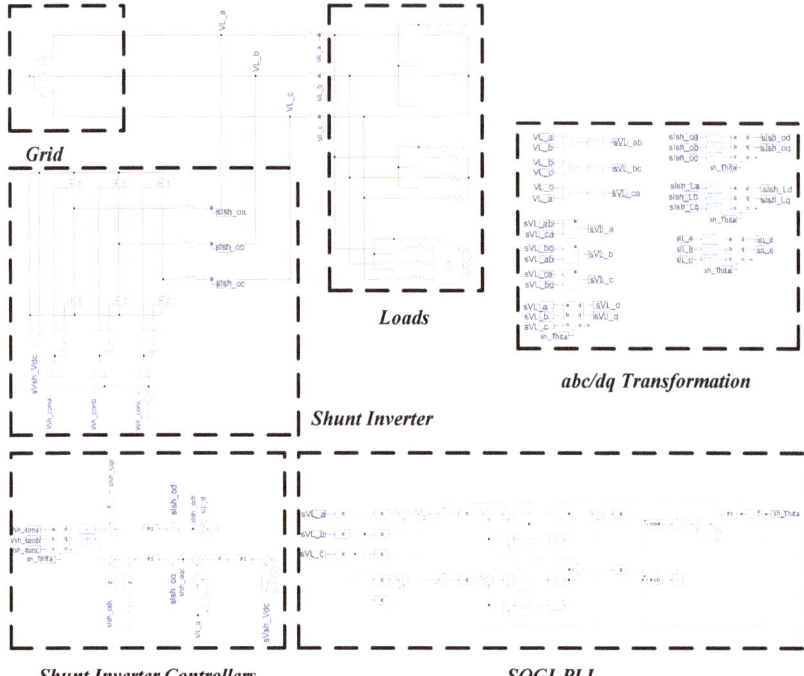

Figure 10. Powersim (PSIM) simulation model of the proposed APF system.

4.1. Simulation Scenario

To demonstrate the performance of the proposed controllers, the integrated compensation for multiple load current quality problems with APF is simulated. In this case, the three-phase load bank consists of a balanced reactive load, an unbalance resistive load, and a non-linear load, as shown in Figure 11. Table 1 shows the detailed values of the loads used. At first (t_0–t_1), the shunt APF, connected to a three-phase power grid with the line to line voltage of 110 V, 60 Hz, adjusts the DC link voltage to 200 V and the compensation function is not activated; at t_1, compensation is activated to achieve a set of balanced grid currents, zero distortion, and unit power factor (PF), as shown in Figure 12. Figures 13–17 show the corresponding simulation results, and Table 2 shows root-mean-square (RMS) currents and total harmonic distortion (THD) data before and after compensation.

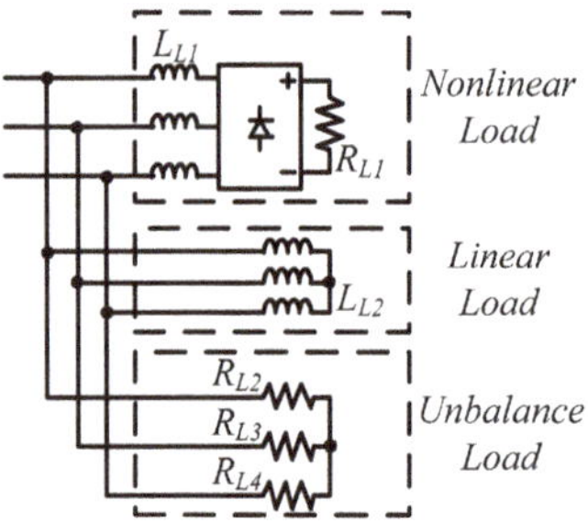

Figure 11. Load condition in simulated scenario.

Table 1. Load parameters for simulation scenario.

Nonlinear Load		Linear Load	Unbalanced Load	
R_{L1}	L_{L1}	L_{L2}	R_{L2}	R_{L3}, R_{L4}
50 Ω	6 mH × 3	152 mH × 3	∞ Ω	50 Ω

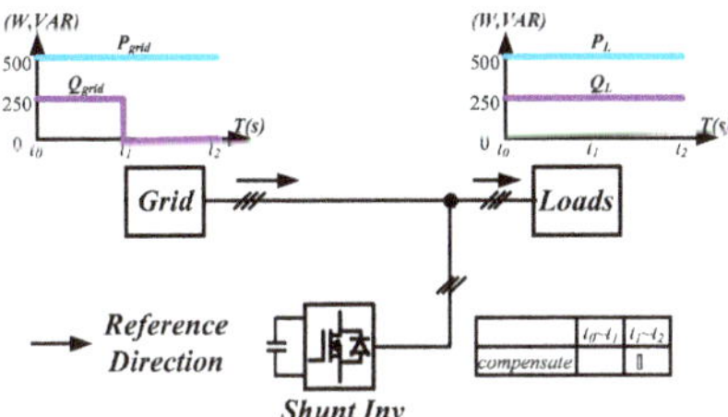

Figure 12. Schematic diagram of simulated scenario.

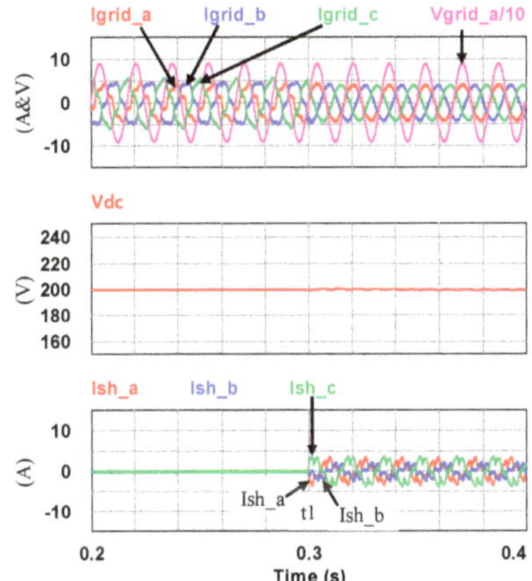

Figure 13. The grid-side phase-a voltage and three-phase currents/DC link voltage/shunt APF three-phase currents (t_0–t_2).

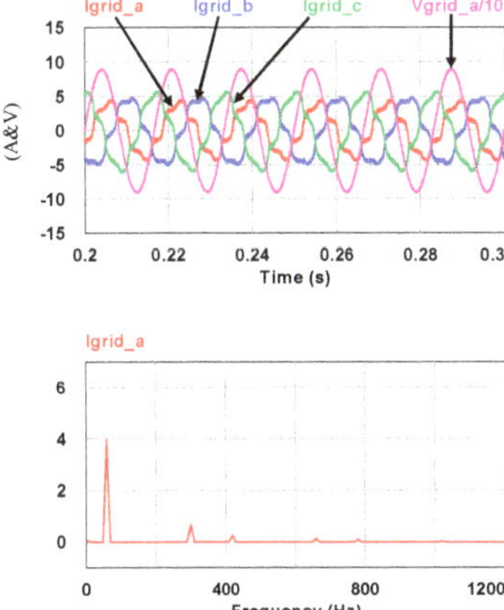

Figure 14. Before t_1: the grid phase-a voltage and three-phase currents/the fast Fourier transform (FFT) waveform of the grid phase-a current.

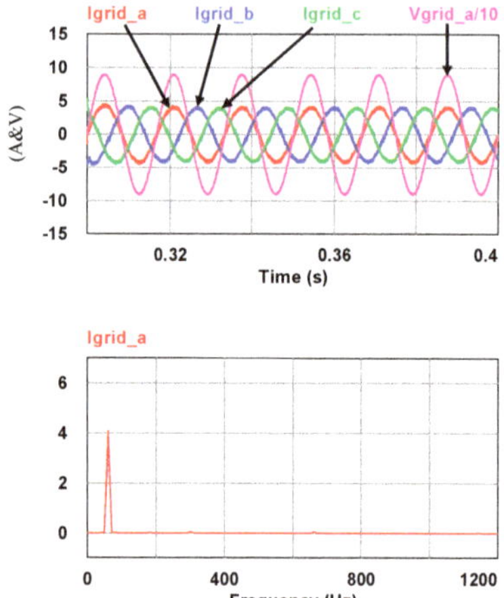

Figure 15. After t_1: the grid phase-a voltage and three-phase currents/FFT waveform of the grid phase-a current.

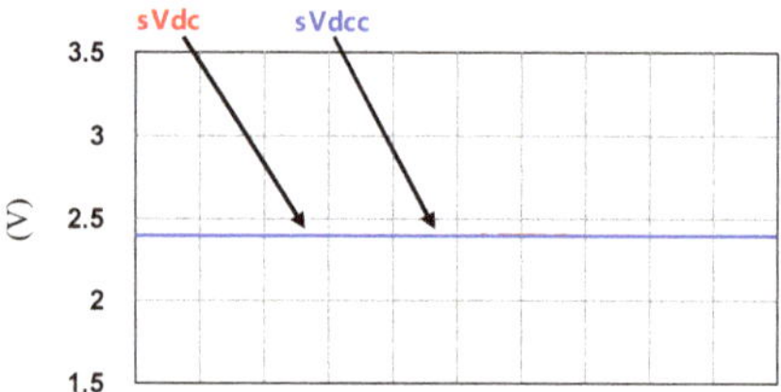

Figure 16. Shunt APF DC link voltage command and feedback signals (t_0–t_2).

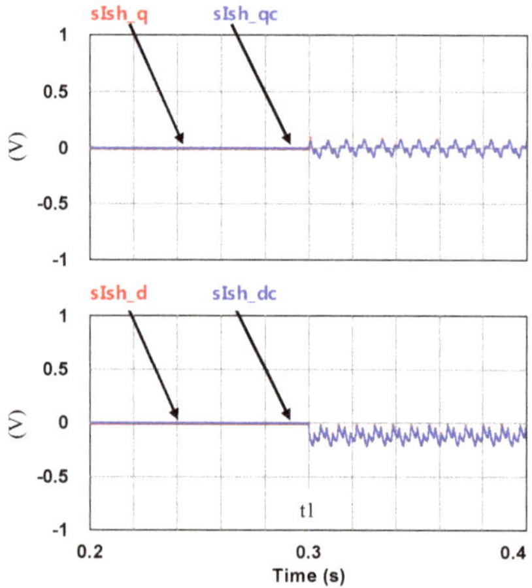

Figure 17. Shunt APF dq-axis current commands and feedbacks (t_0–t_2).

Table 2. Root-mean-square (RMS) currents and total harmonic distortion (THD).

Power Grid Currents and Unbalance Ratio (UR)		
Item	**Without APF (t_0–t_1)**	**With APF (t_1–t_2)**
i_{grid_a} (A)	2.9	2.84
i_{grid_b} (A)	3.42	2.87
i_{grid_c} (A)	3.87	2.89
UR (%)	14.91	0.94
THD		
Item	**Without APF (t_0–t_1)**	**With APF (t_1–t_2)**
THD (i_{grid_a}) (%)	19.18	3.91
THD (i_{grid_b}) (%)	15.56	3.94
THD (i_{grid_c}) (%)	13.89	3.94

5. Hardware Implementation and Test Results

To verify the performance of the proposed GaN-based APF, this section presents the implementation of APF hardware prototype for verification and analysis based on the scenario arranged in the simulation case stated in the previous section. The photograph of the constructed GaN-based APF prototype is shown in Figure 18, where the numbered devices are listed in Table 3. A programmable three-phase AC power supply is adopted to emulate the grid voltage. The Texas Instruments (TI) microcontroller, TMS320F28335 (Texas Instruments, Dallas, TX, USA), is used to provide efficiency and flexibility in controller design. The system parameters and conditions of the experimental tests and measurement scenarios are the same as that used in the previous simulation case presented in Section 4.1. Figures 19–24 show a set of test results; Figure 19 shows the waveforms of measured phase-a voltage and three-phase currents of the grid from t_0 to t_2. Figure 20 shows the DC link voltage and the output three-phase currents of the shunt APF from t_0 to t_2. The related waveforms of grid phase-a voltage and three-phase currents and the fast Fourier transform (FFT) of the grid phase-a current before the before and after the APF is activated are shown in Figures 21 and 22, respectively. As can be seen in Figure 22, after the APF is activated the unbalanced and distorted currents have been well compensated and the current is in phase with the grid voltage achieving the control objective of unity power factor. To demonstrate the performance of the designed controllers, Figure 23 shows the command and feedback signals of DC link voltage and the PI controller output signals. The dq-axis current commands and feedback signals are shown in Figure 24. To provide a set of quantitative results, Table 4 shows the measured RMS currents and calculated THD data before and after compensation. In the stage of hardware construction and tests, the system efficiencies at different switching frequencies are also explored. The arrangement of the test scenario and the detailed results are presented in the next section.

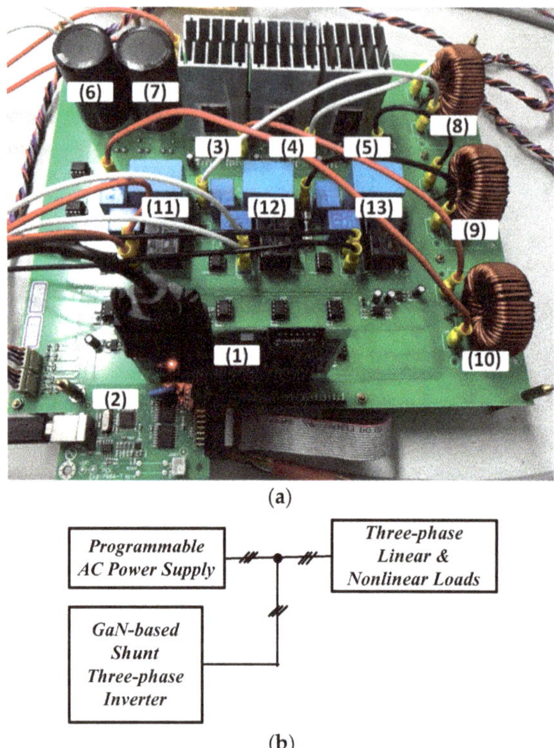

Figure 18. (**a**) Photo of GaN-based three-phase APF hardware; (**b**) schematic of the hardware test and system.

Table 3. Devices in Figure 18a.

Number	Device	Value/Part Number
(1)	Microcontroller	TMS320F28335
(2)	Interface circuit of the microcontroller	N/A
(3)–(5)	GaN HEMT pairs	TPH3207
(6) and (7)	DC link capacitors	680 mF/450 V
(8)–(10)	Filter inductors	0.5 mH
(11)–(13)	Filter capacitors	10 μF/300 V

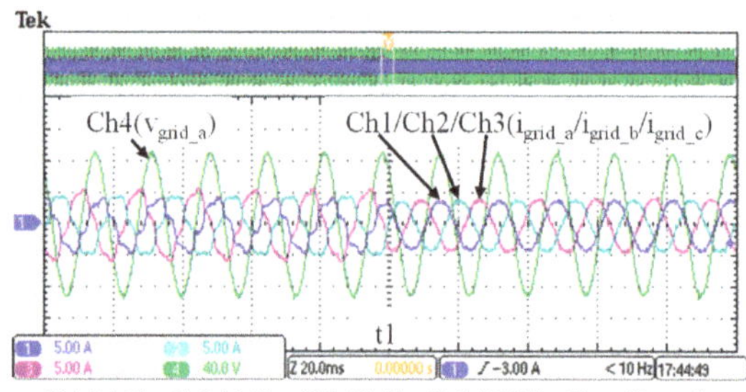

Figure 19. Grid phase-a voltage and three-phase currents (t_0–t_2).

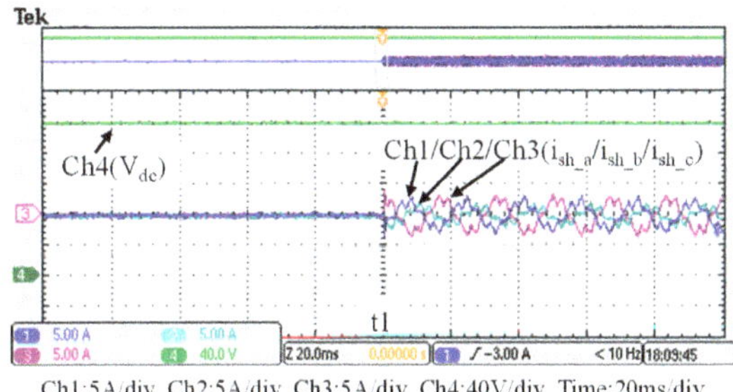

Figure 20. DC link voltage and the shunt APF three-phase currents (t_0–t_2).

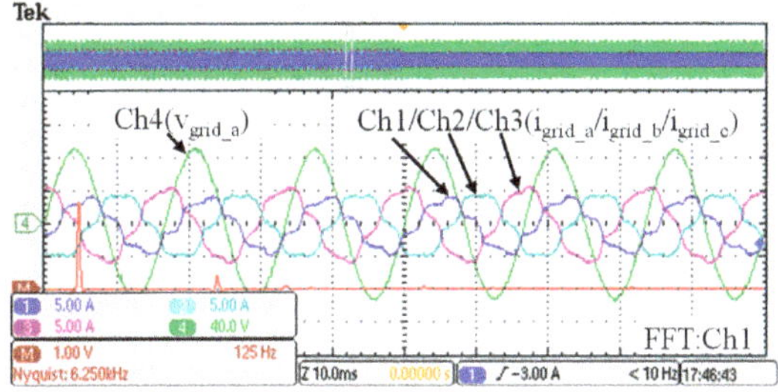

Figure 21. Before t_1: the grid phase-a voltage and three-phase currents and the fast Fourier transform (FFT) waveform of the grid phase-a current.

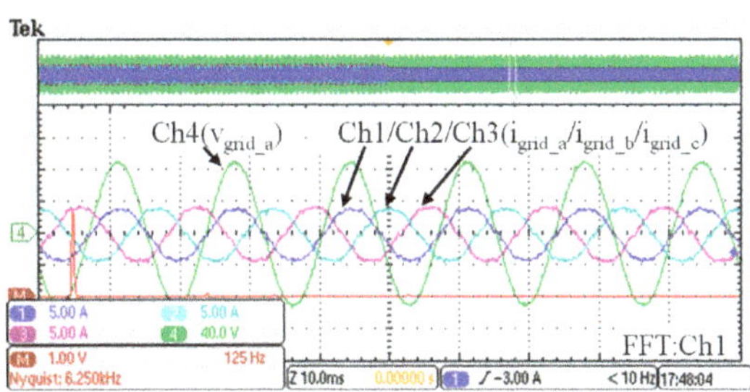

Figure 22. After t_1: the grid phase-a voltage and three-phase currents and FFT waveform of the grid phase-a current.

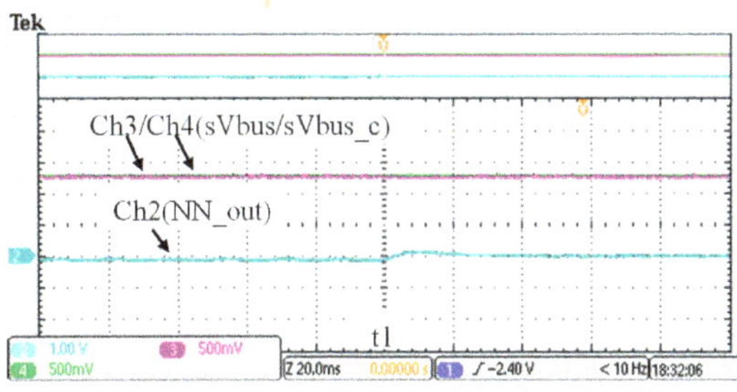

Figure 23. The command and feedback signals of DC link voltage and the proportional-integral (PI) controller output signal (t_0–t_2).

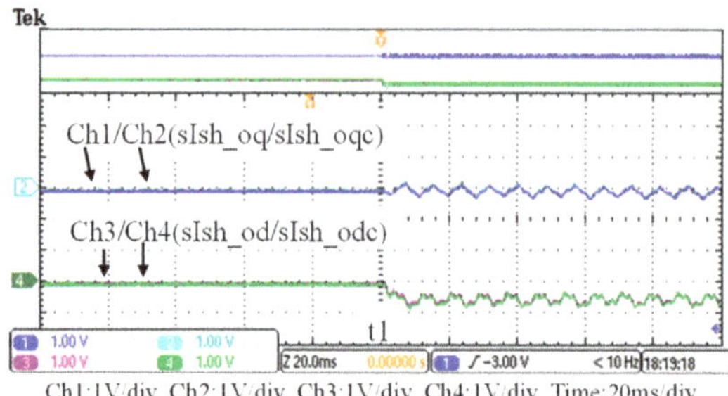

Figure 24. The shunt APF dq-axis current commands and feedbacks (t_0–t_2).

Table 4. RMS currents and THD.

Power Grid Currents and Unbalance Ratio (UR)		
Item	**Without APF (t_0–t_1)**	**With APF (t_1–t_2)**
i_{grid_a} (A)	2.93	2.91
i_{grid_b} (A)	3.43	2.94
i_{grid_c} (A)	3.95	3.02
UR (%)	14.83	2.13
	THD	
Item	**Without APF (t_0–t_1)**	**With APF (t_1–t_2)**
THD (i_{grid_a}) (%)	20.23	4.15
THD (i_{grid_b}) (%)	16.57	4.08
THD (i_{grid_c}) (%)	15.48	4.05

6. Discussion

6.1. The Analysis of System Efficiency

In this paper, the system efficiencies at different switching frequencies (50 kHz, 80 kHz, and 100 kHz) and under different load levels are explored. Figure 25 shows the system block diagram of the efficiency tests. In this test, the AC terminals of the proposed three-phase GaN-based APF are connected to the three-phase power grid having the line to line voltage of 110 V and its DC terminal voltage is regulated at the designed 200 V by the proposed voltage controller of the APF. For testing the APF efficiencies under different load levels, a programmable electronic load is connected to the DC terminal of the APF. By setting different P_{dc} and measuring the corresponding P_{ac}, the efficiency at a specific power level and switching frequency can be readily calculated. In this paper, three switching frequencies, i.e., 50, 80, and 100 kHz were tested. The calculated results are graphically shown in Figure 26. As can be seen in Figure 26, the maximum efficiency appears at about 50% of the rated load and it is found that when the switching frequency increases the efficiency decreases. This is mainly due to the increase in switching losses.

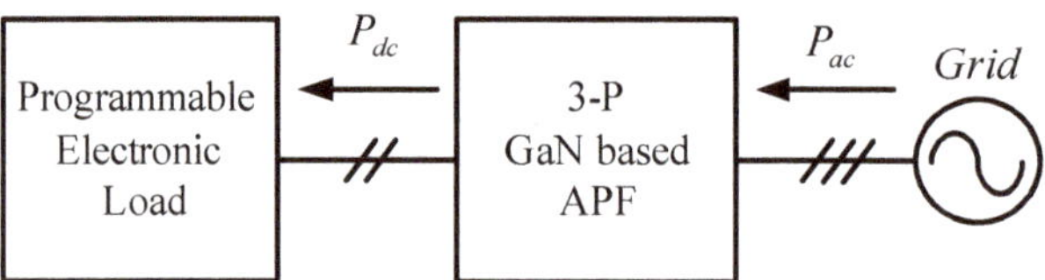

Figure 25. The system block diagram of the efficiency tests.

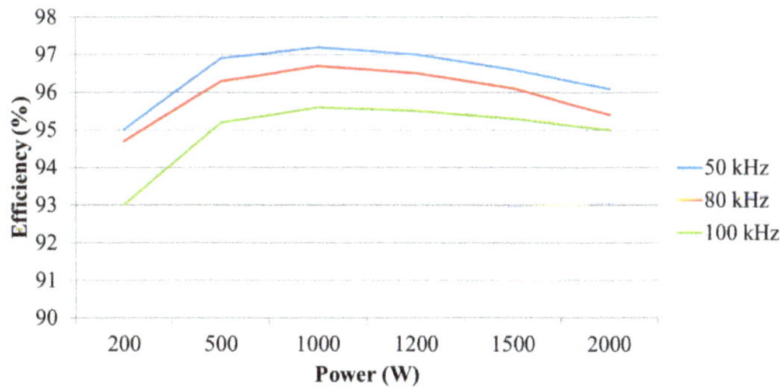

Figure 26. Efficiencies of GaN-based shunt APF system at different switching frequencies.

In the aspect of the efficiency comparison with different technologies, it is indeed difficult to establish a fair comparative basis due to some system constrains, e.g., the switching technique used, system capacities, quality of components used, and the control functions designed for the circuits. To provide a set comparative result, three recently published technical papers [16–18] using different technologies are summarized in Table 5.

Table 5. Performance comparison of different technologies.

Paper	Switching Device	Function	Power	Switching Frequency	Efficiency
[16]	IGBT	Motor drive	8 kW	20 kHz	95.5%
[17]	MOSFET	Motor drive	1.5kW	15 kHz	92%
[18]	SiC	Electric vehicle	8.8 kW	50 kHz	97%
proposed	GaN	Active power filter	2 kVA	50 kHz	97.2%
				80 kHz	96.7%
				100 kHz	95.6%

6.2. The Thermographic Analysis of the System

Figure 27 shows a set of thermographic photos (using FLIER-E63900, T198547, E4) of the proposed GaN-based three-phase APF prototype operating under the rated capacity of 2 kW with different switching frequencies. As can be seen in the photos, the temperature of the switching devices increases as the switching frequency increases and the temperature of capacitors and the circuit board remain under 25 °C. Table 6 shows the summary of the recorded temperature data gathered from the presented thermographic photos. Based on the results of thermographic analysis, it has been found that the greatest losses are located at the six power-switching devices and the three relays which are designed for ensuring a successful synchronizing control with the power grid to which the proposed APF is connected. It should be noted that in practice, these relays can be removed after the overall control system of the APF has been properly tuned. This means that the maximum efficiency of the proposed GaN-based APF can be further improved.

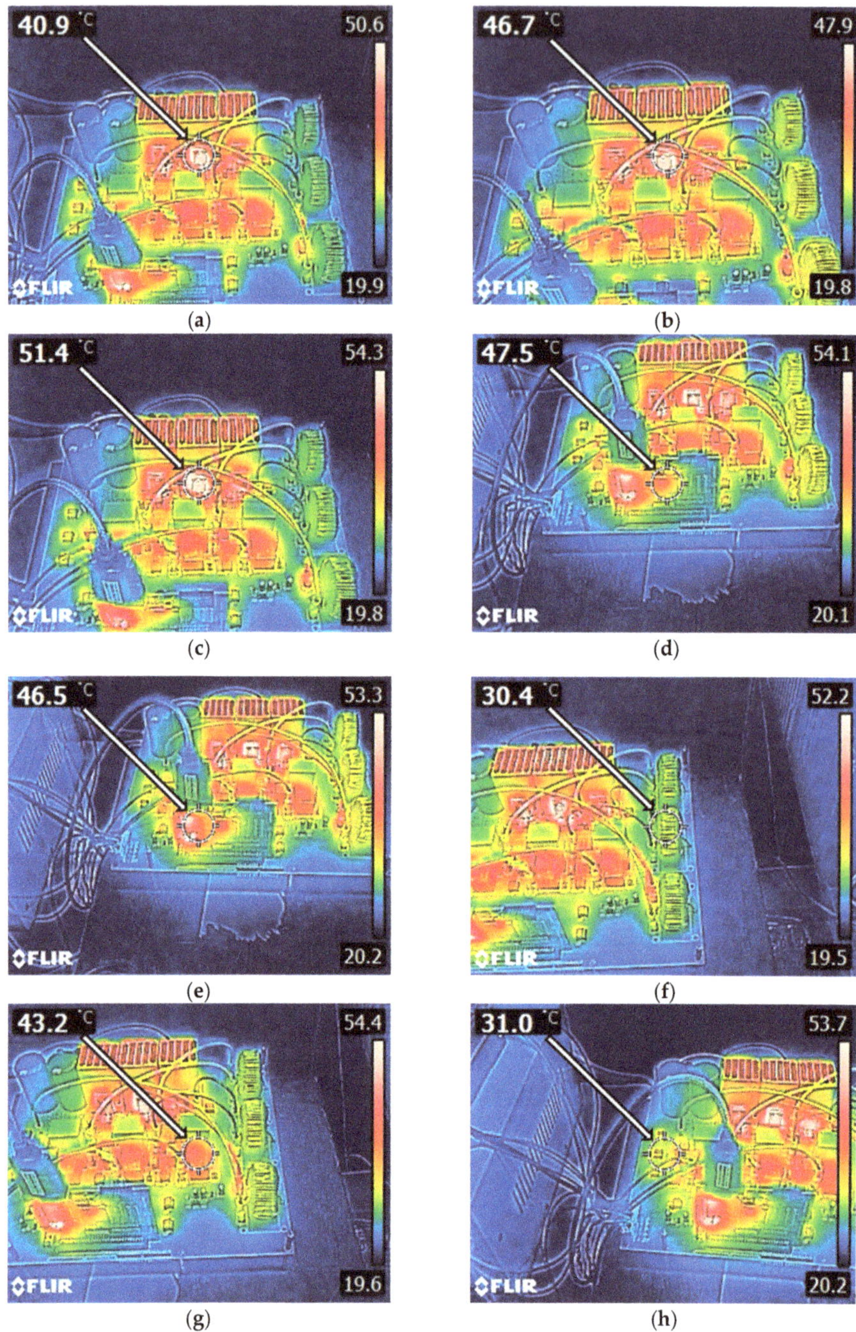

Figure 27. Thermographic photos of the proposed GaN-based three-phase APF prototype: (**a**) TPH3207 device switching at 50 kHz; (**b**) TPH3207 device switching at 80 kHz; (**c**) TPH3207 device switching at 100 kHz; (**d**) DSP; (**e**) communication interface; (**f**) inductors; (**g**) relay; (**h**) signal processing integrated circuits (ICs).

Table 6. Summary of operating temperatures of individual devices.

Sensed Object	Operating Temperature (°C)
TPH3207 @ 50kHz switching frequency	40.9
TPH3207 @ 80kHz switching frequency	46.7
TPH3207 @ 100kHz switching frequency	51.4
DSP	47.5
Communication interface	46.5
Inductors	30.4
Relays	43.2
Signal processing integrated circuits (ICs)	31.0

6.3. Related Technical Issues

As mentioned previously, the proposed GaN-based three-phase APF circuit prototype is demonstrated for the first time. There are some technical issues to be further improved. These include: (1) the driving circuits can be improved to achieve higher switching frequency and to reduce the size of inductors; (2) the three relays can be removed or replaced with new devices having better quality in conduction losses to increase the overall system efficiency; (3) the layout of the circuit can be improved to reduce the noise level in current- and voltage-sensing mechanisms. It is important to note that the noise level in sensing signals constitutes the operating limits of the switching frequency of the proposed GaN-based APF circuit.

7. Conclusions

It has been expected that the mass application of RE-based distributed generation (DG) microgrids or smart grids, and various static and dynamic nonlinear loads is the future trend of development in power systems. To ensure an acceptable PQ level, the development of advanced PQ compensation schemes using new technologies in terms of power-switching devices and state-of-the-art control algorithms is a significant task. In this aspect, this paper has demonstrated, for the first time, a shunt-type GaN HEMTs-based three-phase APF controlled by DSP systems and type II controllers to achieve simultaneous compensation for current harmonics, load imbalance, and reactive currents. Based on the results obtained from simulation and the hardware tests, the proposed 2-kVA GaN-based three-phase shunt APF prototype with digitally integrated control scheme is able to achieve satisfactory compensation results in improving system-wide load current quality of a complex load network consisting of distorted, non-linear and unbalanced loads. In this study, the TPH3207 power switching devices and Si8271 driving integrated circuits (ICs) are successfully adopted. It has been found that GaN HEMTs provide superior performance to conventional Si-based power switches in terms of switching frequency, temperature feature and system efficiency. To further evaluate the system performance of the constructed GaN-based circuit prototype, in terms of efficiency, hotspot distribution and power losses in components, a thermographic analysis has been carried out. Results and discussions for improving the proposed implementation scheme have been presented. With the proposed APF operating at 50 kHz switching frequency, the THD and UR of the three-phase grid currents can be greatly reduced. The best THD improvement recorded is from 20.23% to 4.15% and UR is from 14.83% to 2.13% and the highest system efficiency of 97.2% has been achieved. For future research works, better circuit components and GaN HEMT driving methods based on a bootstrap design can be used for achieving higher switching frequency and better system performance.

Author Contributions: The corresponding author, C.-T.M. conducted the research work, proposed the design methods, designed the simulation and hardware test scenarios, verified the technical contents and polished the final manuscript. Z.-H.G., a postgraduate student in the department of electrical engineering, national united university, performed the paper search, managed figures and checked the related data. All authors have read and agreed to the published version of the manuscript.

Funding: This research was funded by MOST, Taiwan, with grant numbers, MOST 108-2221-E-239-007, MOST 107-2221-E-239-036 and MOST 106-2221-E-239-026 and the APC was funded by MOST 108-2221-E-239-007.

Acknowledgments: The author would like to thank the Ministry of Science and Technology (MOST) of Taiwan for financially support the energy-related research regarding key technologies and the design of advanced power and energy systems.

Conflicts of Interest: The authors declare no conflict of interest.

References

1. Gayatri, M.T.L.; Parimi, A.M.; Pavan Kumar, A.V. A review of reactive power compensation techniques in microgrids. *Renew. Sustain. Energy Rev.* **2018**, *81*, 1030–1036. [CrossRef]
2. Naderi, Y.; Hosseini, S.H.; Zadeh, S.G.; Mohammadi-Ivatloo, B.; Vasquez, J.C.; Guerrero, J.M. An overview of power quality enhancement techniques applied to distributed generation in electrical distribution networks. *Renew Sustain. Energy Rev.* **2018**, *93*, 201–214. [CrossRef]
3. Kumar, R.; Bansal, H.O. Shunt active power filter: Current status of control techniques and its integration to renewable energy sources. *Sustain. Cities Soc.* **2018**, *42*, 574–592. [CrossRef]
4. Baimel, D. Implementation of DQ0 control methods in high power electronics devices for renewable energy sources, energy storage and FACTS. *Sustain. Energy Grids* **2019**, *18*, 100218. [CrossRef]
5. Swain, S.; Subudhi, B. Grid integration scheme for a three phase single stage photovoltaic system employing EKF. *IFAC-PapersOnLine* **2018**, *51*, 279–282. [CrossRef]
6. Chauhan, S.; Singh, B. Grid-interfaced solar PV powered electric vehicle battery system with novel adaptive digital control algorithm. *IET Power Electron.* **2019**, *12*, 3470–3478. [CrossRef]
7. Zhai, H.; Zhuo, F.; Zhu, C.; Yi, H.; Wang, Z.; Tao, R.; Wei, T. An optimal compensation method of shunt active power filters for system-wide voltage quality improvement. *IEEE Trans. Ind. Electron.* **2020**, *67*, 1270–1281. [CrossRef]
8. Badoni, M.; Singh, A.; Singh, V.P.; Tripathi, R.N. Grid interfaced solar photovoltaic system using ZA-LMS based control algorithm. *Electr Power Syst. Res.* **2018**, *160*, 261–272. [CrossRef]
9. Prasad, V.; Jayasree, P.R.; Sruthy, V. Active power sharing and reactive power compensation in a grid-tied photovoltaic system. *Mater. Today: Proc.* **2018**, *5*, 1537–1544. [CrossRef]
10. Viswan, V. A review of silicon carbide and gallium nitride power semiconductor devices. *IJRESM* **2018**, *1*, 224–225.
11. Jones, E.A.; Wang, F.; Costinett, D. Review of commercial GaN power devices and GaN-based converter design challenges. *IEEE J. Emerg. Sel. Topics Power Electron.* **2016**, *4*, 707–719. [CrossRef]
12. Spaziani, L.; Lu, L. Silicon, GaN and SiC: There's room for all: An application space overview of device considerations. In Proceedings of the 2018 IEEE 30th International Symposium on Power Semiconductor Devices and ICs (ISPSD), Chicago, IL, USA, 13–17 May 2018; pp. 8–11.
13. Ma, C.T.; Gu, Z.H. Review of GaN HEMT Applications in Power Converters over 500 W. *Electronics* **2019**, *8*, 1401. [CrossRef]
14. Gwóždž, M. Active power filter with sigma-delta modulator in control section. In Proceedings of the 2017 19th International Conference on Electrical Drives and Power Electronics (EDPE), Dubrovnik, Croatia, 4–6 October 2017; pp. 128–132.
15. Otero-De-Leon, R.; Liu, L.; Bala, S.; Manchia, G. Hybrid active power filter with GaN power stage for 5kW single phase inverter. In Proceedings of the 2018 IEEE Applied Power Electronics Conference and Exposition (APEC), San Antonio, TX, USA, 4–8 March 2018; pp. 692–697.
16. Uğur, M.; Saraç, H.; Keysan, O. Comparison of Inverter Topologies Suited for Integrated Modular Motor Drive Applications. In Proceedings of the 2018 IEEE 18th International Power Electronics and Motion Control Conference (PEMC), Budapest, Hungary, 26–30 August 2018; pp. 524–530.
17. Mahmoudian, M.; Gitizadeh, M.; Rajaei, A.H.; Tehrani, V.M. Common mode voltage suppression in three-phase voltage source inverters with dynamic load. *IET Power Electron.* **2019**, *12*, 3141–3148. [CrossRef]
18. Fu, Y.; Li, Y.; Huang, Y.; Bai, H.; Zou, K.; Lu, X.; Chen, C. Design methodology of a three-phase four-wire EV charger operated at the autonomous mode. *IEEE Trans. Transp. Electr.* **2019**, *5*, 1169–1181. [CrossRef]

Article

Vertical Leakage in GaN-on-Si Stacks Investigated by a Buffer Decomposition Experiment

Alaleh Tajalli [1], Matteo Borga [1], Matteo Meneghini [1,*], Carlo De Santi [1], Davide Benazzi [1], Sven Besendörfer [2], Roland Püsche [3], Joff Derluyn [3], Stefan Degroote [3], Marianne Germain [3], Riad Kabouche [4], Idriss Abid [4], Elke Meissner [2], Enrico Zanoni [1], Farid Medjdoub [4] and Gaudenzio Meneghesso [1,*]

1. Department of Information Engineering, University of Padova, 35151 Padova, Italy; tajalli@dei.unipd.it (A.T.); borgamat@dei.unipd.it (M.B.); desantic@dei.unipd.it (C.D.S.); davide.benazzi@studenti.unipd.it (D.B.); zanoni@dei.unipd.it (E.Z.)
2. Fraunhofer Institute for Integrated Systems and Device Technology IISB, 91058 Erlangen, Germany; Sven.Besendoerfer@iisb.fraunhofer.de (S.B.); elke.meissner@iisb.fraunhofer.de (E.M.)
3. EpiGaN, 3500 Hasselt, Belgium; Roland.Puesche@epigan.com (R.P.); joff.derluyn@epigan.com (J.D.); stefan.degroote@epigan.com (S.D.); Marianne.Germain@epigan.com (M.G.)
4. IEMN-CNRS, 59652 Villeneuve d'Ascq, France; riad.kabouche@ed.univ-lille1.fr (R.K.); idriss.abid.etu@univ-lille.fr (I.A.); farid.medjdoub@iemn.univ-lille1.fr (F.M.)

* Correspondence: matteo.meneghini@unipd.it (M.M.); gauss@dei.unipd.it (G.M.)

Received: 19 December 2019; Accepted: 12 January 2020; Published: 17 January 2020

Abstract: We investigated the origin of vertical leakage and breakdown in GaN-on-Si epitaxial structures. In order to understand the role of the nucleation layer, AlGaN buffer, and C-doped GaN, we designed a sequential growth experiment. Specifically, we analyzed three different structures grown on silicon substrates: AlN/Si, AlGaN/AlN/Si, C:GaN/AlGaN/AlN/Si. The results demonstrate that: (i) the AlN layer grown on silicon has a breakdown field of 3.25 MV/cm, which further decreases with temperature. This value is much lower than that of highly-crystalline AlN, and the difference can be ascribed to the high density of vertical leakage paths like V-pits or threading dislocations. (ii) the AlN/Si structures show negative charge trapping, due to the injection of electrons from silicon to deep traps in AlN. (iii) adding AlGaN on top of AlN significantly reduces the defect density, thus resulting in a more uniform sample-to-sample leakage. (iv) a substantial increase in breakdown voltage is obtained only in the C:GaN/AlGaN/AlN/Si structure, that allows it to reach V_{BD} > 800 V. (v) remarkably, during a vertical I–V sweep, the C:GaN/AlGaN/AlN/Si stack shows evidence for positive charge trapping. Holes from C:GaN are trapped at the GaN/AlGaN interface, thus bringing a positive charge storage in the buffer. For the first time, the results summarized in this paper clarify the contribution of each buffer layer to vertical leakage and breakdown.

Keywords: Gallium nitride (GaN) high-electron-mobility transistors (HEMTs); vertical breakdown voltage; buffer trapping effect

---

1. Introduction

Gallium nitride (GaN) high-electron-mobility transistors (HEMTs) on silicon (Si) substrate are excellent devices for power applications, and are expected to find wide application in switching power converters [1], owing to the high breakdown field of GaN (3.3 MV/cm) [2,3]. To grow GaN on Si substrates, it is necessary to develop complex buffer and transition layer architectures; these allow to obtain a good crystalline quality, which in turn leads to an improved electrical performance of the devices. However, the vertical leakage current (Figure 1) in OFF-state conditions may limit the breakdown voltage of power GaN-HEMTs. Recently, different efforts have been developed in order to improve the vertical robustness of such devices [4–7].

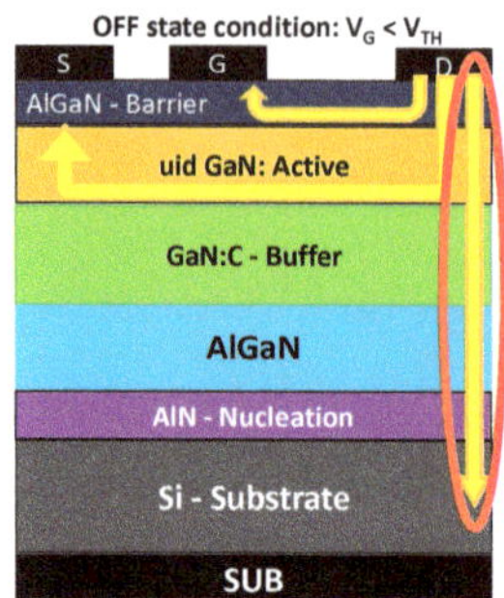

Figure 1. Schematic representation of GaN high-electron-mobility transistors (HEMTs), showing vertical leakage current between drain and substrate in the OFF-state condition.

In this work, we investigated the vertical leakage current and breakdown of GaN-on-Si HEMTs by a sequential growth experiment. Specifically, we fabricated three tailored samples in order to analyze the contribution of the aluminum nitride (AlN) nucleation, the AlGaN buffer and the C-doped GaN layer (GaN:C) on the vertical leakage current and breakdown. Corresponding devices were characterized at room and high temperatures.

Furthermore, the physical origin of I–V hysteresis and charge trapping were analyzed; finally, UV LED irradiation was used to investigate the charge trapping phenomena in the Al-rich AlGaN buffer layers. Evidence for negative charge storage is found for the AlGaN and AlN layers, while we demonstrate that the GaN:C layer can lead to significant positive charge storage.

2. Experimental Details

Three different structures were grown using metal organic chemical vapor deposition (MOCVD) under the same growth conditions until growth stopped (Figure 2). The first structure consisted of a 200 nm thick layer of AlN, epitaxially grown on a conductive 1×10^{19} cm^{-3} boron-doped silicon substrate, referred to as structure A. Structure B had an additional 350 nm thick $Al_{0.70}Ga_{0.30}N$ layer grown on top of the AlN layer. Lastly, structure C was made of a silicon substrate, a 200 nm AlN layer, a step-graded stack of AlGaN layers with a total thickness of 2350 nm, and a 2450 nm thick carbon doped GaN layer; the carbon doping concentration was 2×10^{19} cm^{-3}. Ohmic Ti (100 nm)/Au (400 nm) contacts were defined by photolithography; 95×95 µm^2 patterns that had been isolated by N-implantation were used for electrical measurements [8]. The wafer diameter of all structures was 200 mm.

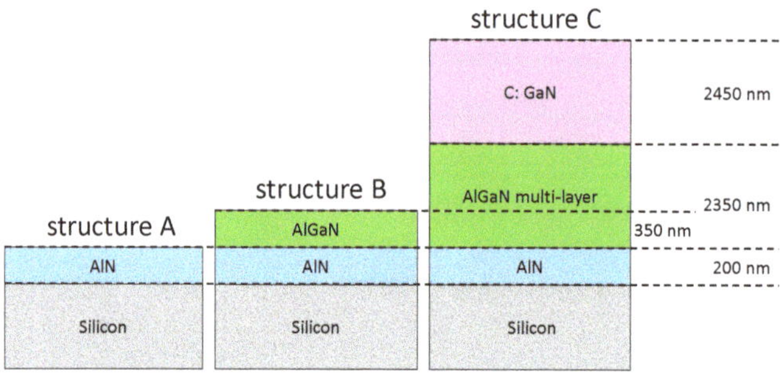

Figure 2. Schematic representation of the three investigated structures.

Current-voltage characterization was carried out in order to investigate leakage current, breakdown, and its physical origins at room and high temperatures as well as charge trapping before and after UV irradiation. These measurements were done with a Keysight B1505A high-voltage parameter analyzer, equipped with a high-voltage source measurement unit. The breakdown voltage was extrapolated as the voltage at which the current of 8 mA was reached. Epitaxial quality in terms of surface morphology and threading dislocation (TD) density were investigated using a JSM-7500F scanning electron microscope (SEM) (JEOL Ltd., Tokyo, Japan), a Bruker Dimension Icon atomic force microscope (AFM) in tapping mode, and a JEM-2200FS transmission electron microscope (TEM) (JEOL Ltd., Tokyo, Japan).

3. Results and Discussion

3.1. Material Quality

To investigate the material quality of the various layers, we used several experimental techniques. Using cross-sectioning and TEM-analysis, the TD-densities were estimated to be 3.2×10^{10} cm^{-2} for the AlN nucleation layers of all three structures, 2.3×10^{10} cm^{-2} for the $Al_{70}Ga_{30}N$ layers of structures B and C, and 3.1×10^{9} cm^{-2} for the GaN:C of structure C. Surface morphologies were analyzed by SEM and AFM. The results observed by SEM are shown in Figure 3. Structure A exhibits a strongly V-pitted surface due to non-optimized nucleation conditions [9]. The V-pit density was ~1.5×10^{10} cm^{-2} with a typical V-pit diameter of ~60 nm. A smaller V-pit density of ~5×10^{9} cm^{-2}, but with a larger typical V-pit diameter of ~120 nm was observed for structure B. Finally, no V-pits were observed for structure C, which was due to successful overgrowth by or below GaN:C. The corresponding rms-roughness values obtained on a scale of 5×5 μm^2 by AFM are 2.35 nm, 7.55 nm, and 0.25 nm for structures A, B, and C respectively.

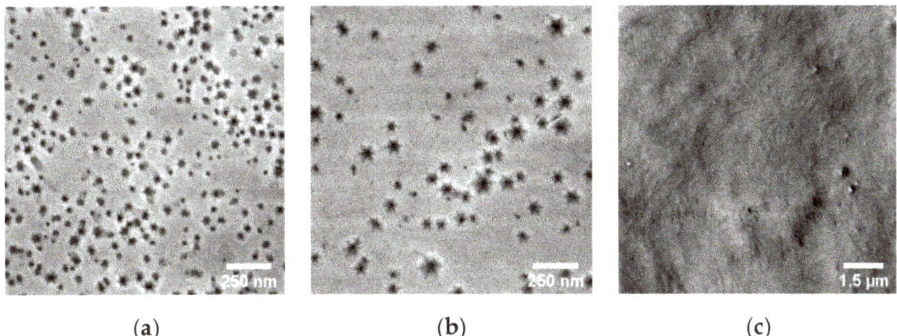

(**a**) (**b**) (**c**)

Figure 3. Representative scanning electron microscope (SEM) images of the as-grown surfaces of structure A (**a**), structure B (**b**), and structure C (**c**).

A non-optimal coalescence of AlN islands on Si during the 3D growth mode gave a rough surface decorated by V-pits. Overgrowth of such a V-pitted AlN-layer led to enlarging of V-pits in subsequent layers due to a slower growth rate of V-pit side planes compared to the c-plane. Impurities, especially oxygen, can be trapped within the V-pit trace [9]. The presence of a large density of V-pits in wafers A and B resulted in a high variability in the vertical leakage characteristics, as shown in Figure 4 (notice that apart from a couple of outliers, the I–V curves of structure B were already much more uniform compared to those of structure A). On the other hand, the use of a thick GaN:C layer results in a much narrower distribution of the leakage curves due to the substantial reduction in V-pit and dislocation density obtained through the growth of a thick C-doped buffer layer.

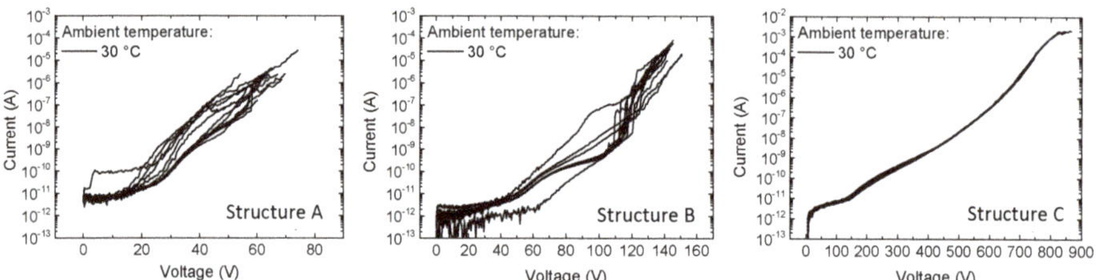

Figure 4. Vertical current-voltage (I–V) characteristics until breakdown for ten devices on structure A (**left**), structure B (**center**) and structure C (**right**).

3.2. Vertical Leakage

Vertical leakage characterizations with a grounded substrate and an ohmic contact sweep from 0 V up to catastrophic breakdown were performed on several samples of each structures. Figure 5 shows the typical leakage characteristics of the three heterostructures at room temperature and 170 °C. The vertical leakage current of sample A was not significantly thermally activated. This suggests that conduction through AlN is dominated by a field-driven tunneling of electrons from an electron inversion layer at the Si/AlN interface into the AlN layer (in agreement with [10]), possibly with the contribution of deep levels in AlN (Figure 6b). This process strongly depends on the availability of deep states within the AlN layer close to the interface with Si, i.e., on the local density of defects such as those described above.

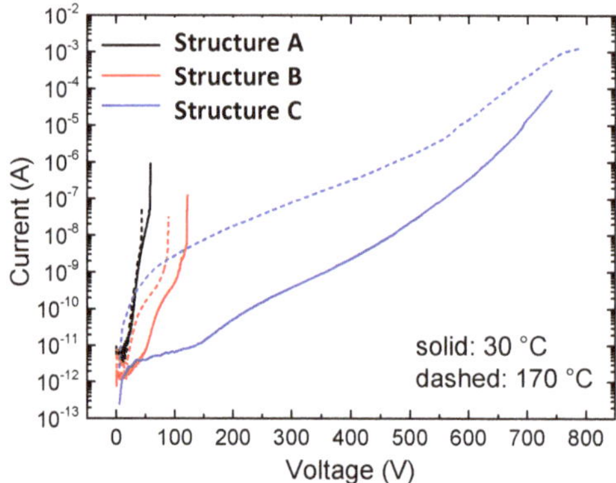

Figure 5. Current–voltage characteristics carried out at both low and high temperatures in the three structures under analysis.

Contrary to structure A, the leakage current of structure B is temperature dependent. The carriers in AlGaN moved towards the top ohmic contact by means of a thermally-activated defect-assisted conduction mechanism (Figure 7b), which became the bottleneck for conduction and therefore dominated the overall behavior. This might be because the AlGaN layer showed a lower defect density as compared to AlN in terms of V-pits and threading dislocations TDs.

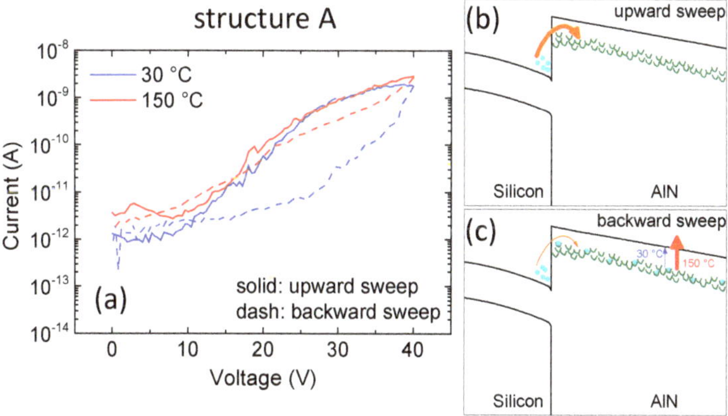

Figure 6. (**a**) Double sweep (upward: solid, backward: dashed) current-voltage characterization on sample A at both room and high temperature. (**b**,**c**) represent a schematic band diagram during the upward sweep and the backward sweep, respectively.

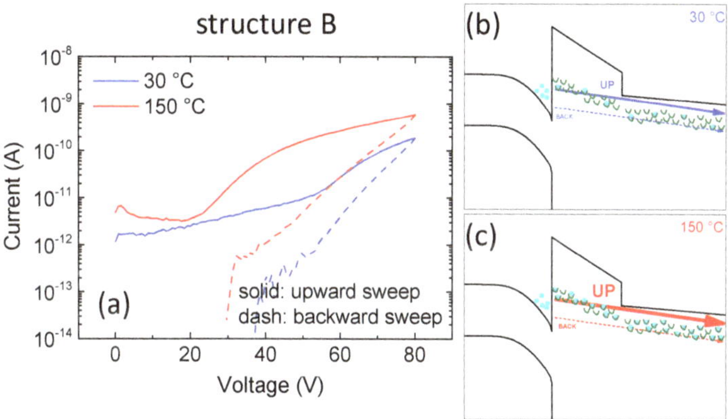

Figure 7. (**a**) Double sweep (upward: solid, backward: dashed) current-voltage characterization on the structure B at both room and high temperature. (**b**,**c**) represent a schematic band diagram at room and high temperature, respectively.

Blue lines in Figure 5 show the strong impact of temperature on the vertical leakage of structure C, which was compatible to the presence of a acceptor level like C_N within GaN:C. Thus, for this sample, conduction was likely not only related to electrons as in structures A and B, but also to holes generated by the 0.9 eV deep acceptor C_N. Holes in GaN:C can easily flow towards the bottom of this layer and, especially at high temperature, towards the Si substrate. Hole generation was stronger at high temperatures (consistent with [3,11]).

3.3. Charge Trapping

In order to study the presence of charge trapping, we analyzed the hysteresis in the vertical leakage characteristics. To this end, the three structures under analysis were submitted to a double voltage sweep (upward and backward) at both room temperature and 150 °C. A large hysteresis is indicative of strong trapping effects: if negative charge is stored in the epitaxial stack during the upward sweep, the current during the backward sweep is significantly reduced [12]. The voltage step used was 1 V, while the integration time was 20 ms. Contrary to vertical leakage measurements, a maximum voltage was chosen in order to avoid catastrophic breakdown of the devices. This voltage was 40 V, 80 V, and 600 V for structures A, B, and C, respectively.

The results obtained in this study can be explained as follows:

For structure A (Figure 6a), the upward current was higher than the backward current (positive shift). This indicates that electrons that were injected from Si into AlN were trapped at defects in the AlN nucleation layer (Figure 6b). However, this effect was weaker at high temperature, leading to a considerably reduced hysteresis, possibly due to a thermally-assisted de-trapping process [13] compatible with the presence of a donor-like trap like oxygen in V-pit traces. In other words, at high temperature there was a lower amount of trapped charges during the backward sweep with respect to the room temperature condition (Figure 6b,c).

Structure B showed positive shifts for both temperatures as well. However, the shifts were much more prominent and, contrary to structure A, the shift at high temperature was stronger. As discussed above, leakage current in structure B was strongly temperature dependent due to a thermally-activated defect-assisted conduction mechanism (Figure 7b). At high temperature the contribution of trapped electrons to conduction was higher and therefore a larger hysteresis was observed.

In contrast to all other samples, sample C showed a negative shift (Figure 8a), i.e., a higher current was measured for the backward sweep, especially at high temperature. This effect was ascribed to the presence of positive charges (holes) trapped in the buffer. We suggest that holes flowed from GaN:C to the GaN:C/AlGaN heterointerface. Due to the valence band discontinuity, part of these holes may have been trapped there, thus modifying the band diagram as schematically represented in Figure 8b. This resulted in a higher current during the backward sweep. A high temperature enhanced this phenomenon, thanks to the higher number of holes available.

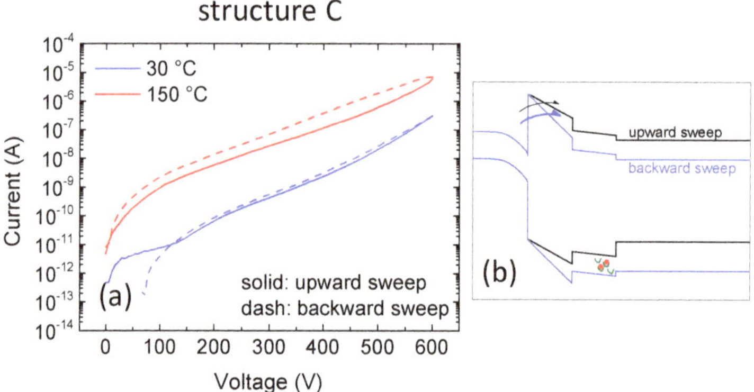

Figure 8. (**a**) Double sweep (upward: solid, backward: dashed) current-voltage characterization on structure C at both room and high temperature. (**b**) represents a schematic band diagram where the electrostatic effect of the positive trapped charges is highlighted.

3.4. The Effect of UV Light Irradiation on Current–Voltage Charactrization

Figures 6 and 7 show the presence of considerable electron trapping in the AlN and AlGaN layers. To characterize the traps which caused the experimentally observed hysteresis in the IV characteristic of structure B, we analyzed the effect of UV light with a wavelength of 385 nm on vertical leakage and charge trapping. For this, we performed I–V characterization before and after exposing the sample to UV light for 30 min.

Figure 9 shows the result for the current-voltage characterization upward and backward sweep from 0 V to 80 V. Three consecutive I–V measurements were carried out on a fresh device without UV irradiation. A considerable decrease of both the current (at 80 V) and the hysteresis (ΔV at 10 pA) at the second and the third I–V characterization was observed, indicating a slow de-trapping process. Remarkably, after UV irradiation, there was a full recovery of the I–V behavior, which can be explained as follows: during the upward sweep, electrons were injected and trapped into defects of the AlN and AlGaN layers and created a substantial hysteresis. When the sample was illuminated with UV-light corresponding to E = 3.22 eV, the trapped electrons are re-emitted leading to a recovery of I–V behavior. These results confirm the existence of deep-traps, as provided by TDs especially, in the Al-rich layers and explain why the I–V curves show a slowly recovering hysteresis.

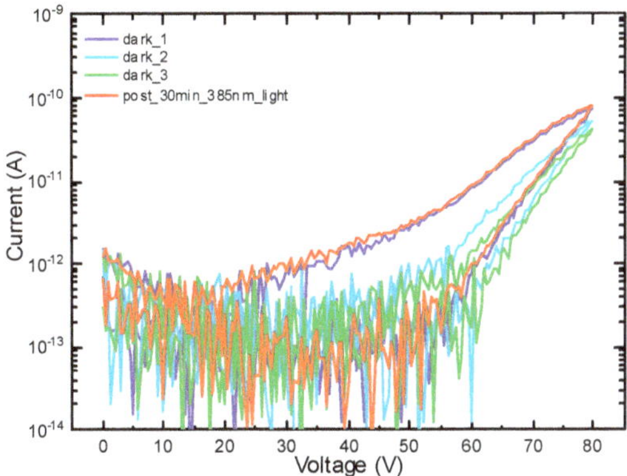

Figure 9. Double sweep current–voltage characterization on structure B, first, second, and third I–V on a fresh sample, then after 30 min irradiation with an LED light at a wavelength of 385 nm.

3.5. Vertical Breakdown

The vertical breakdown voltage of the three stacks was investigated in detail by applying a voltage sweep on the top ohmic contact while keeping the substrate grounded. Several devices were tested for each structure. The mean values obtained for samples A, B, and C were 64 V, 142 V, and 780 V, respectively (Figure 10a). When comparing the breakdown voltages of structures A and B, it is worth noticing that, by increasing the thickness of the epi-layers by 175%, the breakdown voltage increased by 120%. This indicates a non-uniform distribution of the electric field among the epi-layers. This might be ascribed to the different conductivity of the layers, as well as to the piezoelectric and spontaneous polarization charges that originated at the AlN/AlGaN interface. On the other hand, Figure 10a also demonstrates the role of the thick carbon-doped layer in increasing the vertical robustness.

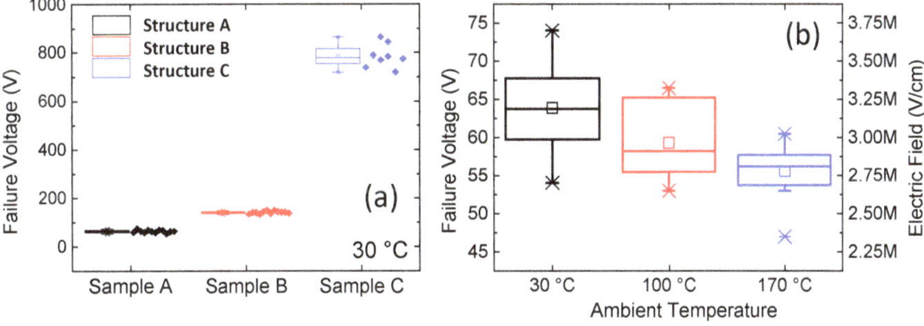

Figure 10. (**a**) Breakdown voltage evaluated on the three wafers under analysis; (**b**) dependence of the breakdown voltage of wafer A on temperature.

For a more detailed analysis, sample A was taken in order to evaluate the temperature dependence of the critical electrical field of the AlN nucleation layer. Considering that the energy gap of AlN at 0 K is 6.15 eV [14], and that such an energy gap decreases at high temperature accordingly with the Bose–Einstein equation [14] and knowing the dependence of the breakdown field on the energy gap [15], the expected (theoretical) breakdown electric field reduction in the range of 30–170 °C was calculated to be equal to 6%. The experimental results show a reduction of the critical electric field between 30 °C and 170 °C, higher than 12% (Figure 10b).

To understand why the breakdown voltage has a stronger temperature dependence than predicted, we plotted the failure voltage versus the leakage current at 30 V of the devices (Figure 11). A bias of 30 V was chosen, because corresponding current is above noise level, but low enough to cause no damage to the sample. As can be noticed, a clear trend can be observed: devices with higher leakage current show a lower breakdown voltage, and consequently a lower breakdown field, of the AlN nucleation layer. This demonstrates that the breakdown process of AlN grown on Si is not only field-dependent, but also current driven. The flow of current at localized defect sites may lead to a premature breakdown of the samples, consistent with the percolation theory mentioned in Borga et al.'s research [8]. This explains why the breakdown field of AlN-on-Si (3.25 MV/cm) is much lower than that of highly crystalline AlN (up to 12 MV/cm, [8]).

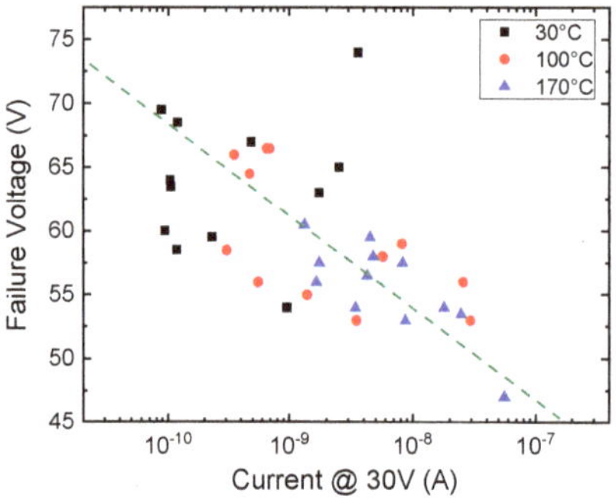

Figure 11. Failure voltage vs. leakage current measured at 30 V on sample A at different ambient temperatures.

4. Conclusions

In this work we compared three structures obtained by sequential epitaxial growth on Si substrate. We were able to separately evaluate the contribution of AlN, AlGaN, and C-doped GaN to the vertical conduction in the GaN-on-Si stack. Also, we described the related trapping processes, showing that in presence of AlN/AlGaN layers trapping is dominated by negative charge, while in the presence of C-doped GaN, positive charge trapping also plays a role.

In addition, we investigated the breakdown voltage of the samples, indicating that a C-doped GaN layer is needed to substantially increase the breakdown strength. This is ascribed to both the insulating properties of C-doped GaN, and to the increased thickness of the vertical stack. In addition, SEM and AFM analysis were used to confirm the substantial absence of extended defects like V-pits at the surface of the C-doped layer. Finally, the temperature dependence of the critical electric field of AlN grown on a Si substrate was studied. Experimental results point out that the failure of the AlN is not related to the intrinsic breakage of the semiconductor crystal, but to a defect-related phenomenon.

Author Contributions: Data curation, D.B., R.P., R.K., S.B. and I.A.; writing—review and editing, A.T., M.B., M.M., C.D.S., S.B., J.D., S.D., M.G., E.M., E.Z., F.M. and G.M. All authors have read and agreed to the published version of the manuscript.

Funding: This work was supported by the project InRel-NPower (Innovative Reliable Nitride based Power Devices and Applications). This project has received funding from the European Union's Horizon 2020 research and innovation programme under grant agreement No. 720527.

Conflicts of Interest: The authors declare no conflict of interest.

References

1. Amano, H.; Baines, Y.; Beam, E.; Borga, M.; Bouchet, T.; Chalker, P.R.; Charles, M.; Chen, K.J.; Chowdhury, N.; Chu, R.; et al. The 2018 GaN power electronics roadmap. *J. Phys. D Appl. Phys.* **2018**, *51*, 163001. [CrossRef]
2. Kaminski, N.; Hilt, O. SiC and GaN devices—Wide bandgap is not all the same. *IET Circuits Devices Syst.* **2014**, *8*, 227–236. [CrossRef]
3. Meneghini, M.; Meneghesso, G.; Zanoni, E. *Power GaN Devices*; Springer: Cham, Switzerland, 2017.
4. Zhou, C.; Jiang, Q.; Huang, S.; Chen, K.J. Vertical leakage/breakdown mechanisms in AlGaN/GaN-on-Si structures. *IEEE Electron Device Lett.* **2012**, *33*, 1132–1134. [CrossRef]
5. Uren, M.J.; Karboyan, S.; Chatterjee, I.; Pooth, A.; Moens, P.; Banerjee, A.; Kuball, M. "Leaky Dielectric" Model for the Suppression of Dynamic Ron in Carbon-Doped AlGaN/GaN HEMTs. *IEEE Trans. Electron Devices* **2017**, *64*, 2826–2834. [CrossRef]
6. Moens, P.; Banerjee, A.; Uren, M.J.; Meneghini, M.; Karboyan, S.; Chatterjee, I.; Vanmeerbeek, P.; Cäsar, M.; Liu, C.; Salih, A.; et al. Impact of buffer leakage on intrinsic reliability of 650V AlGaN/GaN HEMTs. In *2015 IEEE International Electron Devices Meeting (IEDM)*; IEEE: Piscataway, NJ, USA, 2015. [CrossRef]
7. Umeda, H.; Suzuki, A.; Anda, Y.; Ishida, M.; Ueda, T.; Tanaka, T.; Ueda, D. Blocking-voltage boosting technology for GaN transistors by widening depletion layer in Si substrates. In *2010 International Electron Devices Meeting*; IEEE: Piscataway, NJ, USA, 2010. [CrossRef]
8. Borga, M.; Meneghini, M.; Benazzi, D.; Canato, E.; Püsche, R.; Derluyn, J.; Abid, I.; Medjdoub, F.; Meneghsso, G.; Zanoni, E. Buffer breakdown in GaN-on-Si HEMTs: A comprehensive study based on a sequential growth experiment. *Microelectron. Reliab.* **2019**, *100*, 113461. [CrossRef]
9. Besendörfer, S.; Meissner, E.; Tajalli, A.; Meneghini, M.; Freitas, J.A., Jr.; Derluyn, J.; Medjdoub, F.; Meneghesso, G.; Friedrich, J.; Erlbacher, T. Vertical breakdown of GaN on Si due to V-pits. *J. Appl. Phys.* **2020**, *127*, 015701. [CrossRef]
10. Yacoub, H.; Fahle, D.; Finken, M.; Hahn, H.; Blumberg, C.; Prost, W.; Kalisch, H.; Herken, M.; Vescan, A. The effect of the inversion channel at the AlN/Si interface on the vertical breakdown characteristics of GaN-based devices. *Semicond. Sci. Technol.* **2014**, *29*, 115012. [CrossRef]
11. Meneghini, M.; Tajalli, A.; Moens, P.; Banerjee, A.; Zanoni, E.; Meneghesso, G. Trapping phenomena and degradation mechanisms in GaN-based power HEMTs. *Mater. Sci. Semicond. Process.* **2017**, *78*, 118–126. [CrossRef]
12. Meneghini, M.; Tajalli, A.; Moens, P.; Banerjee, A.; Stockman, A.; Tack, M.; Gerardin, S.; Bagatin, M.; Paccagnella, A.; Zanoni, E.; et al. Total suppression of dynamic-ron in AlGaN/GaN-HEMTs through proton irradiation. In *2017 IEEE International Electron Devices Meeting (IEDM)*; IEEE: Piscataway, NJ, USA, 2017; pp. 33.5.1–33.5.4. [CrossRef]
13. Stockman, A.; Uren, M.; Tajalli, A.; Meneghini, M.; Bakeroot, B.; Moens, P. Temperature dependent substrate trapping in AlGaN/GaN power devices and the impact on dynamic ron. In *2017 47th European Solid-State Device Research Conference (ESSDERC)*; IEEE: Piscataway, NJ, USA, 2017; pp. 130–133. [CrossRef]
14. Guo, Q.; Yoshida, A. Temperature Dependence of Band Gap Change in InN and AlN. *Jpn. J. Appl. Phys.* **1994**, *33*, 2453–2456. [CrossRef]
15. Hudgins, J.L.; Simin, G.S.; Santi, E.; Khan, M.A. An assessment of wide bandgap semiconductors for power devices. *IEEE Trans. Power Electron.* **2003**, *18*, 907–914. [CrossRef]

Article

Physical-Based Simulation of the GaN-Based Grooved-Anode Planar Gunn Diode

Ying Wang [1,2], Liu-An Li [3], Jin-Ping Ao [2,*] and Yue Hao [2]

1. School of Electronic Information, Northwestern Polytechnical University, Xi'an 710072, China; yingwang@nwpu.edu.cn
2. The State Key Discipline Laboratory of Wide Band Gap Semiconductor Technology, School of Microelectronics, Xidian University, Xi'an 710071, China; yhao@xidian.edu.cn
3. School of Electronics and Information Technology, Sun Yat-Sen University, Guangzhou 510275, China; liliuan@mail.sysu.edu.cn

* Correspondence: jpao@xidian.edu.cn; Tel.: +86-029-8843-1212

Received: 7 December 2019; Accepted: 9 January 2020; Published: 16 January 2020

Abstract: In this paper, a novel gallium nitride (GaN)-based heterostructure Gunn diode is proposed for the first time to enhance the output characteristics of Gunn oscillation waveforms. A well-designed grooved anode contact is adopted to separate the long-channel diode into two short-channel diodes in parallel. If the grooved anode contact is positioned in the middle of the device, the output power nearly doubles in the grooved-anode diode compared with the single-channel ones, as does the output frequency. Based on the numerical results, the best output characteristics are obtained at the 2.0-μm symmetrical grooved-anode diode, which produces nearly 5.48 mW of power at the fundamental frequency of 172.81 GHz, with 3.13% efficiency of power conversion. If the grooved anode contact is not positioned in the middle of the diode, the harmonic frequency would be enhanced. The GaN heterostructure grooved-anode Gunn diode has been demonstrated to be an excellent solid-state source of terahertz oscillator.

Keywords: wide band gap semiconductors; numerical simulation; terahertz Gunn diode; grooved-anode diode

1. Introduction

Terahertz (THz) waves (300 GHz–10 THz) have been extensively studied in recent years due to their potential applications in the fields of communication, imaging, radar, spectroscopy and security screening [1–3]. Terahertz has been a "research gap" for a long time. Indeed, no powerful radiation sources have been available until the last few years. From a practical point of view, solid-state devices show excellent potential as terahertz sources, which can be integrated with other electronic or optoelectronic devices within a single chip [4]. GaN-based Gunn diodes are one of the most excellent solid-state terahertz oscillators [4] and attract much interest thanks to the unique properties of gallium nitride, such as its wide band gap (3.42 eV) [5], high electron mobility [6], high breakdown field (3.3 MV/cm) [5], high thermal conductivity [7] and so on. In recent years, the research has been mainly concentrated on heterostructural planar Gunn diodes rather than the traditional vertical ones [8–12]. Compared with the traditional vertical Gunn diode, the heterostructural planar Gunn diode has great advantages. First of all, it allows for the easy integration with other devices in terahertz monolithic integrated circuits, as all the contacts of planar diodes are fabricated on one plane [10]. Secondly, its oscillation frequency is controllable by determining the contacts' distance. Lastly, due to the excellent electron transport properties of the two-dimensional electron gas (2DEG), the planar Gunn diode generates a higher oscillation frequency than bulk ones [9–12]. However, on one hand, the radio frequency (RF) output power of the planar Gunn diode has been predicted to be much lower than the vertical ones [13–15]. On the other hand, the fundamental oscillation of Gunn diodes with a channel length of 1–2 μm reported so far have been far from the terahertz regime [8–11]. Achieving a Gunn diode with a terahertz oscillation and higher output characteristics is a worldwide problem that should be solved with great urgency. High RF power and high operation frequency seem to be two contradictory pursuits which are difficult to satisfy simultaneously. The shorter the transit region length, the higher the oscillation frequency is [16]. However, the operation bias for short-transit-region diodes is relatively low, which limits their RF output characteristics. In addition, the small-size devices also face a complicated process. Some reported planar Gunn diodes, like nanowire slot diodes and self-switching diodes (SSD) produce very high frequency oscillation, even as high as several THz [17–21]. Nevertheless, all of these emerging devices face the same serious problem–that is, how to generate sufficient RF power. Some theoretical work also shows other solutions, like the harmonic Gunn diode reported in [22,23] and the multi-channel Gunn diode in [24], which either suffers from low RF output power or a complicated process. In order to achieve higher frequency and higher output power simultaneously, for the first time, we propose this new-type grooved-anode Gunn diode, which is realized by simply etching a rectangle groove onto the semiconductor layer at one lithographic step. The anode contact is deposited in the rectangle groove and two cathodes are defined as being adjacent to terminals of AlGaN/GaN heterostructural channel. Therefore, one diode actually turns into two diodes placed in parallel. In the symmetric grooved-anode Gunn diode, where the length of the left and right channel is equal, the output power and oscillation frequency approximately doubles in the grooved-anode diodes compared with the single-channel ones. In the asymmetric grooved-anode Gunn diode, the harmonic frequency is greatly enhanced. In this paper, we present a detailed study into the GaN-based heterostructural grooved-anode Gunn diode based on the Silvaco simulator. We have demonstrated that it effectively improves the RF output power and operation frequency of the Gunn diode simultaneously as compared with many other structures.

The structure of the grooved-anode diode and the simulation method are described in Section 2. The numerical Results and theoretical analysis are given in Section 3. Important conclusions are given in Section 4.

2. Device Structure and Simulation Method

The GaN-based Gunn diode studied in this paper is illustrated in Figure 1. Figure 1a shows the structure of the grooved-anode diode, in which all the structural dimension parameters are labeled clearly. The doping levels of all the material layers are set to be 1×10^{15} cm^{-3}. The adoption of the $Al_{0.1}Ga_{0.9}N$ back barrier layer leads to the enhancing confinement of the 2DEG under a high electric field. A rectangular groove is etched through the GaN layer, which ensures that electrons have individual paths in the left and right channels. To achieve the dual-channel diode, a groove-anode area should be defined, and the groove area is etched using a controllable low-damage chlorine-based Inductively Coupled Plasma- Reactive Ion Etching (ICP-RIE) process [25,26]. The definition of the specific technology parameters should be explored further in the actual manufacturing. The anode ohmic contact is deposited in the rectangle grooves and the two cathode ohmic contacts are defined as vertical contacts instead of surface contacts. On one hand, the vertical contacts introduce the lowest parasitic resistance; on the other hand, electric field peaks are easily formed by the surface contact, which results in the premature breakdown of the devices. The total length of the Gunn diode is set to be L. Meanwhile, in the grooved-anode diode, the length of each channel is set to be L_a and L_b separately, where $L_a + L_b = L$. The width of the device is defined as a default of 1 μm in this 2-D simulator. The vertical ohmic contacts can be realized by Molecular beam epitaxy (MBE) regrown technology. In order to simplify the calculation, we performed the simulations under ideal conditions. We assumed that the groove separates the longer channel completely, and the substrate is totally insulated. Therefore, we do not discuss the coupling effect through the substrate here. An energy balance (EB) model with higher order solution of the general Boltzman transport equation was adopted instead of the drift diffusion (DD) model. The energy relaxation time τ_ε and momentum relaxation time τ_m for GaN 2DEG are defined as 500 and 4 fs, respectively [27–29]. The temperature is set to be 300 K, ideally. In order to calculate the RF output characteristics of the Gunn diode, we use the same method as explained in detail in [22–24].

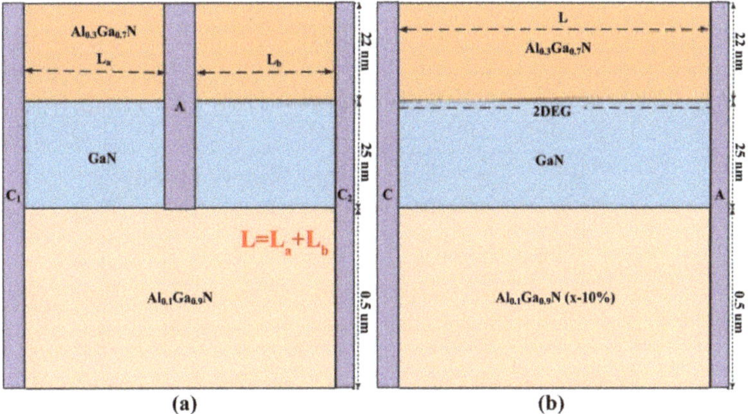

Figure 1. Schematic structures of GaN-based heterostructure Gunn diodes: (**a**) Grooved-anode diode, (**b**) Single-channel diode.

In order to calculate the electrical characteristics of the diode, we put a single-tone sinusoidal voltage of form $V_{DC} + V_{AC}\sin(2\pi ft)$ across the diode instead of embedding it to a resonant circuit, as the external circuit adds complexity to the calculation and easily results in non-convergence. This method is very popular in the analysis of the RF performance of Gunn diodes [30–37] and its validity has been proved in previous publications [33–37]. The applied DC voltage V_{DC} has to be above a critical value so that the device is biased in the negative differential mobility regime. The DC voltage V_{DC} is proportional to length of the diode and $V_{AC} = 1/4V_{DC}$. For example, for the 0.6-μm single channel Gunn diode, $V_{DC} = 16$ V, $V_{AC} = 4$ V; for the 1.2-μm single-channel diode, $V_{DC} = 32$ V, $V_{AC} = 8$ V; for the 0.6-0.6-μm grooved-anode Gunn diode, $V_{DC1} = V_{DC2} = 16$ V, $V_{AC1} = V_{AC2} = 4$ V. The DC-to-AC conversion efficiency η is defined as $\eta = -P_{AC}/P_{DC}$ (P_{AC} is the time-average AC power; P_{DC} is the dissipated DC power).

The AC power delivered is given by [30]

$$P_{AC} = \frac{V_{AC}}{T_R}\int_0^{T_R} I(t)\sin(2\pi ft)dt \tag{1}$$

$$\text{where } T_R = \frac{1}{f} \tag{2}$$

Similarly, the DC power dissipation in the device is [30]

$$P_{DC} = \frac{V_{DC}}{T_R}\int_0^{T_R} i(t)dt \tag{3}$$

3. Results and Discussion

Both the traditional GaN-based heterostructural Gunn diode with only one channel and the grooved-anode diode were studied as a comparison. In the tradition planar Gunn diode, as the channel length L ranges from 1.2 to 3.6 μm, the frequency f ranges from 143.38 to 45.63 GHz, as shown in Figure 2. The f–L curve of the grooved-anode diode nearly doubles as compared with that of the single-channel diode, as the real channel length of grooved-anode diode is only half of the single-channel diode. The curves also show that f varies inversely to L, which almost matches the formula f = v_{sat}/L. In grooved-anode diode, the best output characteristics are achieved at an L of 2.0 μm. The 2.0-μm-grooved-anode diode operates at a frequency f of 172.81 GHz with a DC-to-AC efficiency η of 3.13% and RF output power P_{RF} of 5.48 mW. Meanwhile, the 2.0-μm-single-channel diode generates a frequency of 86.47 GHz with η of 2.09% and P_{AC} of 3.15 mW, as shown in Figures 2 and 3. Therefore, we conclude that the operating frequency and RF output power nearly doubles as compared with the single-channel diode with the same channel length. In addition, the DC-to-AC efficiency η is also greatly enhanced in the grooved-anode diode.

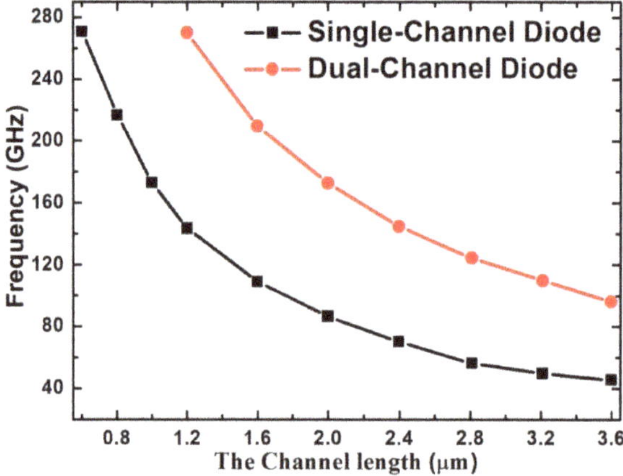

Figure 2. Variation of frequency f with the channel length L in the grooved-anode diode and single-channel diode.

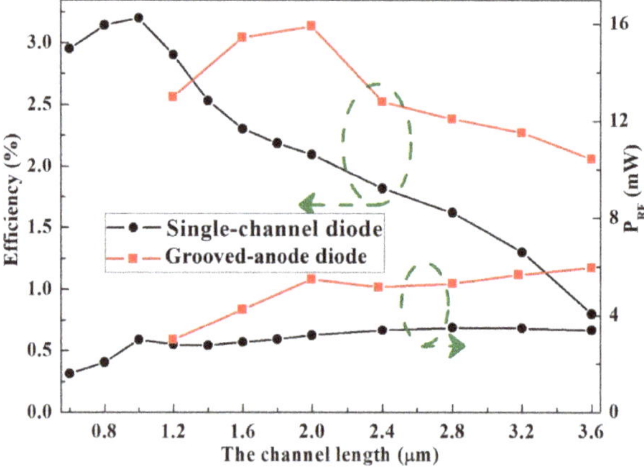

Figure 3. Variation of efficiency η with the channel length L and RF output power P_{AC} with L in the grooved-anode diode and the variation of the single-channel diode.

Figure 4 gives the I-V characteristics for the grooved-anode diode and single-channel diode of 2.0 μm, from which we can see the grooved-anode diode provides practically the same current, twice the value obtained in the single diode. The grooved-anode diode is equal to two shorter diodes in parallel connection, as the grooved anode is designed in the middle of the diode and divides the long channel into two shorter channels; therefore, the anode current doubles. Perturbation occurs in the I-V characteristics as the applied voltage increases above 17 V in the grooved-anode diode and nearly 34 V in the single-channel diode, which means there are electron domains coming into being in the 2DEG channel. When the anode voltage of the 2.0-μm -grooved-anode diodes increases up to 28 V, prefect stable dipole domains come into being. Figure 5a,b separately give the electron concentration profiles and the corresponding electric field profiles during one oscillating period extracted from the 2.0-μm-grooved-anode diode which show the electron periodic movement during one oscillation period. From Figure 5, we can observe the distinct formation of dipole domains rather than accumulation layers. Theoretically, the dipole domain mode is the most stable mode and generates the highest RF output power and efficiency as compared with other operation modes of the Gunn diode. Based on [13], in order to make the GaN Gunn diode work at a stable dipole domain mode, some conditions should be satisfied [16]:

$$N_{AC} \times L_{AC} > 5 \times 10^{13}\ cm^{-2} \tag{4}$$

$$N_{AC} \times d > 2 \times 10^{11}\ cm^{-2} \tag{5}$$

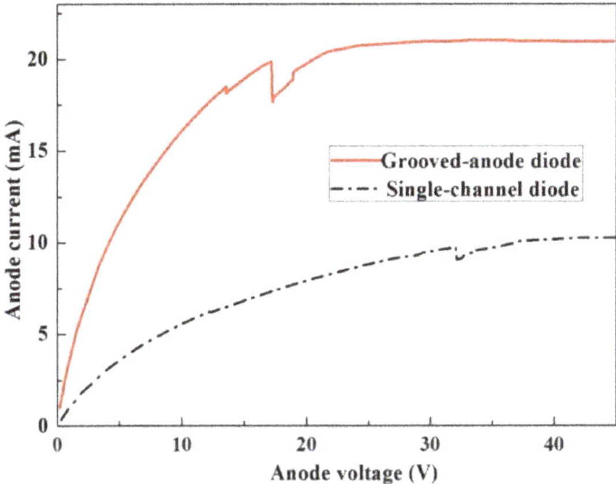

Figure 4. DC I-V output for the grooved-anode diode and single-channel diode as L = 2.0 μm.

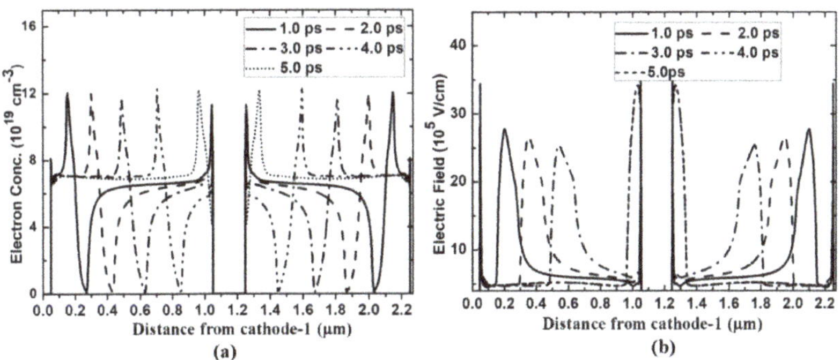

Figure 5. (**a**) Electron concentration and (**b**) Electric field profiles in one oscillation period in the grooved-anode diode as L = 2.0 μm.

N_{AC} is the electron concentration of the active channel, L_{AC} is channel length and d is the channel thickness. These conditions are easily satisfied in GaN-based heterostructure Gunn diode, as in the AlGaN/GaN/AlGaN heterojunction, the electron concentration of 2DEG is up to 10^{19} cm^{-3}. Such a high 2DEG is induced by a strong polarization effect of AlGaN/GaN without any doping and well confined in the quantum well. Therefore, 2DEG is well away from the ionized impurity scattering, ensuring the easier formation of the dipole domain in a short channel length. Even when the length of the GaN planar Gunn diode is reduced to several hundred nanometers or the channel thickness is reduced to several nanometers, the diode can still generate stable dipole-domain mode oscillation, as shown in Figure 6. However, a shorter channel length results in a smaller electron domain. In the 1.2-μm-grooved-anode diode, each channel length is 0.6 μm, which is not long enough for the electron dipole to grow mature before exiting from the anode side. Therefore, comparing Figures 5 and 6, especially the Figures 5b and 6b, from the electric field distribution for the 2.0-μm and 1.2-μm diodes we can conclude that a bigger domain forms in the 2.0-μm-grooved-anode diode. Therefore, higher DC-to-AC efficiency is obtained in the 2.0-μm diode.

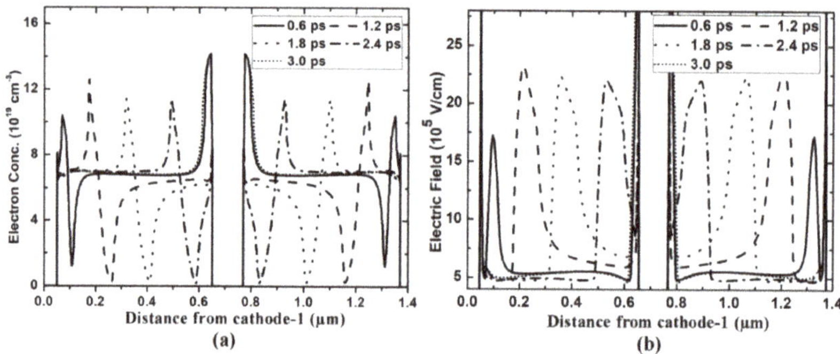

Figure 6. (**a**) Electron concentration and (**b**) Electric field profiles in one oscillation period in the grooved-anode diode as L = 1.2 μm.

Meanwhile, if the channel length is too long, the electron domain grows to its mature size before reaching the anode contact. The remaining channel will be regarded as invalid growth space for the electron domain, which results in the decrease of the efficiency of the Gunn diode. The formation of the electron domain results in the current decline. When the electron domain grows to its full size, the current will drop to its lowest value. If L is too long, the lowest current will last for a period until the electron domain begins to disappear from the anode side. As shown in Figure 7, the lowest-value part of the current wave obviously extends as the channel length L increases, which aggravates the nonlinearity of the oscillation wave, and enhances the harmonic component of the oscillation wave. It is worth noting that two current peaks occur in the 3.6-μm diode, as shown in Figure 7. Figure 8 gives the electron movement tracks in the 3.6-μm diode, which shows that two dipole domains form simultaneously inside each channel; while one forms at near the cathode side, one forms near the middle of the channel. As the electron concentration of 2DEG is very high, therefore, the effective channel lengths for both nucleating points satisfy the condition of the dipole domain. The formation of two domains inside one oscillation circle also aggravates the nonlinearity of the oscillation wave, and weakens the fundamental frequency. In addition, as the two domains restrict each other, neither of them are able to grow to their full size. Therefore, the fundamental and harmonic components of both are reduced. Figure 9 gives the frequency spectrum diagram of the grooved-anode diode for different L from 1.2 to 3.6 μm, which demonstrates that the harmonic component enhances and fundamental component decreases with L. As a result, the noise performance deteriorates with L. In conclusion, in order to avoid the harmonic component enhancement and increase the fundamental component, suitable channel length is of great importance.

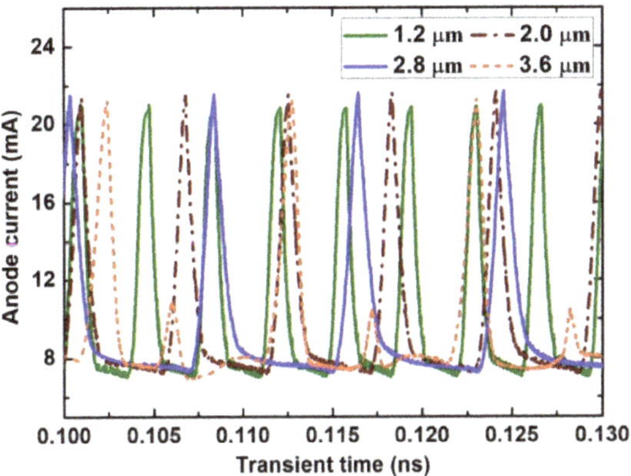

Figure 7. The current oscillation wave in the symmetric grooved-anode diode.

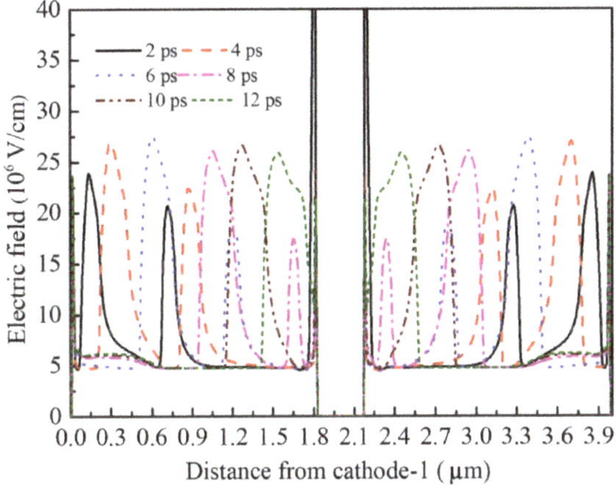

Figure 8. The electric field profiles in one oscillation period in the grooved-anode diode as L = 1.8 μm.

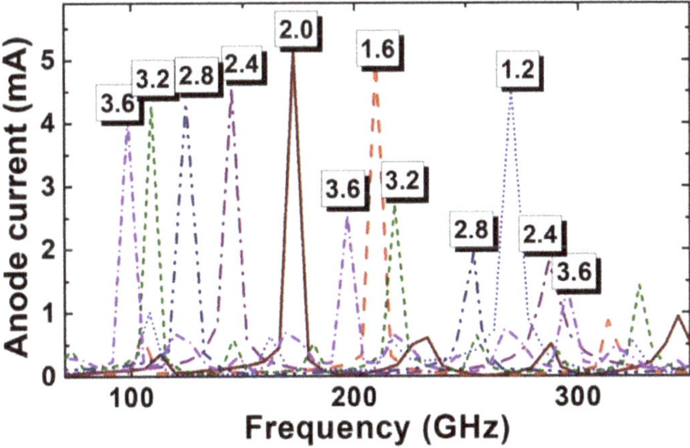

Figure 9. The frequency spectrum diagram of the grooved-anode diode for different L.

We also study the asymmetric grooved-anode Gunn diode, where $L_b = 2L_a = 2.0$ μm. In order to ensure that the electric field of each channel is in an appropriate range to generate stable Gunn oscillations, we set $V_{DC2} = 2V_{DC1} = 53.4$ V, $V_{AC2} = 2V_{AC1} = 13.4$ V. As the length of the right channel is twice as long as the length of the left channel, the movement period of the electron domain in the right channel should be twice as long as that in the left channel. This is verified by Figure 10, which shows the electric field profiles in one oscillation period derived in the 1.0-2.0-μm-grooved-anode diode. When the dipole domain in the right channel disappears from the anode contact, the dipole domain in the left channel completes two circles, which results in the two current peaks in the current oscillation wave, as shown in Figure 11a. The small current peak is generated as the domain exits from the shorter channel, and the larger current peak is generated as the domain exits from the longer channel. Therefore, two frequencies are obtained, and the second harmonic frequency is greatly enhanced, as shown in Figure 11b. As the two channels are independent of each other, the fundamental and harmonic frequencies are enhanced at the same time. The frequency of the second harmonic is about 175.47 GHz, η is about 1.85%, and P_{AC} is about 4.45 mW. The asymmetric structure realizes the modulation of the operation frequency via a single diode. By changing the length proportion of the two channels, the size and the number of the harmonic wave are able to be controlled.

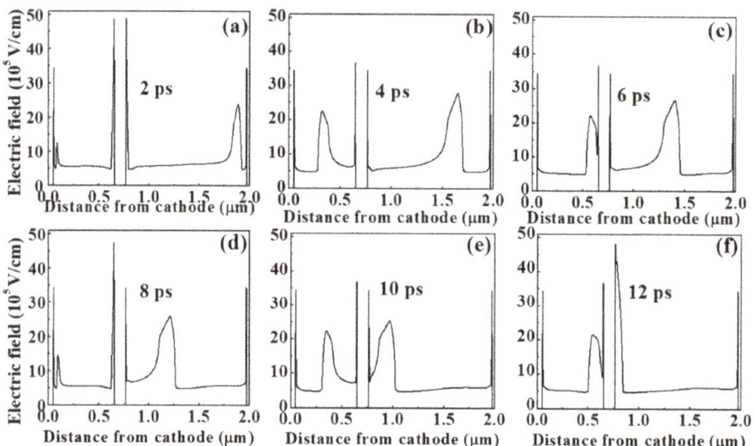

Figure 10. (**a–f**) The electric field profiles derived at the same time step in one oscillation period in the grooved-anode diode as $L_a = 1.0$ μm and $L_b = 2.0$ μm.

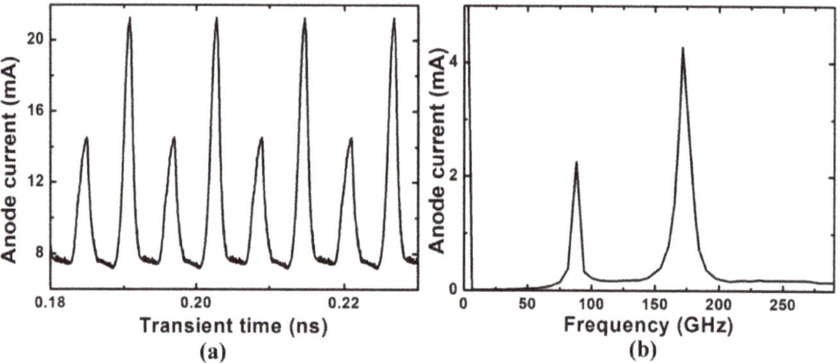

Figure 11. (**a**) The current oscillation wave and (**b**) the frequency spectrum diagram of the grooved-anode diode, where $L_a = 1.0$ μm and $L_b = 2.0$ μm.

4. Conclusions

In this paper, for the first time we propose the grooved-anode planar Gunn diode to greatly enhance the RF output power at a high operation frequency. We present an explicit numerical study into its working principle and output characteristics based on a simulation method. The grooved-anode diode is equal to two shorter diodes in parallel connection, as the grooved anode divides the long channel into two shorter channels. In the symmetric grooved-anode diode, the RF output power is almost doubled in the grooved-anode diode as compared with the single-channel diode. The 1.0-1.0-μm grooved-anode diode shows the best output characteristics. It operates at a fundamental frequency of 172.81 GHz and the corresponding DC-to-AC conversion efficiency is about 3.13%. It produces over 5.48 mW of power, nearly twice as high as that of the 1.0-μm single-channel diode. This novel grooved-anode diode realizes the enhancement of the frequency and RF output power simultaneously, by simply etching a rectangular grooved anode onto the semiconductor layer at one lithographic step, which provides good design ideas in improving the output characteristic of the terahertz sources and other power devices. In the asymmetric 1.0-2.0-μm grooved-anode diode, two frequencies are obtained, and the second harmonic is enhanced as compared with the fundamental wave. The harmonic-enhanced Gunn diode shows its potentials as a mixer or frequency multiplier. Furthermore, it will provide a fast conversion between two different frequencies without connecting with other terahertz oscillators. We have demonstrated that the proposed GaN heterostructure grooved-anode planar Gunn diode is an excellent candidate as a solid-state terahertz device.

Author Contributions: Conceptualization, Y.W. and L.-A.L.; Methodology, J.-P.A.; Software, Y.W.; Validation, Y.W., L.-A.L.; Formal Analysis, Y.W.; Investigation, Y.H.; Resources, J.-P.A.; Data Curation, Y.W.; Writing-Original Draft Preparation, Y.W.; Writing-Review & Editing, L.-A.L.; Visualization, L.-A.L.; Supervision, J.-P.A.; Project Administration, Y.H.; Funding Acquisition, Y.H. All authors have read and agreed to the published version of the manuscript.

Funding: This work was partially supported by the National Key Research and Development Program (Grant No. 2017 YFB043000), the National Natural Science Foundation of China (Grant No. 61274092) and the China Postdoctoral Science Foundation Grant (2019M653554).

Conflicts of Interest: The authors declare no conflict of interest.

References

1. Rostami, A.; Rasooli, H.; Baghban, H. Terahertz and Infrared Quantum Cascade Lasers. In *Terahertz Technology Fundamentals and Applications*; Springer: Berlin/Heidelberg, Germany; New York, NY, USA, 2011; Volume 77.
2. Wallance, V.P.; Fitzgerald, A.J.; Shankar, S.; Flanagan, N.; Pye, R.; Cluff, J.; Arnone, D.D. Terahertz pulsed imaging of basal cell carcinoma ex vivo and in vivo. *Br. J. Dermatol.* **2004**, *151*, 424–432. [CrossRef]
3. Cohen, I.; Cahan, R.; Shani, G.; Cohen, E.; Abramovich, A. Effect of 99 GHz continuous millimeter wave electro-magnetic radiation on *E. coli* viability and metabolic activity. *Int. J. Radiat. Biol.* **2010**, *86*, 390–399. [CrossRef]
4. Hao, Y.; Yang, L.A.; Zhang, J.C. GaN-based Semiconductor Devices for Terahertz Technology. *Terahertz Sci. Technol.* **2008**, *1*, 51–364.
5. Ambacher, O.; Smart, J.; Shealy, J.R.; Weimann, N.G.; Chu, K.; Murphy, M.; Schaff, W.J.; Eastman, L.F.; Dimitrov, R.; Wittmer, L.; et al. Two-dimensional electron gases induced by spontaneous and piezoelectric polarization charges in N- and Ga-face AlGaN/GaN heterostructures. *J. Appl. Phys.* **1999**, *85*, 3222–3233. [CrossRef]
6. Piprek, J. *Nitride Semiconductors: Principles and Simulation*; Wiley: Hoboken, NJ, USA, 2007.
7. Yang, L.A.; Long, S.; Guo, X.; Hao, Y. A comparative investigation on sub-micrometer InN and GaN Gun diodes working at terahertz frequency. *J. Appl. Phys.* **2012**, *111*, 104514. [CrossRef]
8. Khalid, A.; Pilgrim, N.J.; Dunn, G.M.; Holland, M.C.; Stanley, C.R.; Thayne, I.G.; Cunmming, D.R.S. A Planar Gunn Diode Operating Above 100 GHz. *IEEE Electron Devices Lett.* **2007**, *28*, 849–851. [CrossRef]
9. Pilgrim, N.J.; Khalid, A.; Dunn, G.M.; Cumming, D.R.S. Gunn oscillations in planar heterostructure diodes. *Semicond. Sci. Technol.* **2008**, *23*, 075013. [CrossRef]

10. Li, C.; Khalid, A.; Caldwell, S.H.P.; Holland, M.C.; Dunn, G.M.; Thayne, I.G.; Cumming, D.R.S. Design, fabrication and characterization of In0.23Ga0.77As-channel planar Gunn diodes for millimeter wave applications. *Solid State Electron.* **2011**, *64*, 67–72. [CrossRef]
11. Song, A.M.; Missous, M.; Omling, P.; Peaker, A.R.; Samuelson, L.; Seifert, W. Operation and high-frequency performance of nanoscale unipolar rectifying diodes. *Appl. Phys. Lett.* **2003**, *83*, 1881. [CrossRef]
12. Chowdhury, S.; Mishra, U.K. Lateral and Vertical Transistors Using the AlGaN/GaN Heterostructure. *IEEE Trans. Electron Devices* **2013**, *60*, 3060. [CrossRef]
13. Zhang, Y.; Sun, M.; Liu, Z.; Piedra, D.; Pan, M.; Gao, X.; Lin, Y.; Zubair, A.; Yu, L.; Palacios, T. Novel GaN trench MIS barrier Schottky rectifiers with implanted field rings. In Proceedings of the 2016 IEEE International Electron Devices Meeting (IEDM), San Francisco, CA, USA, 3–7 December 2016.
14. Zou, X.; Zhang, X.; Chong, W.C.; Tang, C.W.; Lau, K.M. Vertical LEDs on Rigid and Flexible Substrates Using GaN-on-Si Epilayers and Au-Free Bonding. *IEEE Trans. Electron Devices* **2016**, *63*, 1587–1593. [CrossRef]
15. Millithaler, J.F.; Íñiguez-de-la-Torre, I.; Íñiguez-de-la-Torre, A.I.; González, T.; Sangare, P.; Ducournau, G.; Gaquiere, C.; Mateos, J. Optimized V-shape design of GaN nanodiodes for the generation of Gunn oscillations. *Appl. Phys. Lett.* **2014**, *104*, 073509. [CrossRef]
16. Sze, S.M.; Ng, K.K. *Physics of Semiconductor Devices*, 3rd ed.; Wiley: New York, NY, USA, 2007.
17. González, T.; Íñiguez-de-la-Torre, A.; Pardo, D.; Mateos, J.; Song, A.M.J. Monte Carlo analysis of Gunn oscillations in narrow and wide band-gap asymmetric nanodiodes. *Phys. Conf. Ser.* **2009**, *193*, 012018. [CrossRef]
18. Xu, K.Y.; Li, J.; Xiong, J.W.; Wang, G. Mutual phase-locking of planar nano-oscillators. *AIP Adv.* **2014**, *4*, 067108. [CrossRef]
19. Íñiguez-de-la-Torre, I.; Mateos, J.; González, T.; Sangaré, P.; Faucher, M.; Grimbert, B.; Brandli, V.; Ducournau, G.; Gaquière, C. Searching for THz Gunn oscillations in GaN planar nanodiodes. *J. Appl. Phys.* **2012**, *111*, 113705. [CrossRef]
20. Balocco, C.; Halsall, M.; Vinh, N.Q.; Song, A.M. THz operation of asymmetric-nanochannel devices. *J. Phys. Condens. Matter* **2008**, *20*, 384203. [CrossRef]
21. Balocco, C.; Kasjoo, S.R.; Lu, X.F.; Zhang, L.Q.; Alimi, Y.; Winnerl, Y.S.; Song, A.M. Room-temperature operation of a unipolar nanodiode at Terahertz frequencies. *Appl. Phys. Lett.* **2011**, *98*, 223501. [CrossRef]
22. Wang, Y.; Wang, Z.; Ao, J.; Hao, Y. The modulation of multi-domain and harmonic wave in a GaN planar Gunn diode by recess layer. *Semicond. Sci. Technol.* **2016**, *31*, 025001. [CrossRef]
23. Wang, Y.; Yang, L.; Wang, Z. Ultra-short channel GaN high electron mobility transistor-like Gunn diode with composite contact. *J. Appl. Phys.* **2014**, *116*, 849. [CrossRef]
24. Wang, Y.; Yang, L.; Wang, Z. The enhancement of the output characteristics in the GaN based multiple-channel planar Gunn diode. *Phys. Status Solidi* **2016**, *213*, 1252–1258. [CrossRef]
25. Lanford, W.B.; Tanaka, T.; Otoki, Y.; Adesida, I. Recessed-gate enhancement-mode GaN HEMT with high threshold voltage. *Electron. Lett.* **2015**, *41*, 449–450. [CrossRef]
26. Wang, H.; Wang, Y.; Zhu, G.; Wang, Q.; Xin, Q.; Han, L.; Song, A. A novel thermally evaporated etching mask for low-damage dry etching. *IEEE Trans. Nanotechnol.* **2017**, *16*, 290–295. [CrossRef]
27. Balkan, N.; Arikan, M.C.; Gokden, S.; Tilak, V.; Schaff, B.; Shealy, R.J. Energy and momentum relaxation of hot electrons in GaN/AlGaN. *J. Phys. Condens. Matter* **2002**, *14*, 3457. [CrossRef]
28. Matulionis, A.; Liberis, J. Microwave noise in AlGaN/GaN channels. *IEE Proc. Circuits Devices Syst.* **2004**, *151*, 148–154. [CrossRef]
29. Matulionis, A.; Morkoç, H. Hot phonons in InAlN/AlN/GaN heterostructure 2DEG channels. In Proceedings of the SPIE OPTO: Integrated Optoelectronic Devices, San Jose, CA, USA, 18 February 2009.
30. Dunn, G.; Kearney, M.J. A theoretical study of differing active region doping profiles for W-band (75–110 GHz) InP Gunn diodes. *Semicond. Sci. Technol.* **2003**, *18*, 794. [CrossRef]
31. Sevik, C.; Bulutay, C. Gunn oscillations in GaN channels. *Semicond. Sci. Technol.* **2004**, *19*, 190. [CrossRef]
32. Tech, Y.P.; Dunn, G.M.; Priestley, N.; Carr, M. Monte Carlo simulations of asymmetry multiple transit region Gunn diodes. *Semicond. Sci. Technol.* **2005**, *20*, 418.
33. García, S.; Vasallo, B.G.; Mateos, J.; González, T. Time-domain Monte Carlo simulations of resonant-circuit operation of GaN Gunn diodes. In Proceedings of the 9th Spanish Conference on Electronic Devices (CDE 2013), Valladolid, Spain, 12–14 February 2013; pp. 79–82.

34. Kamoua, R.; Eisle, H.; Haddad, G.I. D-band (110–170 GHz), InP gunn devices. *Solide State Electron.* **1993**, *36*, 1547.
35. Judaschke, R. Comparison of modulated impurity-concentration inp transferred electron devices for power generation at frequencies above 130 GHz. *IEEE Trans. Microw. Theory Tech.* **2000**, *48*, 719–724. [CrossRef]
36. Zybura, M.F.; Jones, S.H.; Tait, G.; Duva, J.M. Efficient computer aided design of GaAs and InP millimeter wave transferred electron devices including detailed thermal analysis. *Solid State Electron.* **1995**, *38*, 873–880. [CrossRef]
37. García, S.; Íñiguez-de-la-Torre, I.; Pérez, S.; Mateos, J.; González, T. Numerical study of sub-millimeter Gunn oscillations in InP and GaN vertical diodes: Dependence on bias, doping, and length. *J. Appl. Phys.* **2013**, *114*, 074503. [CrossRef]

 *micromachines* 

Article

Improving Output Power of InGaN Laser Diode Using Asymmetric $In_{0.15}Ga_{0.85}N/In_{0.02}Ga_{0.98}N$ Multiple Quantum Wells

Wenjie Wang [1,2,*], Wuze Xie [1,2], Zejia Deng [1,2] and Mingle Liao [1,2]

1 Microsystem & Terahertz Research Center, China Academy of Engineering Physics, Chengdu 610200, China; xiewuze@mtrc.ac.cn (W.X.); dengzejia@mtrc.ac.cn (Z.D.); liaomingle@mtrc.ac.cn (M.L.)
2 Institute of Electronic Engineering, China Academy of Engineering Physics, Mianyang 621999, China
* Correspondence: wangwenjie@mtrc.ac.cn

Received: 14 November 2019; Accepted: 11 December 2019; Published: 13 December 2019

Abstract: Herein, the optical field distribution and electrical property improvements of the InGaN laser diode with an emission wavelength around 416 nm are theoretically investigated by adjusting the relative thickness of the first or last barrier layer in the three $In_{0.15}Ga_{0.85}N/In_{0.02}Ga_{0.98}N$ quantum wells, which is achieved with the simulation program Crosslight. It was found that the thickness of the first or last InGaN barrier has strong effects on the threshold currents and output powers of the laser diodes. The optimal thickness of the first quantum barrier layer (FQB) and last quantum barrier layer (LQB) were found to be 225 nm and 300 nm, respectively. The thickness of LQB layer predominantly affects the output power compared to that of the FQB layer, and the highest output power achieved 3.87 times that of the reference structure (symmetric quantum well), which is attributed to reduced optical absorption loss as well as the reduced vertical electron leakage current leaking from the quantum wells to the p-type region. Our result proves that an appropriate LQB layer thickness is advantageous for achieving low threshold current and high output power lasers.

Keywords: asymmetric multiple quantum wells; barrier thickness; InGaN laser diodes; optical absorption loss; electron leakage current

---

1. Introduction

InGaN-based multi-quantum well (MQW) laser diodes (LDs) have drawn much attention in recent years due to their potential as the light sources in the applications of high-density optical storage systems, laser printing, full-color displays, small portable projector, among the others [1–10]. To date, although the InGaN based blue LDs are already commercialized and the material growth technology has been significantly improved, the factors to achieve optimal optoelectronic performance have not been fully developed. There are still many places that need to be improved. In general, high output power and low threshold current are two key indicators to attain superior laser diode performances. The emission mechanism of the InGaN/(In)GaN multiple-quantum-well structure is important and necessary for further improving the performance of LD. The material composition, number, and thickness of the multiple quantum well (WQW) in the active region are crucial structural parameters for optimizing the structure of LD because they directly affect the optical and electrical properties of the device. Among these parameters, the composition fluctuation of the quantum well mainly affects the wavelength of the laser emission, because the band gap of the quantum well varies with composition. Meanwhile, the effect of QW number on the performance of MQW LDs has also been studied both theoretically and experimentally [11–13], which indicates that the blue-violet laser (emission wavelength between 392 nm and 420 nm) with two InGaN wells could obtain the lowest threshold current when the band gap ratio is 7/3. But the QW number is highly dependent on other parameters such as laser material, output wavelength, structural design, and so on [14]. When the emission wavelength is around 416 nm, our simulation results and experiments show that the performance of the laser device obtained from the three quantum wells structure is slightly better than that of the two quantum wells structure. Thereby, the InGaN lasers with wavelengths around 416 nm generally employ two or three quantum wells structure [15]. Meanwhile, Alahyarizadeh's research shows that the thickness of QW can also manipulate the emission characteristic of the InGaN QW [16]. In addition, the experimental data shows that the threshold current is the smallest when the width of the quantum well is around 3 nm, which is consistent with the high-quality sample results [17]. Despite all these findings, the influence of barrier parameters is less studied, especially the thick barrier. The current research mainly focused on improving the carrier distribution by the special composite barrier layer [15], and few on the distribution of the optical field. The main reason is that the common barrier layer is too thin relative to the waveguide layer and the cladding layer, and the effect on the distribution of the optical field is relatively small.

In principle, the barrier layer of MQW structure affects the asymmetric distribution of carrier concentration in the p-doped and n-doped regions and the optical field distribution over the whole LD structure. Due to the huge difference between electron (600 cm^2/Vs) and hole mobility (10 cm^2/Vs) [18], the carrier distribution in the MQW region is non-uniform. The concentration of holes in the n-doped region is significantly lower than the concentration of electrons in the p-doped region. The asymmetry of the carrier distribution in the MQW and the optical field distribution deviating from the center of the quantum well are thus enhanced, and the luminous efficiency would be reduced [19]. Since the barrier thicknesses of the conventional InGaN/GaN MQW LDs are usually thin (8~20 nm), it is difficult to achieve effective carrier transport into the MQW near the n-layer. So, the actual carrier density will not be uniform in every QW, and it is hard to affect optical field distribution and the associated absorption loss. Many researchers have improved the performance of LD by inserting an unintentionally doped (In) GaN or AlGaN thin layer between LQB and EBL, or thickening of the upper and lower waveguide layers to reduce optical absorption loss and expand the distribution of the optical field [20–22]. However, it is difficult to further reduce optical absorption loss and improve the optical field distribution in these device structures, because the upper waveguide layer is always p-type with a high absorption coefficient. It is more convenient and effective to improve the optical field distribution by adjusting the parameters of the quantum well region with a low absorption coefficient. For the sake of eliminating the unfavorable effect of the asymmetric carrier and optical field distribution mentioned above, the MQW active region needs to be redesigned and optimized.

In this letter, the relative thickness of the first or last barrier layers in the three $In_{0.15}Ga_{0.85}N/In_{0.02}Ga_{0.98}N$ quantum wells laser diode, are designed and optimized to enhance the optical field distribution into the MQW while reducing the optical absorption loss. Simultaneously, by redesigning the barrier layer, the comparison between their optical and electrical characteristics (especially leakage current) are analyzed and compared with the theoretical simulation results.

2. Device Structure and Simulation Setup

For the illustration purpose, a structure of three $In_{0.15}Ga_{0.85}N/In_{0.02}Ga_{0.98}N$ quantum wells laser diodes with various first or last barrier layers are designed to increase the optical field distribution in the active region, and to reduce the optical absorption loss and the vertical electron leakage current. For comparison with symmetric quantum well structure, there are series of structures with various first or last barrier layers of $In_{0.15}Ga_{0.85}N/In_{0.02}Ga_{0.98}N$ quantum wells laser diodes shown in Figure 1. These laser structures with varying thicknesses of quantum barrier layers consist of a GaN free-standing substrate, a 1 μm thick GaN buffer layer, an n-$Al_{0.08}Ga_{0.92}N$ confinement layer with a thickness of 1 μm and the doping concentration of 3×10^{18} cm^{-3}, an n-type GaN lower waveguide (LWG) layer with a thickness of 100 nm and the doping concentration of 1×10^{18} cm^{-3}, a three-period 3 nm $In_{0.15}Ga_{0.85}N$ well/15 nm $In_{0.02}Ga_{0.98}N$ barrier quantum well with background concentration of 5×10^{16} cm^{-3}, a p-$Al_{0.2}Ga_{0.8}N$ electron blocking layer (EBL) with 20 nm thickness, a p- GaN upper WG

(UWG) layer with 100 nm thickness, a p-AlGaN confinement layer with 500 nm thickness, and a p-GaN contact layer with 80 nm thickness. The doping concentration of all p-type layers is 5×10^{18} cm^{-3}.

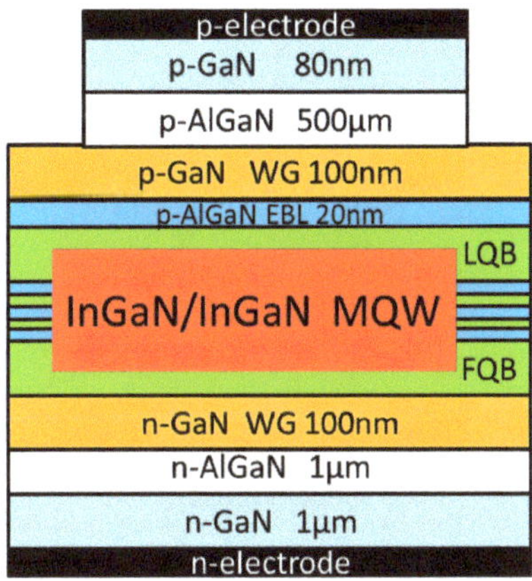

Figure 1. Schematic of $In_{0.15}Ga_{0.85}N/In_{0.02}Ga_{0.98}N$ quantum well laser diodes.

The differences in these laser structures are only the first or last barrier thickness of the $In_{0.15}Ga_{0.85}N/In_{0.02}Ga_{0.98}N$ quantum wells. In the reference structure (marked as FQB-15 or LQB-15), the active region of MQW is symmetrical and the barrier layer thicknesses is 15 nm. In the first barrier series of the LD structures (marked as FQB-XX) and the last barrier series of the LD structures (marked as LQB-XX), the first or last barrier is an $In_{0.02}Ga_{0.98}N$ single layer with different thickness.

Asymmetric $In_{0.15}Ga_{0.85}N/In_{0.02}Ga_{0.98}N$ MQW samples with various barrier layer thicknesses, photoelectric properties of these laser structures, including the optical filed distributions, optical confinement factors, peak modal gains, electron leakage currents and output powers, are theoretically simulated by a powerful semiconductor laser simulation tool Crosslight Device Simulation Software (Crosslight Software Inc., Vancouver, Canada) [23,24]. In order to eliminate the influence of the contact, both the p and n electrodes are set as an ideal Ohmic contact type. The cavity length and ridge width are set to be 800 µm and 2 µm, respectively. The reflectivity of both sides of the laser cavity is set to 19%. The screening factor is set to be 0.25 [25], and the band offset ratio ($\Delta E_c/\Delta E_g$) is set to be 0.67 [26]. Meanwhile, for the n-type and p-type layers, their absorption coefficients are set as 5 cm^{-1} and 50 cm^{-1} [27], respectively. The absorption coefficient of highly doped EBL is 100 cm^{-1}.

3. Results and Discussions

First, the effect of the layer thickness of the first quantum barrier layer is investigated. The optical characteristics of the four FQB thicknesses of 15 nm, 45 nm, 100 nm and 225 nm are simulated, respectively. Then, three $In_{0.15}Ga_{0.85}N/In_{0.02}Ga_{0.98}N$ quantum wells with LQB thicknesses of 100 nm, 300 nm and 500 nm, respectively, are calculated. The refractive index profile of the symmetric quantum well LD structure (red line) and the optical field distributions of the FQB-15 and FQB-100 structures are shown in Figure 2. The refractive indices of InN, GaN and AlN are set as 3.4167, 2.5067, and 2.0767, respectively. The refractive indices of $In_xGa_{1-x}N$ and $Al_xGa_{1-x}N$ are calculated using an approximate method as follows:

$$n(In_xGa_{1-x}N) = [n(InN) - n(GaN)] \cdot x + n(GaN) \quad (1)$$

$$n(Al_xGa_{1-x}N) = [n(AlN) - n(GaN)] \cdot x + n(GaN) \quad (2)$$

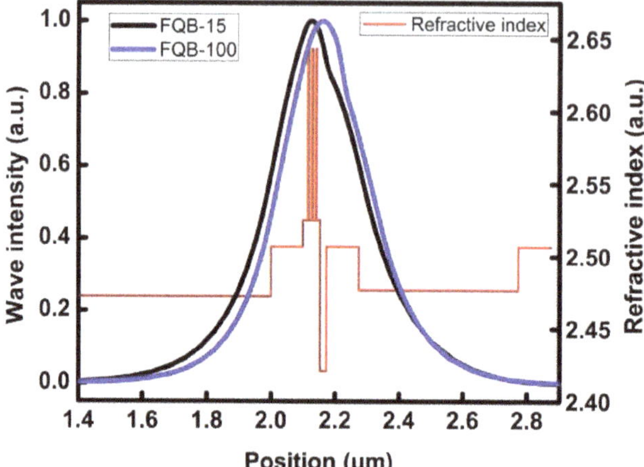

Figure 2. The refractive index profile of the symmetric quantum well LD structure (red line) and the optical field distributions of the FQB-15 and FQB-100 structures.

According Equations (1) and (2), the refractive indices of the $Al_{0.2}Ga_{0.8}N$, $Al_{0.07}Ga_{0.93}N$, $Al_{0.08}Ga_{0.92}N$, $In_{0.15}Ga_{0.85}N$ and $In_{0.02}Ga_{0.98}N$ are calculated to be 2.4207, 2.4766, 2.4723, 2.6432 and 2.5249, respectively. Due to the large refractive index difference between the GaN waveguide layer and the upper and lower confinement layers, most of the optical field is located to the layer between the upper and lower GaN waveguides. However, the appearance of the low refractive index $Al_{0.2}Ga_{0.8}N$ EBL decreases the mean value of the p-type refractive index, resulting in an asymmetric optical field distribution on the left and right sides of the quantum wells, and the peak is not situated at the middle of the quantum wells. The peak position shifted slightly to the n-type side, and the optical limit factor (OCF) did not reach the maximum value. In addition, a large part of the optical field is distributed in the p region with a large absorption coefficient, which causes a large optical absorption loss and the laser's threshold current rises.

When the FQB or LQB layer becomes thicker, the EBL or n-GaN WG will move away from the quantum wells. The optical field peak moves to the p-type region, so that more optical fields are distributed in the quantum well region, and the OCF is enhanced. In addition, since the optical field is close to a Gaussian distribution, when the FQB or LQB layer becomes thicker, the optical field proportion of the p-type region decreases significantly, thereby reducing light absorption loss.

However, the FQB or LQB layer should not be too thick, either. As the thickness of the FQB or LQB layer increases, the optical field is further deviated from the quantum wells. Meanwhile, the dotted line in Figure 3 indicates that the peak position of the light field will pass through the quantum wells region and continue to move to the p region, moving to the position between the quantum wells and EBL. It can be seen that with the increase of the FQB or LQB layer thickness, the optical field is more distributed inside the FQB or LQB layer. However, the optical field peak has deviated from the quantum wells. Even if more optical fields are distributed inside the FQB or LQB layer, the optical absorption loss further decreases, but the OCF may be reduced by the shift of the peak position of the optical field [28]. This is consistent with the trend of OFC in Figure 4 as the thickness of the barrier layer changes. Therefore, with the increase of the FQB or LQB layer thickness, the peak position of the optical field moves from the n-type regions to the p-type regions, and the optical confinement factor slowly grows to the maximum and then decreases rapidly.

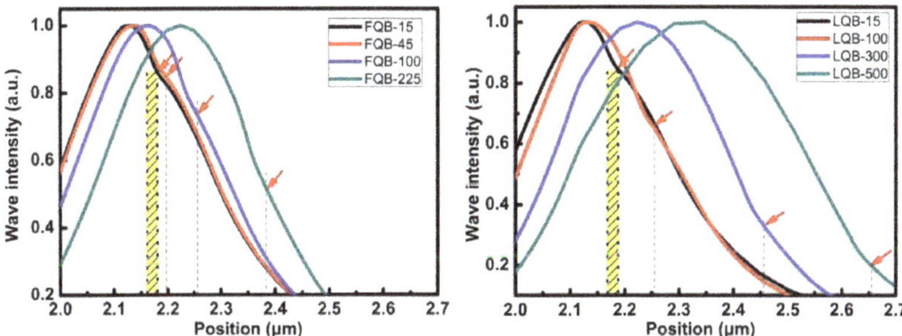

Figure 3. Optical field distribution of FQB or LQB structures (dashed lines and red arrows indicate the position of the electron blocking layer.).

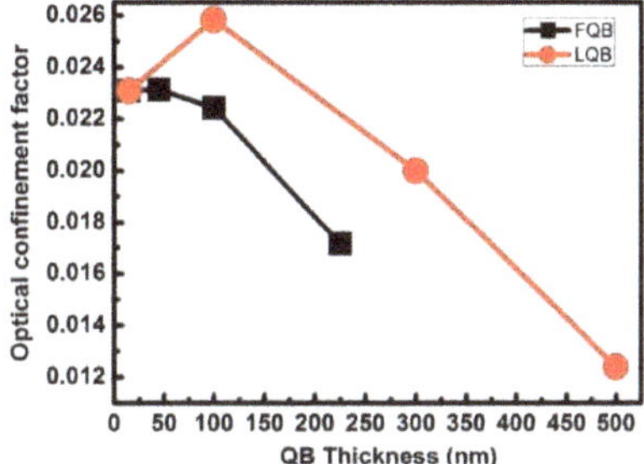

Figure 4. Optical confinement factor of LDs with different FQB or LQB layer thicknesses.

The peak modal gain of LDs with different FQB or LQB layer thicknesses are shown in Figure 5. When the optical confinement effect in the quantum well region is enhanced, the modal gain will increase accordingly, because the modal gain is the product of OCF Γ_0 and the material gain g of the quantum well. Taking a specific injection current value of 35 mA as an example, the calculated peak modal gain of four lasers with varying FQB or LQB layer thicknesses are compared, as shown by the red dotted line in Figure 5. When these structures are injected with the same current, the peak modal gain grows from 4 to 20 m^{-1} and the corresponding LQB layer thickness increases to 100 nm. When the thickness of the LQB layer continues to increase to 500 nm, the peak modal gain decreases from 20 to -3 m^{-1}. In addition, when the thickness of the FQB layer exceeds 45 nm, the peak modal gain will decrease, which confirms that the optical field shift will not be beneficial to OCF. This is consistent with the trend of the OFC in Figure 4.

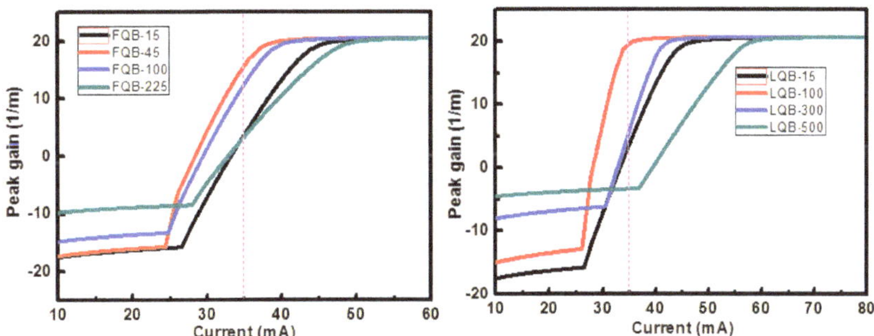

Figure 5. Peak modal gain of LDs with different FQB or LQB layer thicknesses.

The changes in OCF and optical absorption loss (OAL) have a large effect on the electrical characteristics of the $In_{0.15}Ga_{0.85}N/In_{0.02}Ga_{0.98}N$ QW LD. Figure 6 shows a completely different V-I characteristic of lasers with varying FQB and LQB thickness. In the laser structure of this paper, the quantum wells region is close to the electron blocking layer. Increasing the thickness of LQB makes the distance of the electrons reflected by the EBL to the well layer larger, resulting in greater resistance. However, increasing the thickness of FQB has little effect on the reflection of EBL on electrons, and the resistance is basically unchanged. This is quite different from the electron-overflow-suppression (EOS) layer [22] above the upper waveguide layer. Reference [22] indicates that the EOS layer can eliminate the voltage rise only when the EOS layer is located between the upper waveguide layer and the p-doped region. L-I characteristics of lasers with varying FQB or LQB layer thicknesses are shown in Figure 7. When the modal gain and loss are completely balanced, the modal gain tends to saturate and the laser reaches the threshold condition [29]:

$$\Gamma_0 g = \alpha_i + \alpha_m \quad (3)$$

where α_i is the optical absorption loss, α_m is the mirror loss. As the thickness of the FQB and LQB layers increases, the threshold current will increase first and then decrease. The threshold currents of lasers with FQB thicknesses of 15, 45, 100, and 225 nm are 48, 39, 42, and 50 mA, respectively. Similar trends also occur when the thickness of the LQB changes. When the thickness of LQB is 100, 300, and 500 nm, the corresponding threshold currents are 36 mA, 42 mA, and 57 mA, respectively. In particular, LQB-300 has a lower threshold current density than the reference structure when the OFC value is smaller than that of the reference structure, due a small α_i. The change trend of the laser output power in Figure 7 is similar to the threshold current trend in Figure 5. When the thickness of LQB is 300 nm, the LD device achieves the highest output power of 240 mW.

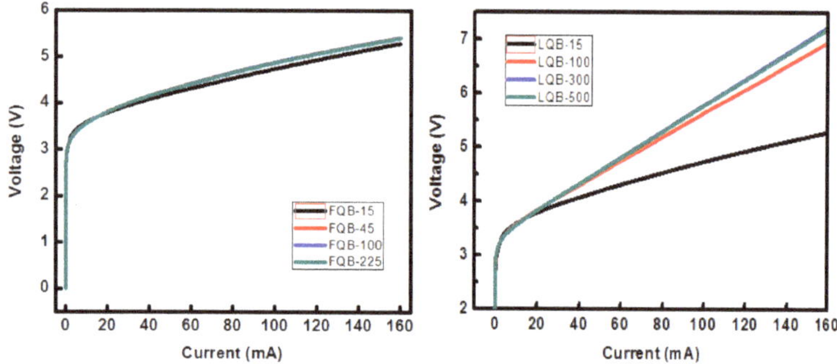

Figure 6. V-I characteristic of lasers with varying FQB and LQB thickness.

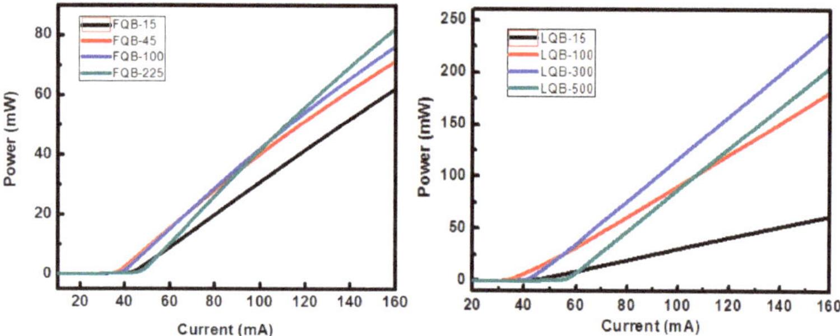

Figure 7. L-I characteristics of lasers with varying FQB or LQB layer thicknesses.

The optimum thickness value of the FQB or LQB layer can be determined based on the changes in the OAL and OCF. The L-I characteristic curves of asymmetric quantum well LDs with FQB or LQB layers of various thicknesses are calculated. Figure 8 depicts the variation in L-I characteristics with different FQB or LQB layer laser structure under an injection current of 160 mA. It can be seen that the laser output power increase evidently as the thickness of FQB (or LQB) layer increases from 15 nm to 225 nm (or 300 nm). That is, when the FQB (or LQB) layer gradually increases from the initial thickness of 15 nm, it will bring two factors that increase the optical confinement factor and decrease the absorption loss, which are conducive to increasing the output power. For the asymmetric quantum well LDs with an FQB (or LQB) layer thicker than 300 nm (or 400 nm), the deviation of the peak position of the optical field from the quantum wells caused a sharp decrease in OCF, which led to a rapid decrease in the output power of the laser. There is a relatively stable output power area between the two regions where the output power changes rapidly, that is, 200 nm to 250 nm for the LBQ layer and 250 nm to 350 nm for the FBQ layer. This is the result of the comprehensive effects of the disadvantage of OCF reduction and the positive compensation of optical absorption loss decrease. When the thickness of the FQB or LQB layer is further increased, the output power drops, which is caused by the rapid decline of OCF is not enough to be compensated by the reduction of the optical absorption loss, so the output power of the asymmetric quantum well laser decreases. The optimum thickness of the FQB (or LQB) layer in the laser structure is around 225 nm (or 300 nm). However, the thickness of the LQB layer is more important than the FQB layer. At 160 mA, the maximum output power of the LQB-300 is 240 mW, which is 2.92 times that of the FQB-225 layer, and 3.87 times that of the reference structure.

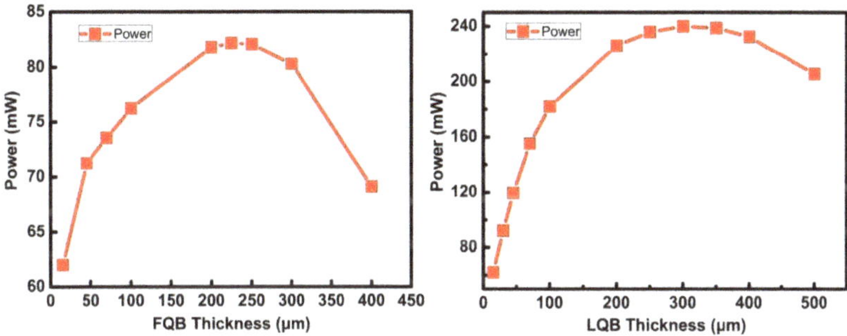

Figure 8. Variation in L-I characteristics with different FQB or LQB layer laser structure under an injection current of 160 mA.

In order to compare the impact of the first and last barrier layer thickness on the output power, the vertical electron current density of the different FQB and LQB layer structures are simulated, as indicated in Figure 9. It indicates that in the normal working state of the laser, electrons enter the quantum wells from the n-type region, resulting in radiative and non-radiative recombination of electrons and holes in each quantum well, so that the electron current density gradually decreases in the quantum wells. It is noted that the $Al_{0.2}Ga_{0.8}N$ EBL can effectively confine electrons in the quantum wells, and at the same time, it can reflect electrons back into the quantum wells and recombine with holes, thereby reducing the electron leakage current. The percentage of electron leakage can be calculated by dividing the electron current entering the p-type region from the quantum wells by the electron current injected into the quantum wells. For the 15, 45, 100, and 225 nm FQB layers, the percentages of electron leakage are 67.9%, 52.7%, 56.9%, and 62.8%, respectively. In a series of laser structures with varying FQB thicknesses, when the thickness of the FQB layer exceeds 45 nm, the decrease in electron current density from the bottom quantum well to the top quantum well becomes greater, which indicates that the degree of recombination of electrons and holes is inconsistent in each quantum well. In addition, the blocking ability of $Al_{0.2}Ga_{0.8}N$ EBL becomes weaker as the FQB thickens.

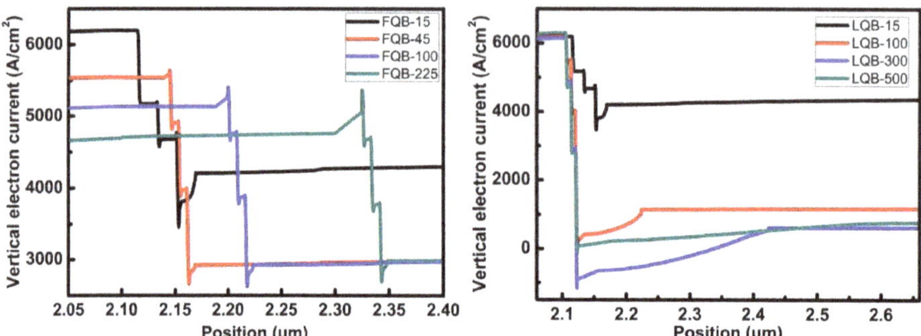

Figure 9. Vertical electron current density of LDs with different FQB or LQB layer thickness.

The electron current densities of LQB structures are significantly different from that of FQB structures. For the 15, 100, 300, and 500 nm FQB layers, the percentages of electron leakage in the LQB layer changed structure are 67.9%, 18.3%, 9.9%, and 12.1%, respectively. It can be clearly seen that compared with the FQB structure, the electron current density of the LQB structure decreases more, and the change of the electron current density is more obvious in the upper quantum well, indicating that more carriers are recombined in the LQB structure than that of reference structure. This is consistent with the results of the energy band diagrams in Figure 10. Now, the largest effective EBL barrier for electrons in the LQB series is ΔEn = 215 meV in LQB-300, which is much greater than that of the reference LD (ΔEn = 184 meV). A large increase in the effective EBL barrier for electrons results in a significant reduction in electron leakage and improves device performance. The electrons can be effectively reflected to the quantum wells region to increase the effective recombination probability of the two quantum wells near the n-type region, so that the electron concentration distribution in the quantum wells region is more even. This is more advantageous to carrier recombination and device efficiency. When the thickness of the barrier layer is increased to more than 300 nm, the electron leakage current remains substantially unchanged or even slightly increased. It means that, the increase in the thickness of the barrier layer does not effectively reduce the electron leakage, but increases the resistance of the device, so that the threshold current becomes larger, which is consistent with Figure 7. The LQB-300 structure has the smallest electron leakage current in the p region, which indicates that greatly reduced electron leakage is critical to improve the output power of InGaN-based LDs. Therefore, by decreasing optical absorption loss and greatly reducing the electron leakage, combining these two factors, we obtain the optimal structure LQB-300 which produces the maximum power of 3.87 times that of the reference structure power.

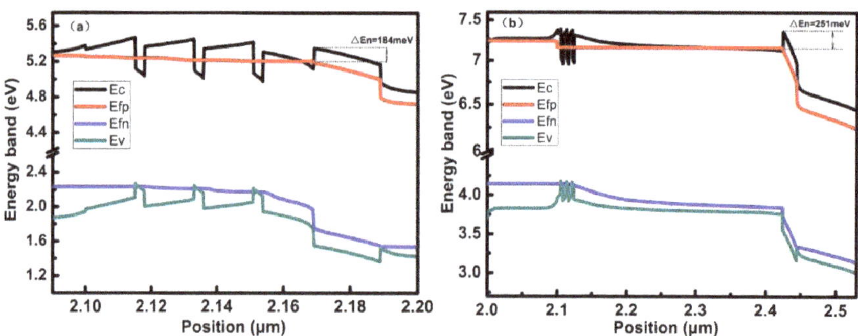

Figure 10. Energy band diagrams of reference LD structure (**a**) and LQB-300 structure (**b**) under an injection current of 160 mA.

4. Conclusions

A series of $In_{0.15}Ga_{0.85}N/In_{0.02}Ga_{0.98}N$ MQW laser diodes (LDs) with different barrier layers are investigated with the simulator Crosslight. It is found that when the first (or last) $In_{0.15}Ga_{0.85}N$ barrier layer is no more than 45 nm (or 100 nm), due to the increase in the thickness of the first (or last) barrier layer, the optical field is limited better in the MQW, and the optical absorption loss is reduced. Subsequently, a low threshold current and high output power are achieved. As the thickness of the barrier layer becomes larger, the output powers of the lasers gradually increase, and the positive effects of the reduced optical absorption loss are partially compensated by the negative effects of the OCF reduction. However, when the FQB (or LQB) layer is thicker than 225 nm (or 300 nm), the photoelectric performance of LDs become worse. It is due to the rapid decrement of the OCF, which is not enough to be compensated by the reduction of the optical absorption loss. Nevertheless, compared to the FQB structure, the thick LQB layer will significantly reduce the vertical electron leakage current leaking from quantum wells to p-type region, especially the LQB-300 structure. As a result, the maximum output power of the LQB-300 is 3.87 times that of the reference structure.

Author Contributions: Conceptualization, W.W.; Methodology, W.W.; Validation, W.W.; Formal Analysis, W.X.; Investigation, W.W., W.X., M.L. and Z.D.; Resources, W.W.; Data Curation, W.W.; Writing—Original Draft Preparation, W.W.; Writing—Review and Editing, W.W.

Funding: This research is funded by the Science Challenge Project (No. TZ2016003-2-1), National Key R&D Program of China (No. 2017YFB0403103), and National Natural Science Foundation of China (No. 61504126).

Acknowledgments: The authors would like to thank Liang Feng and Sun Song for their helpful suggestions and comments.

Conflicts of Interest: The authors declare no conflict of interest.

References

1. Nakamura, S.; Senoh, M.; Nagahama, S.; Iwasa, N.; Yamada, T.; Matsushita, T.; Sugimoto, Y.; Kiyoku, H. Room-temperature continuous-wave operation of InGaN multi-quantum-well structure laser diodes. *Appl. Phys. Lett.* **1996**, *69*, 4056–4058. [CrossRef]
2. Nakamura, S. III–V nitride based light-emitting devices. *Solid State Commun.* **1997**, *102*, 237–248. [CrossRef]
3. Schmidt, M.C.; Kim, K.C.; Farrell, R.M.; Feezell, D.F.; Cohen, D.A.; Saito, M.; Fujito, K.; Speck, J.S.; DenBaars, S.P.; Nakamura, S. Demonstration of nonpolar m-plane InGaN/GaN laser diodes. *Jpn. J. Appl. Phys.* **2007**, *46*, L190–L191. [CrossRef]
4. Sizov, D.; Bhat, R.; Zah, C.E. Gallium indium nitride-based green lasers. *J. Lightwave Technol.* **2012**, *30*, 679–699. [CrossRef]
5. Zhao, D.G.; Yang, J.; Liu, Z.S.; Chen, P.; Zhu, J.J.; Jiang, D.S.; Shi, Y.S.; Wang, H.; Duan, L.H.; Zhang, L.Q.; et al. Fabrication of room temperature continuous-wave operation GaN-based ultraviolet laser diodes. *J. Semicond.* **2017**, *38*, 051001. [CrossRef]
6. Kim, K.S.; Son, J.K.; Lee, S.N.; Sung, Y.J.; Paek, H.S.; Kim, H.K.; Kim, M.Y.; Ha, K.H.; Ryu, H.Y.; Nam, O.H.; et al. Characteristics of long wavelength InGaN quantum well laser diodes. *Appl. Phys. Lett.* **2008**, *92*, 101103. [CrossRef]
7. Uchida, S.; Takeya, M.; Ikeda, S.; Mizuno, T.; Fujimoto, T.; Matsumoto, O.; Goto, S.; Tojyo, T.; Ikeda, M. Recent progress in high-power blue-violet lasers. *IEEE J. Sel. Top. Quantum Electron.* **2003**, *9*, 1252–1259. [CrossRef]
8. Thahab, S.M.; Abu Hassan, H.; Hassan, Z. Performance and optical characteristic of InGaN MQWs laser diodes. *Opt. Express* **2007**, *15*, 2380–2390. [CrossRef]
9. Alahyarizadeh, G.; Amirhoseiny, M.; Hassan, Z. Effect of different EBL structures on deep violet InGaN laser diodes performance. *Opt. Laser Technol.* **2016**, *76*, 106–112. [CrossRef]
10. Stanczyk, S.; Czyszanowski, T.; Kafar, A.; Czernecki, R.; Targowski, G.; Leszczynski, M.; Suski, T.; Kucharski, R.; Perlin, P. InGaN laser diodes with reduced AlGaN cladding thickness fabricated on GaN plasmonic substrate. *Appl. Phys. Lett.* **2013**, *102*, 151102. [CrossRef]
11. Nakamura, S.; Senoh, M.; Nagahama, S.I.; Iwasa, N.; Yamada, T.; Matsushita, T.; Kiyoku, H.; Sugimoto, Y. InGaN-based multi-quantum-well structure laser diodes. *Jpn. J. Appl. Phys.* **1996**, *35*, L74–L76. [CrossRef]

12. Nakamura, S.; Senoh, M.; Nagahama, S.I.; Iwasa, N.; Matsushita, T.; Mukai, T. Blue InGaN-based laser diodes with an emission wavelength of 450 nm. *Appl. Phys. Lett.* **2000**, *76*, 22–24. [CrossRef]
13. Kuo, Y.K.; Liou, B.T.; Chen, M.L.; Yen, S.H.; Lin, C.Y. Effect band offset ratio violet-blue InGaN laser characteristics. *Opt. Commun.* **2004**, *231*, 395–402. [CrossRef]
14. Martín, J.A.; Sánchez, M. Comparison between a graded and step-index optical cavity in InGaN MQW laser diodes. *Semicond. Sci. Technol.* **2005**, *20*, 290–295. [CrossRef]
15. Chang, S.H.; Chen, J.R.; Lee, C.H.; Yang, C.H. Effects of built-in polarization and carrier overflow on InGaN quantum-well lasers with AlGaN or AlInGaN electronic blocking layers. *Proc. SPIE Optoelectron. Devices Phys. Fabr. Appl. III* **2006**, *6368*, 636813.
16. Chen, P.; Zhao, D.G.; Jiang, D.S.; Zhu, J.J.; Liu, Z.S.; Le, L.C.; Yang, J.; Li, X.; Zhang, L.Q.; Liu, J.P.; et al. The effect of composite GaN/InGaN last barrier layer on electron leakage current and modal gain of InGaN-based multiple quantum well laser diodes. *Phys. Stat. Sol. A* **2015**, *212*, 2936–2943. [CrossRef]
17. Chow, W.W.; Amano, H.; Takeuchi, T.; Han, J. Well width dependence of threshold in InGan/AlGaN quantum well lasers. In Proceedings of the Technical Digest. Summaries of Papers Presented at the Conference on Lasers and Electro-Optics. Postconference Edition. CLEO'99. Conference on Lasers and Electro-Optics (IEEE Cat. No.99CH37013), Baltimore, MD, USA, 28 May 1999.
18. Nakamura, S.; Mukai, T.; Senoh, M.; Iwasa, N. Thermal Annealing Effects on P-Type Mg-Doped GaN Films. *Jpn. J. Appl. Phys.* **1992**, *31*, L139–L142. [CrossRef]
19. Meyaard, D.S.; Lin, G.B.; Shan, Q.F.; Cho, J.; Schubert, E.F.; Shim, H.; Kim, M.H.; Sone, C. Asymmetry of carrier transport leading to efficiency droop in GaInN based light-emitting diodes. *Appl. Phys. Lett.* **2011**, *99*, 251115. [CrossRef]
20. Takeya, M.; Tojyo, T.; Asano, T.; Ikeda, S.; Mizuno, T.; Matsumoto, O.; Goto, S.; Yabuki, Y.; Uchida, S.; Ikeda, M. High-power AlGaInN lasers. *Phys. Stat. Sol. A* **2002**, *192*, 269–276. [CrossRef]
21. Ikeda, M.; Uchida, S. Blue-Violet Laser Diodes Suitable for Blu-ray Disk. *Phys. Stat. Sol. A* **2002**, *192*, 407–413. [CrossRef]
22. Kawaguchi, M.; Imafuji, O.; Nozaki, S.; Hagino, H.; Takigawa, S.; Katayama, T.; Tanaka, T. Optical-loss suppressed InGaN laser diodes using undoped thick waveguide structure. *Proc. SPIE* **2016**, *9748*, 974818.
23. Li, X.; Zhao, D.G.; Jiang, D.S.; Chen, P.; Liu, Z.S.; Zhu, J.J.; Shi, M.; Zhao, D.M.; Liu, W. Suppression of electron leakage in 808 nm laser diodes with asymmetric waveguide layer. *J. Semicond.* **2016**, *37*, 014007. [CrossRef]
24. Le, L.C.; Zhao, D.G.; Jiang, D.S.; Chen, P.; Liu, Z.S.; Yang, J.; He, X.G.; Li, X.J.; Liu, J.P.; Zhu, J.J.; et al. Suppression of electron leakage by inserting a thin undoped InGaN layer prior to electron blocking layer in InGaN-based blue-violet laser diodes. *Opt. Express* **2014**, *22*, 11392–11398. [CrossRef] [PubMed]
25. Fiorentini, V.; Bernardini, F.; Ambacher, O. Evidence for nonlinear macroscopic polarization in III–V nitride alloy heterostructures. *Appl. Phys. Lett.* **2002**, *80*, 1204–1206. [CrossRef]
26. Bernardini, F.; Fiorentini, V. Nonlinear macroscopic polarization in III-V nitride alloys. *Phys. Rev. B* **2001**, *64*, 085207. [CrossRef]
27. Yang, J.; Zhao, D.G.; Liu, Z.S.; Jiang, D.S.; Zhu, J.J.; Chen, P.; Liang, F.; Liu, S.T.; Liu, W.; Xing, Y.; et al. Suppression the leakage of optical field and carriers in GaN-based laser diodes by using InGaN barrier layers. *IEEE Photonics J.* **2018**, *10*, 1–8. [CrossRef]
28. Lee, S.N.; Son, J.K.; Paek, H.S.; Sung, Y.J.; Kim, K.S.; Kim, H.K.; Kim, H.; Sakong, T.; Park, Y.; Ha, K.H.; et al. High-power AlInGaN-based violet laser diodes with InGaN optical confinement layers. *Appl. Phys. Lett.* **2008**, *93*, 091109. [CrossRef]
29. Melo, T.; Hu, Y.L.; Weisbuch, C.; Schmidt, M.C.; David, A.; Ellis, B.; Poblenz, C.; Lin, Y.D.; Krames, M.R.; Raring, J.W. Gain comparison in polar and nonpolar/semipolar gallium-nitride-based laser diodes. *Semicond. Sci. Technol.* **2012**, *27*, 024015. [CrossRef]

Article

A Novel GaN Metal-Insulator-Semiconductor High Electron Mobility Transistor Featuring Vertical Gate Structure

Zhonghao Sun [1], Huolin Huang [1,*], Nan Sun [1], Pengcheng Tao [1], Cezhou Zhao [2] and Yung C. Liang [3]

[1] School of Optoelectronic Engineering and Instrumentation Science & School of Microelectronics, Dalian University of Technology, Dalian 116024, China; sunzhonghao@mail.dlut.edu.cn (Z.S.); dgsunnan@mail.dlut.edu.cn (N.S.); pctao@dlut.edu.cn (P.T.)

[2] Department of Electrical and Electronic Engineering, Xi'an Jiaotong-Liverpool University, Suzhou 215123, China; cezhou.zhao@xjtlu.edu.cn

[3] Department of Electrical and Computer Engineering, National University of Singapore, Singapore 119260, Singapore; chii@nus.edu.sg

* Correspondence: hlhuang@dlut.edu.cn

Received: 3 November 2019; Accepted: 29 November 2019; Published: 5 December 2019

Abstract: A novel structure scheme by transposing the gate channel orientation from a long horizontal one to a short vertical one is proposed and verified by technology computer-aided design (TCAD) simulations to achieve GaN-based normally-off high electron mobility transistors (HEMTs) with reduced on-resistance and improved threshold voltage. The proposed devices exhibit high threshold voltage of 3.1 V, high peak transconductance of 213 mS, and much lower on-resistance of 0.53 mΩ·cm^2 while displaying better off-state characteristics owing to more uniform electric field distribution around the recessed gate edge in comparison to the conventional lateral HEMTs. The proposed scheme provides a new technical approach to realize high-performance normally-off HEMTs.

Keywords: wide-bandgap semiconductor; high electron mobility transistors; vertical gate structure; normally-off operation; gallium nitride

1. Introduction

Wide-bandgap GaN-based high electron mobility transistors (HEMTs) are promising candidates in the applications of high-frequency and high-power electronics owing to their superior material properties such as large bandgap, high critical breakdown field, and high-density carriers in the form of two-dimensional electron gas (2DEG) with high mobility over 2000 cm^2/V·s [1–5]. Nowadays, much progress has been achieved in GaN-based HEMTs with the development of the material quality and the innovation of the device structure [6,7]. However, there are still several important issues unaddressed, among which, the normally-off operation with a large threshold voltage (V_{th}) is a big concern when the chip products are pushed toward the market [8–10]. Several device architectures, such as p-GaN cap, barrier layer-recessed, fluorinated-gate, and cascode-connected metal-insulator-semiconductor high electron mobility transistors (MIS-HEMTs), have been reported to shift the threshold voltage to be positive [11–16]. Indeed, the device performances have been improved significantly in the past ten years. However, these normally-off HEMTs still suffer from either large on-resistance (R_{on}) or low V_{th} which increases the risk of device switching failure. Basically, the enhancement of V_{th} comes at the expense of increasing the R_{on} of the devices. A trade-off, hence, has to be made between them [17]. In recent years, several novel normally-off schemes such as tri-gate, Fin-gate, and trench etching in the SiO_2 buried layer have also been proposed and developed [18–20].

However, a truly effective device structure to gain both high V_{th} and low R_{on} simultaneously should be developed to resolve the mentioned issues effectively.

In this work, a novel scheme featuring a vertical short gate channel is proposed and demonstrated in AlGaN/GaN MIS-HEMTs to realize the normally-off operation. The source electrode is located at the trenched GaN bulk region with gate dielectric layer covering the side wall. The proposed vertical gate HEMT (VG-HEMT) is able to get a higher V_{th} (>3 V) and at the same time a lower R_{on} due to the short vertical channel (200–500 nm in this work). To verify this design, a conventional HEMT was fabricated and its output characteristics were utilized to extract the physical parameters for the subsequent TCAD simulations for the VG-HEMT design which finally exhibits a much higher output current and transconductance (g_m) while displaying a better off-state electric field profile. The quantitative dependences of saturated current density and R_{on} on scattering factors in the trenched vertical gate channel were also investigated in detail in the work.

2. The Proposed Device and Physical Principle

Figure 1 shows the cross-sectional schematics of GaN-based HEMTs, where the same L_{GD} (distance between gate and drain) and gate length (including the gate overlap) have been employed. In the conventional lateral gate-recessed HEMTs (LG-HEMTs), as shown in Figure 1a, the length of the recessed gate is usually designed to be 1–3 μm, which is limited by the influence of the UV light diffraction in the lithography process. It is hard to get very high pattern resolution just using the common mask aligner. If the pattern resolution below 1 μm is required, a more precise and expensive lithography apparatus such as stepper has to be used. The technology processes, such as gate channel etching and metal lift-off, with resolution below 1 μm are more rigorous and challenging, and hence increase the cost. It is not easy to control the precision around few nanometers for fully etching just the thin AlGaN barrier (~20 nm), and the resulting over-etch leads to a severe degradation of 2DEG channel mobility. Therefore, an unstable V_{th} and large R_{on} are still the typical issues unaddressed till now.

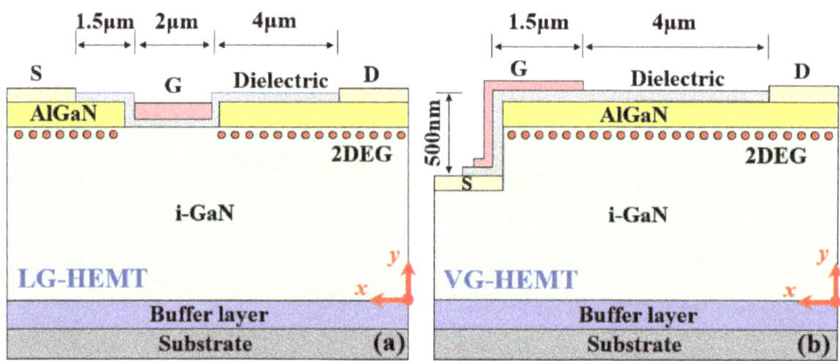

Figure 1. Cross-sectional schematics of (**a**) the lateral gate-recessed-high electron mobility transistors (LG-HEMT) and (**b**) the vertical gate HEMT (VG-HEMT).

In contrast to this, the conductive 2DEG channel is removed completely by fine etching in the VG-HEMTs and the source electrode is placed at the wafer body, as shown in Figure 1b. A decent source contact can be formed by Si ion implantation beneath the source region and post-annealing treatment [21]. To address the mismatch issue, the self-aligned technology will be employed in the future to precisely adjust the source metal edge to the edge of the recessed region. To make a better isolation between gate and source, a multiple dielectric stack or a relative thick dielectric film (e.g., 30 nm) will be employed. The vertical gate on the sidewall can modulate the electric field in the vertical channel for electron accumulation and hence control the device on/off switching. Considering the large G-to-S capacitance, the VG-HEMT devices will not be used in high-frequency or microwave applications. But, for the conventional switching applications at the frequency less than 1.0 MHz, they should be able to handle. Moreover, more schemes such as increasing the dielectric thickness between the gate and source and reducing the overlap dimension of the gate electrode will be employed to reduce the capacitance. To improve the gate breakdown endurance, the bi-layer dielectric and slanted gate schemes are considered and the gate overdrive circuit will also be designed to protect the device. Compared with the long recessed-gate channel in the LG-HEMTs, the main benefits in the VG-HEMTs include a shorter gate channel originated from the low-mobility trenched region based on a simple lithography technique and greater tolerance on etching depth error. Therefore, the device is able to achieve both high V_{th} and low R_{on}.

3. Fabrication Work and Parameter Calibration for TCAD simulation

The heterostructure comprising 25 nm $Al_{0.2}Ga_{0.8}N$ barrier layer and 4 μm undoped GaN channel and buffer layers was grown on a 6-inch p-Si (111) substrate by metal-organic chemical vapor deposition (MOCVD) technique. The conventional GaN-based HEMTs were fabricated and characterized. The fabrication work started with the device isolation by employing Cl^--based plasma etching. Source and drain contacts were formed by depositing Ti/Al/Ni/Au stack using E-beam evaporation system followed by a rapid thermal annealing (RTA) at 875 °C for 30 s in N_2 atmosphere. The gate electrode was deposited by evaporating Ni/Au alloy. Then, 300 nm SiO_2 layer was deposited on the device surface at 300 °C by plasma enhanced chemical vapor deposition (PECVD) system for passivation purpose. Figure 2a indicates the structural schematic of the GaN HEMT. Figure 2b shows the microscopy image of the fabricated device and the I–V characteristics measured using Agilent B1505A system and benchmarked by the simulation data.

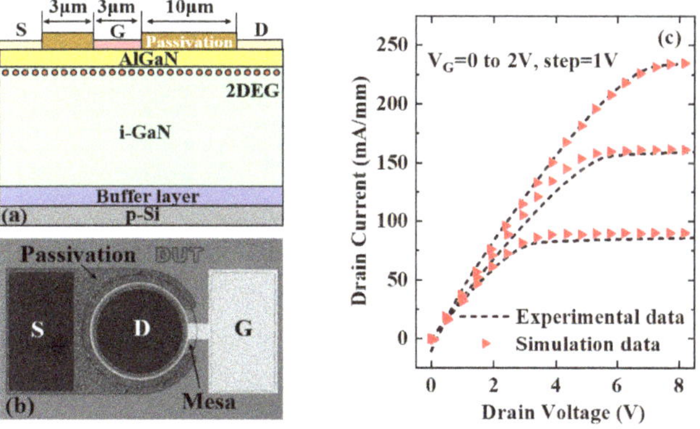

Figure 2. (**a**) Schematic and (**b**) microscopy image of the fabricated normally-on AlGaN/GaN HEMTs, and (**c**) comparisons of output I–V characteristics of the devices for physical parameter calibration.

In TCAD simulations, Auger recombination, Shockley–Read–Hall (SRH) recombination, and Van Overstraeten–De Man models were used to simulate the device behaviors. To investigate the dry etching effects on the carrier mobility of gate channel in the normally-off devices, several mobility models were selected. The Arora model was employed to determine the doping-dependent mobility in the low-field case, and the transferred electron model was used to describe the effect of a transfer of electrons into an energetically higher side valley with a much larger effective mass in the high-field case [22]. The mobility degradations caused by high field and interface scattering are given by Meinerzhagen–Engl model and Lombardi model, respectively [23,24]. The comparison of simulated output I–V characteristics with the measurement data is shown in Figure 2c. The displayed I–V curves match well and the average mismatch for the current density is within 5%. This confirms the validity of the parameters used in the simulations. The physical parameters used in simulation are listed in Table 1 [3]. Especially, in the VG-HEMT, it is also of great importance to know the the mobility degradation at the etched sidewall. Thus, the effects of the etched roughness at the gate region on the device performances were investigated in detail in the next section.

Table 1. Physical parameters used in technology computer-aided design (TCAD) simulations after calibration [3].

Physical Parameters	Values
Electron effective mass in GaN	$0.22{\cdot}m_e$
Electron affinity	3.4 eV
Relative dielectric constant in GaN	9.7
Background electron concentration in i-GaN layer	5.0×10^{15} cm^{-3}
Electron mobility in 2DEG channel	1500 cm^2/(V·s)
2DEG sheet density	8.0×10^{12} cm^{-2}
Electron saturation velocity in GaN	1.8×10^7 cm/s

4. Results and Discussions

The output I–V characteristics of the normally-off LG-HEMT and VG-HEMT are illustrated in Figure 3. The V_{th} values are defined by extrapolating the linear section of the I_D–V_G curve to the voltage axis. The normally-off operation with a positive V_{th} is realized in both devices. The VG-HEMT exhibits a V_{th} of 3.1 V and a high peak g_m of 213 mS/mm compared with 3.6 V and 114 mS/mm in the LG-HEMT, respectively. The similar V_{th} values in both devices are found when the influence of the polar plane is ignored considering the fact that usually the gate surface polarization might be damaged severely after a long-time plasma etching in the gate recess process and the polarization charges at the lateral surface are generally compensated by the high-density surface states, such as donor-like traps.

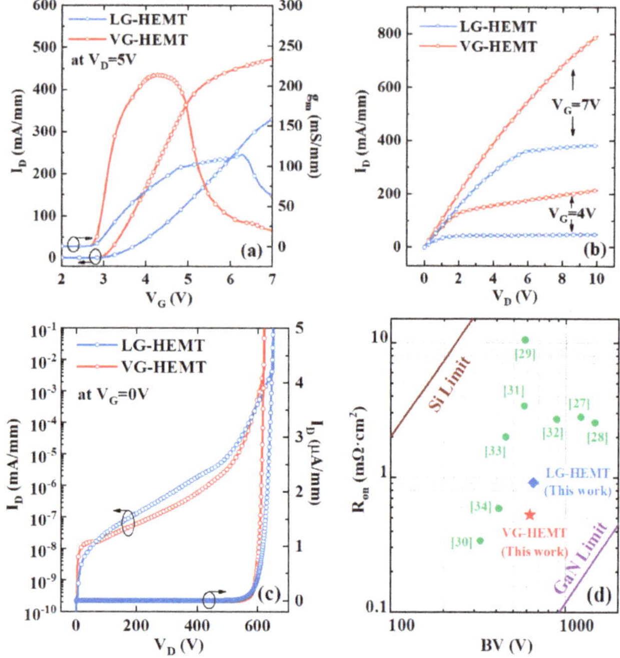

Figure 3. Comparisons of (**a**) transfer characteristics, (**b**) output I_D–V_D curves, and (**c**) off-state breakdown characteristics between the LG-HEMT and VG-HEMT devices, and (**d**) performance comparisons among the reported normally-off HEMTs including the VG-HEMT in this work (the red star).

Theoretically, the V_{th} model for the MIS-HEMT can be expressed in Equation (1) [25]:

$$V_{th} = \frac{\varphi_b}{q} - \frac{\Delta E_c}{q} - \frac{\varphi_f}{q} - \frac{q \cdot \Delta Q_{it} t_{ox}}{\varepsilon_{ox}} \quad (1)$$

where φ_b is the barrier height for gate metal on the dielectric (3.2 eV for Ni on Si_3N_4), ΔE_c the conduction band offset between Si_3N_4 and GaN (1.5 eV), φ_f the conduction band distance from the Fermi-level in GaN (0.2 eV), t_{ox} the dielectric thickness (20 nm), ε_{ox} the permittivity of dielectric (6.6×10^{-13} F/cm), and ΔQ_{it} the net charge density at the interface (-3.4×10^{12} cm^{-2} in this work which is reasonable and reported widely for the III–V group GaN-based materials) [26,27]. The calculated V_{th} of the VG-HEMT is found to be 3.2 V, which is roughly consistent with the simulation data at 3.1 V. Furthermore, the slight difference of the V_{th} value between the two devices is mainly caused by the increased drain current and transconductance values in the VG-HEMT considering that the effective vertical gate length is much less than that in the LG-HEMT. The maximum current density and R_{on} are found to be 793 mA/mm and 0.53 mΩ·cm^2 at $V_G = 7$ V in the VG-HEMT, while 381 mA/mm and 0.92 mΩ·cm^2 in LG-HEMT, respectively. The VG-HEMT shows a higher drain current mainly owing to the saving of the L_{GS} length by placing the source electrode under the gate and a shorter gate channel originated from the low-mobility trenched region. The shorter total channel length and narrower trenched-gate originated from the structural feature of the proposed VG-HEMT result in the increase of the drain current. Furthermore, a similar off-state breakdown voltage level (606 V vs. 615 V at $I_{D,off} = 1$ µA/mm) is revealed between the VG-HEMT and LG-HEMT, as shown in Figure 3c. The proposed VG-HEMT exhibits even a lower drain leakage current owing to a more uniform electric field distribution. Figure 3d gives the performance comparisons with other reported normally-off HEMTs, which clearly indicates the advantage of the proposed scheme by employing a vertical short gate structure [27–34].

To further investigate the electrical behaviors of the VG-HEMT, its energy band and electron density distributions at different biases are calculated and compared, as shown in Figure 4a,b. The conduction band of vertical channel is pulled down to the Fermi level with the increasing gate and drain bias and an electron density peak is formed in the conduction channel. The surface potential of the semiconductor can be affected by the different dielectric deposition processes due to the influence of the high-density dielectric/GaN interface traps (10^{12}–10^{13} cm^{-2}) and the short distance to the conduction channel. As a result, the conduction band bending occurs at the interface, as shown in Figure 4a. The concentration of the accumulated electrons is given by Equation (2) [35]:

$$n_e = \frac{C_{ox}}{q} \cdot (V_g - \frac{\varphi_b - \varphi_{it} - \varphi_f - \Delta E_c}{q}) \quad (2)$$

where C_{ox} is the capacitance of the gate dielectric (3.32×10^{-7} F/cm^{-2}), V_g the applied gate voltage, and φ_{it} the impact on the surface potential caused by the interface electrons. Using the ΔQ_{it} value of 3.4×10^{12} cm^{-2}, the calculated peak n_e at $V_g = 7$ V is 7.8×10^{12} cm^{-2}, which is approximately consistent with the data of 7.2×10^{12} cm^{-3} for the blue line after integral calculation in Figure 4b.

Furthermore, the physical mechanisms of output current conduction in the VG-HEMT were analyzed. The I_D–V_D characteristics of these two devices are compared because they have similar electrode distance L_{GD} and gate length L_G. A decent source contact technology will make it easier to obtain a low source resistance and minimize the effect on the on-resistance. In this case, the surface roughness at the etched gate region becomes the major factor that influences the on-resistance of the device. The mobility degradation in the etched vertical channel is strongly dependent on several scattering factors. The sidewall channel mobility μ is modeled by the following expression based on Lombardi model [36]:

$$\frac{1}{\mu} = \frac{1}{\mu_{ac}} + \frac{1}{\mu_b} + \frac{1}{\mu_{sr}} \quad (3)$$

where μ_{ac} is the mobility contribution by acoustic phonon scattering, μ_b the carrier mobility in bulk, and μ_{sr} the mobility contribution by surface roughness scattering. In the simulations, μ_{ac} was determined to be 818 $cm^2/(V \cdot s)$ using the equation $\mu_{ac} = (BT/E_{eff} + C/E_{eff}{}^{(1/3)}) \cdot (1/T)$ [24], where E_{eff} is the effective field controlled by the channel charge, T (= 300 K) is the temperature, and B (= 9×10^7 cm/s) and C (= 5.8×10^2 $cm^{5/3} \cdot V^{-2/3} \cdot s^{-1}$) are the fitting parameters, respectively [37].The μ_b value of 803 $cm^2/(V \cdot s)$ is extracted from the calibration data. Considering that μ_{ac} and μ_b are almost constant for a technically mature GaN wafer at a certain temperature, μ in the vertical channel is mainly modulated by μ_{sr}, which can be expressed as [33]:

$$\mu_{sr} = \frac{\delta}{E_{eff}^{\alpha}} \tag{4}$$

where α is determined by the experimental fitting. And δ is inversely proportional to the root mean square (RMS) value of the surface roughness, and is the main factor resulting in mobility degradation at the etched surface. The μ_{sr} value was then determined to be 1114 $cm^2/(V \cdot s)$ by employing the parameters extracted from the simulation. Therefore, the mobility value of 297 $cm^2/(V \cdot s)$ was achieved for the sidewall channel in VG-HEMT. The quantitative effects of δ values on the drain current density and R_{on} are shown in Figure 4c,d, taking into account the effect from the roughness scattering [37]. The VG-HEMT exhibits much better current characteristics owing to the less scattering amount in the shorter gate channel, as shown in Figure 4c. Moreover, to further investigate the effect of etching damage on the output characteristic, as well as distinguish the specific damage between the sidewall and surface, more experiments will be carried out in future work.

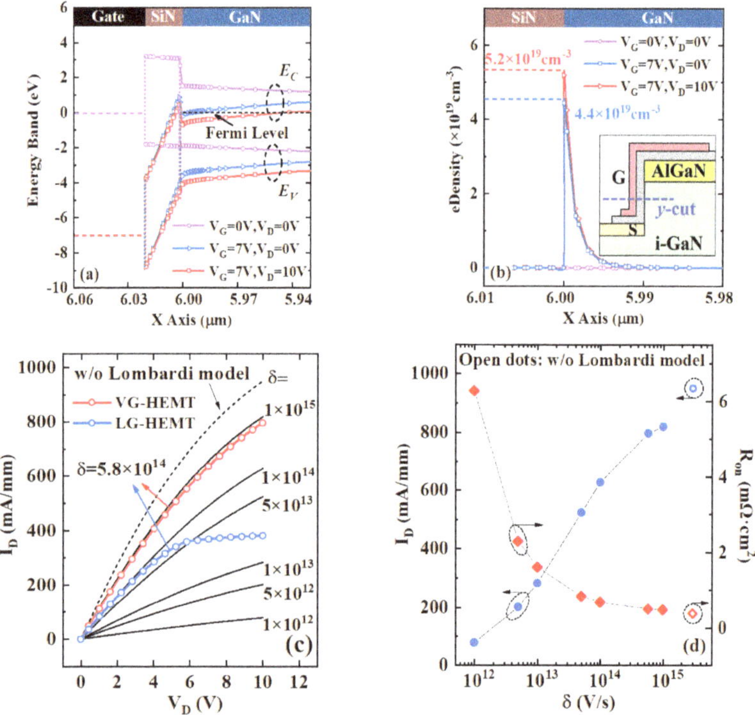

Figure 4. (**a**) Energy band and (**b**) electron concentration diagrams along the horizontal direction across the gate channel ("*y*-cut" marked in the inset). (**c**) I_D–V_D curves affected by different δ values using Lombardi model in the VG-HEMT. (**d**) Dependences of saturated current density and R_{on} on δ values.

Figure 5 shows the 2-D equipotential lines and electric field distributions near the gate region in the same dimension range of the devices at V_D = 400 V and V_G = 0 V. With comparison, the VG-HEMT exhibits more uniform electrostatic potential and electric field distribution at the gate corner toward the drain side where usually the electrical breakdown occurs due to the high field crowding. Without an additional field plate design which requires more lithography and metal deposition steps, the VG-HEMT can get a reduced electric field at the gate corner owing to the presence of a "natural field plate", forming by the overlap of gate metal at the etched edge. The depletion region in the vertical gate structure is expanded to a wider range to sustain a higher voltage and hence reduce the device leakage current. Figure 5c,d show the detailed data plots of the electrostatic potential and electric field extracted in the 2DEG channel of both devices. The electrostatic potential increases more smoothly at the gate corner of the VG-HEMT, while a large field crowding happens in the LG-HEMT. The peak of electric field in the VG-HEMT is reduced to 26% compared with the LG-HEMT case, which indicates a better stability and reliability in the VG-HEMT. Furthermore, an improved breakdown voltage can be hopefully achieved when the length of the "natural field plate" is further optimized. To avoid a punch-through breakdown in the VG-HEMT due to the off-state conducting path formed in GaN bulk, carbon doping in GaN bulk to reduce the background carrier density and design of drain-side field plate will be considered and demonstrated in future work.

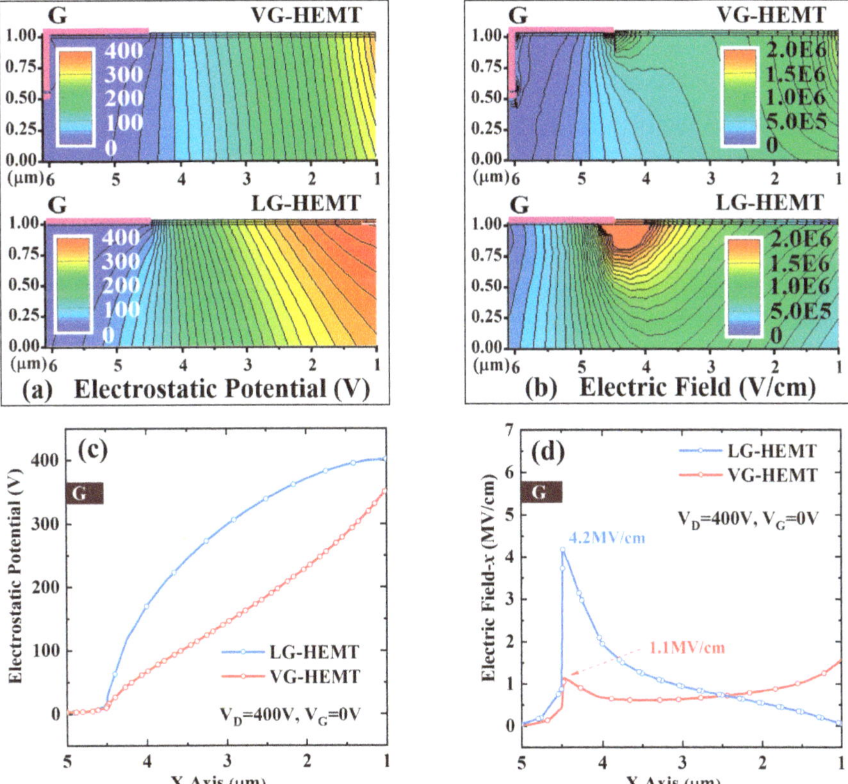

Figure 5. Comparisons of (**a**) electrostatic potential and (**b**) electric field distribution profiles between the LG-HEMT and VG-HEMT devices. The data plots in (**c**) and (**d**) are derived along the two-dimensional electron gas (2DEG) channel within 5 nm near the gate toward the drain side.

5. Conclusion

In summary, an innovative scheme with a vertical short gate structure in GaN-based HEMTs to realize normally-off operation is proposed and verified by TCAD simulations. The VG-HEMT exhibits a large V_{th} of 3.1 V and an improved output current density and R_{on}, which can greatly reduce the power loss of the device. Furthermore, the VG-HEMT displays a lower leakage current during the off state owing to the more uniform electric field distribution by reducing the peak electric field to 26% of the LG-HEMT case. The quantitative dependence of saturated current density and R_{on} on the scattering factors in the trenched vertical gate channel was investigated in detail which concludes the benefit of the proposed short vertical gate VG-HEMT scheme. Although the fabrication process of the VG-HEMT might be complicated and time-consuming, the device structure is proposed for the first time and does have several advantages. More improved performances will be achieved and demonstrated in the following experiments.

Author Contributions: Writing—Original Draft, Z.S.; Investigation, N.S.; Validation, P.T.; Data Curation, C.Z.; Software, Y.C.L.; Writing—Review & Editing, H.H.

Funding: This research was funded by the National Natural Science Foundation of China under Grant Nos. 61971090 and 51607022, the Open Project Program of the Key Laboratory of Semiconductor Materials Science under Grant No. KLSMS-1804 and the Open Project Program of the Key Laboratory of Nanodevices and Applications under Grant No. 18JG02 from the Chinese Academy of Sciences.

Conflicts of Interest: The authors declare no conflict of interest.

References

1. Fukushima, H.; Usami, S.; Ogura, M.; Ando, Y.; Tanaka, A.; Deki, M.; Kushimoto, M.; Nitta, S.; Honda, Y.; Amano, H. Vertical GaN pn diode with deeply etched mesa and capability of avalanche breakdown. *Appl. Phys. Express* **2019**, *12*. [CrossRef]
2. Ando, Y.; Kaneki, S.; Hashizume, T. Improved operation stability of Al2O3/AlGaN/GaN MOS high-electron-mobility transistors grown on GaN substrates. *Appl. Phys. Express* **2019**, *12*, 024002. [CrossRef]
3. Huang, H.; Liang, Y.C.; Samudra, G.S.; Chang, T.F.; Huang, C.F. Effects of gate field plates on the surface state related current collapse in AlGaN/GaN HEMTs. *IEEE Trans. Power Electron.* **2013**, *29*, 2164–2173. [CrossRef]
4. Chen, K.J.; Häberlen, O.; Lidow, A.; lin Tsai, C.; Ueda, T.; Uemoto, Y.; Wu, Y. GaN-on-Si power technology: Devices and applications. *IEEE Trans. Electron Devices* **2017**, *64*, 779–795. [CrossRef]
5. Huang, H.; Sun, Z.; Cao, Y.; Li, F.; Zhang, F.; Wen, Z.; Zhang, Z.; Liang, Y.C.; Hu, L. Investigation of surface traps-induced current collapse phenomenon in AlGaN/GaN high electron mobility transistors with schottky gate structures. *J. Phys. D Appl. Phys.* **2018**, *51*, 345102. [CrossRef]
6. Wei, J.; Lei, J.; Tang, X.; Li, B.; Liu, S.; Chen, K.J. Channel-to-channel coupling in normally-off GaN double-channel MOS-HEMT. *IEEE Electron Device Lett.* **2017**, *39*, 59–62. [CrossRef]
7. Rossetto, I.; Meneghini, M.; De Santi, C.; Pandey, S.; Gajda, M.; Hurkx, G.M.; Croon, J.; Šonský, J.; Meneghesso, G. 2DEG retraction and potential distribution of GaN–on–Si HEMTs investigated through a floating gate terminal. *IEEE Trans. Electron Devices* **2018**, *65*, 1303–1307. [CrossRef]
8. Wang, H.; Wei, J.; Xie, R.; Liu, C.; Tang, G.; Chen, K.J. Maximizing the performance of 650-V p-GaN gate HEMTs: Dynamic RON characterization and circuit design considerations. *IEEE Trans. Power Electron.* **2016**, *32*, 5539–5549. [CrossRef]
9. Sun, R.; Liang, Y.C.; Yeo, Y.C.; Zhao, C. Au-Free AlGaN/GaN MIS-HEMTs With Embedded Current Sensing Structure for Power Switching Applications. *IEEE Trans. Electron Devices* **2017**, *64*, 3515–3518. [CrossRef]
10. Shen, F.; Hao, R.; Song, L.; Chen, F.; Yu, G.; Zhang, X.; Fan, Y.; Lin, F.; Cai, Y.; Zhang, B. Enhancement mode AlGaN/GaN HEMTs by fluorine ion thermal diffusion with high V th stability. *Appl. Phys. Express* **2019**, *12*, 066501. [CrossRef]
11. Wu, T.L.; Marcon, D.; You, S.; Posthuma, N.; Bakeroot, B.; Stoffels, S.; Van Hove, M.; Groeseneken, G.; Decoutere, S. Forward bias gate breakdown mechanism in enhancement-mode p-GaN gate AlGaN/GaN high-electron mobility transistors. *IEEE Electron Device Lett.* **2015**, *36*, 1001–1003. [CrossRef]

12. Wang, H.; Wang, J.; Li, M.; Cao, Q.; Yu, M.; He, Y.; Wu, W. 823-mA/mm Drain Current Density and 945-MW/cm 2 Baliga's Figure-of-Merit Enhancement-Mode GaN MISFETs With a Novel PEALD-AlN/LPCVD-Si 3 N 4 Dual-Gate Dielectric. *IEEE Electron Device Lett.* **2018**, *39*, 1888–1891. [CrossRef]
13. Wang, Y.H.; Liang, Y.C.; Samudra, G.S.; Huang, H.; Huang, B.J.; Huang, S.H.; Chang, T.F.; Huang, C.F.; Kuo, W.H.; Lo, G.Q. 6.5 V high threshold voltage AlGaN/GaN power metal-insulator-semiconductor high electron mobility transistor using multilayer fluorinated gate stack. *IEEE Electron Device Lett.* **2015**, *36*, 381–383. [CrossRef]
14. Huang, H.; Liang, Y.C. Formation of combined partially recessed and multiple fluorinated-dielectric layers gate structures for high threshold voltage GaN-based HEMT power devices. *Solid-State Electron.* **2015**, *114*, 148–154. [CrossRef]
15. Huang, X.; Liu, Z.; Lee, F.C.; Li, Q. Characterization and enhancement of high-voltage cascode GaN devices. *IEEE Trans. Electron Devices.* **2014**, *62*, 270–277. [CrossRef]
16. Ren, J.; Tang, C.W.; Feng, H.; Jiang, H.; Yang, W.; Zhou, X.; Lau, K.M.; Sin, J.K. A Novel 700 V Monolithically Integrated Si-GaN Cascoded Field Effect Transistor. *IEEE Electron Device Lett.* **2018**, *39*, 394–396. [CrossRef]
17. Oka, T.; Nozawa, T. AlGaN/GaN recessed MIS-gate HFET with high-threshold-voltage normally-off operation for power electronics applications. *IEEE Electron Device Lett.* **2008**, *29*, 668–670. [CrossRef]
18. Li, X.; Hove, M.V.; Zhao, M.; Geens, K.; Lempinen, V.P.; Sormunen, J.; Groeseneken, G.; Decoutere, S. 200 V Enhancement-Mode p-GaN HEMTs Fabricated on 200 mm GaN-on-SOI with Trench Isolation for Monolithic Integration. *IEEE Electron Device Lett.* **2017**, *38*, 918–921. [CrossRef]
19. Ma, J.; Matioli, E. High performance tri-gate GaN power MOSHEMTs on silicon substrate. *IEEE Electron Device Lett.* **2017**, *38*, 367–370. [CrossRef]
20. Zhang, Y.; Sun, M.; Piedra, D.; Hu, J.; Liu, Z.; Lin, Y.; Gao, X.; Shepard, K.; Palacios, T. 1200 V GaN vertical fin power field-effect transistors. In Proceedings of the 2017 IEEE International Electron Devices Meeting (IEDM), San Francisco, CA, USA, 2–6 December 2017; pp. 9.2.1–9.2.4.
21. Nishimura, T.; Kasai, T.; Mishima, T.; Kuriyama, K.; Nakamura, T. Reduction in contact resistance and structural evaluation of Al/Ti electrodes on Si-implanted GaN. *NUCL INSTRUM METHODS PHYS RES B* **2019**, *450*, 244–247. [CrossRef]
22. Arora, N.D.; Hauser, J.R.; Roulston, D.J. Electron and hole mobilities in silicon as a function of concentration and temperature. *IEEE Trans. Electron Devices* **1982**, *29*, 292–295. [CrossRef]
23. Barnes, J.J.; Lomax, R.J.; Haddad, G.I. Finite-element simulation of GaAs MESFET's with lateral doping profiles and submicron gates. *IEEE Trans. Electron Devices* **1976**, *23*, 1042–1048. [CrossRef]
24. Lombardi, C.; Manzini, S.; Saporito, A.; Vanzi, M. A physically based mobility model for numerical simulation of nonplanar devices. *IEEE Trans. Comput. Aided Des. Integr. Circuits Syst.* **1988**, *7*, 1164–1171. [CrossRef]
25. Zhang, Y.; Sun, M.; Joglekar, S.J.; Fujishima, T.; Palacios, T. Threshold voltage control by gate oxide thickness in fluorinated GaN metal-oxide-semiconductor high-electron-mobility transistors. *Appl. Phys. Lett.* **2013**, *103*, 033524. [CrossRef]
26. Hua, M.; Zhang, Z.; Wei, J.; Lei, J.; Tang, G.; Fu, K.; Cai, Y.; Zhang, B.; Chen, K.J. Integration of LPCVD-SiN x gate dielectric with recessed-gate E-mode GaN MIS-FETs: Toward high performance, high stability and long TDDB lifetime. In Proceedings of the 2016 IEEE International Electron Devices Meeting (IEDM), San Francisco, CA, USA, 3–7 December 2016; pp. 10.4.1–10.4.4.
27. Yang, C.; Luo, X.; Zhang, A.; Deng, S.; Ouyang, D.; Peng, F.; Wei, J.; Zhang, B.; Li, Z. AlGaN/GaN MIS-HEMT with AlN interface protection layer and trench termination structure. *IEEE Trans. Electron Devices* **2018**, *65*, 5203–5207. [CrossRef]
28. Hao, R.; Li, W.; Fu, K.; Yu, G.; Song, L.; Yuan, J.; Li, J.; Deng, X.; Zhang, X.; Zhou, Q.; et al. Breakdown enhancement and current collapse suppression by high-resistivity GaN cap layer in normally-off AlGaN/GaN HEMTs. *IEEE Electron Device Lett.* **2017**, *38*, 1567–1570. [CrossRef]
29. Hu, Q.; Li, S.; Li, T.; Wang, X.; Li, X.; Wu, Y. Channel Engineering of Normally-OFF AlGaN/GaN MOS-HEMTs by Atomic Layer Etching and High-k Dielectric. *IEEE Electron Device Lett.* **2018**, *39*, 1377–1380. [CrossRef]
30. Huang, H.; Liang, Y.C.; Samudra, G.S.; Ngo, C.L. Au-free normally-off AlGaN/GaN-on-Si MIS-HEMTs using combined partially recessed and fluorinated trap-charge gate structures. *IEEE Electron Device Lett.* **2014**, *35*, 569–571. [CrossRef]

31. Lin, S.; Wang, M.; Sang, F.; Tao, M.; Wen, C.P.; Xie, B.; Yu, M.; Wang, J.; Hao, Y.; Wu, W.; et al. A GaN HEMT structure allowing self-terminated, plasma-free etching for high-uniformity, high-mobility enhancement-mode devices. *IEEE Electron Device Lett.* **2016**, *37*, 377–380. [CrossRef]
32. Zhang, Z.; Fu, K.; Deng, X.; Zhang, X.; Fan, Y.; Sun, S.; Song, L.; Xing, Z.; Huang, W.; Yu, G.; et al. Normally Off AlGaN/GaN MIS-high-electron mobility transistors fabricated by using low pressure chemical vapor deposition Si 3 N 4 gate dielectric and standard fluorine ion implantation. *IEEE Electron Device Lett.* **2015**, *36*, 1128–1131. [CrossRef]
33. Zhang, Z.; Li, W.; Fu, K.; Yu, G.; Zhang, X.; Zhao, Y.; Sun, S.; Song, L.; Deng, X.; Xing, Z.; et al. AlGaN/GaN MIS-HEMTs of Very-Low *V* Hysteresis and Current Collapse With In-Situ Pre-Deposition Plasma Nitridation and LPCVD-Si3N4 Gate Insulator. *IEEE Electron Device Lett.* **2016**, *38*, 236–239. [CrossRef]
34. Xu, Z.; Wang, J.; Liu, J.; Jin, C.; Cai, Y.; Yang, Z.; Wang, M.; Yu, M.; Xie, B.; Wu, W.; et al. Demonstration of normally-off recess-gated AlGaN/GaN MOSFET using GaN cap layer as recess mask. *IEEE Electron Device Lett.* **2014**, *35*, 1197–1199.
35. Sze, S.M.; Ng, K.K. *Physics of semiconductor devices*, 3rd ed.; John wiley & sons: Hoboken, NJ, USA, 2006; pp. 225–316.
36. Rodriguez, N.; Roldan, J.B.; Gamiz, F. An electron mobility model for ultra-thin gate-oxide MOSFETs including the contribution of remote scattering mechanisms. *Semicond. Sci. Technol.* **2007**, *22*, 348. [CrossRef]
37. Du, J.; Yan, H.; Yin, C.; Feng, Z.; Dun, S.; Yu, Q. Simulation and characterization of millimeter-wave InAlN/GaN high electron mobility transistors using Lombardi mobility model. *J. Appl. Phys.* **2014**, *115*, 164510. [CrossRef]

Article

Simulation Study of 4H-SiC Trench Insulated Gate Bipolar Transistor with Low Turn-Off Loss

Hong-kai Mao [1], Ying Wang [1,*], Xue Wu [2] and Fang-wen Su [1]

1 The Key Laboratory of RF Circuits and Systems, Ministry of Education, Hangzhou Dianzi University, Hangzhou 310018, China; 171040042@hdu.edu.cn (H.-k.M.); 171040083@hdu.edu.cn (F.-w.S.)
2 National Key Laboratory of Analog Integrated Circuits, Chongqing 400060, China; xuew86@yeah.net
* Correspondence: wangying7711@yahoo.com

Received: 18 October 2019; Accepted: 15 November 2019; Published: 26 November 2019

Abstract: In this work, an insulated gate bipolar transistor (IGBT) is proposed that introduces a portion of the p-polySi/p-SiC heterojunction on the collector side to reduce the tail current during device turn-offs. By adjusting the doping concentration on both sides of the heterojunction, the turn-off loss is further reduced without sacrificing other characteristics of the device. The electrical characteristics of the device were simulated through the Silvaco ATLAS 2D simulation tool and compared with the traditional structure to verify the design idea. The simulation results show that, compared with the traditional structure, the turn-off loss of the proposed structure was reduced by 58.4%, the breakdown voltage increased by 13.3%, and the forward characteristics sacrificed 8.3%.

Keywords: 4H-SiC; turn-off loss; ON-state voltage; breakdown voltage (BV); IGBT

1. Introduction

In recent years, with the development of semiconductor technology, market demand and power electronics have undergone great changes [1–5]. The market not only requires semiconductor devices that work properly in harsh environments, but devices with small sizes and high integration. Since Si-based devices have reached their material limits, third-generation semiconductor materials represented by SiC are widely used in device design.

There are two mainstreams processing methods for a 4H-SiC insulated gate bipolar transistor (IGBT) at present: one is to make a good compromise between the forward and off characteristics of the device by properly setting the structural parameters of the device such as the n buffer's thickness and doping parameters [6], the minority carrier lifetime in the n^- drift region [7,8], and thickness and doping parameters of the CSL (carrier storage layer) [9,10]; the other is by considering the process conditions and designing a special device structure that can improve by affecting certain characteristics, such as an anode short circuit IGBT [11,12], a super junction IGBT [13], or a collector trench IGBT (CT-IGBT) with an electronic extraction channel [14]. By analyzing the previous research, it can be observed that researchers have been mainly concerned with the compromise between the on-state and breakdown characteristics of IGBT devices, and that relatively little research has been made into the dynamic conversion characteristics of the device.

In this paper, we propose an improved structure of introducing a partial p-polySi/p-SiC heterojunction on the collector side, which is named H-IGBT. By adjusting the doping concentration on both sides of the heterojunction, it is ensured that the heterojunction contributes to electron bleed under the premise of not affecting the forward characteristics, and that the tail current of the device is greatly reduced and turn-off loss is improved.

2. Fabrication Procedure and Parameters

2.1. Device Structure

Figure 1 is a schematic cross-sectional view showing the half-cell structure of a conventional IGBT device (C-IGBT) and the proposed structure of the H-IGBT, respectively. In order to verify the characteristic advantages of the proposed structure, we simulated the electrical characteristics of the device using Silvaco ATLAS two-dimensional simulation software. In the simulation process, we first design the basic structure of a breakdown voltage of 15 kV as the reference structure [15,16], and then apply the design idea to the new structure while keeping most of the parameters unchanged. The relevant parameters of the two structures during the simulation process are listed in Table 1 [17,18]. In this simulation, the carrier lifetime in the n^- drift region is 1 µs, and the carrier lifetime of the n buffer is 0.1 µs.

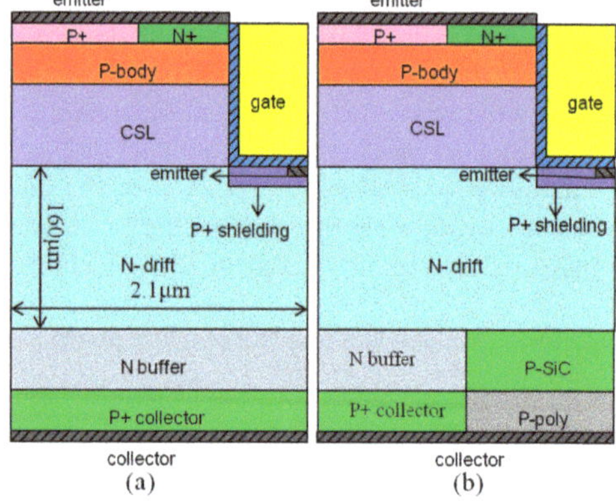

Figure 1. The schematic cross-sectional views of the (**a**) conventional insulated gate bipolar transistor (C-IGBT) and (**b**) heterojunction insulated gate bipolar transistor (H-IGBT).

Table 1. Devices parameters for the simulations.

Parameter	C-IGBT	H-IGBT
Gate oxide wall thickness	0.05 µm	0.05 µm
Gate oxide bottom thickness	0.1 µm	0.1 µm
Half-cell width	2.1 µm	2.1 µm
N^- drift thickness	160 µm	160 µm
P^+ source doping	5×10^{19} cm^{-3}	5×10^{19} cm^{-3}
N^+ source doping	2×10^{19} cm^{-3}	2×10^{19} cm^{-3}
p-body doping	4×10^{17} cm^{-3}	4×10^{17} cm^{-3}
CSL doping	1×10^{15} cm^{-3}	1×10^{15} cm^{-3}
N^- drift doping	4.5×10^{14} cm^{-3}	4.5×10^{14} cm^{-3}
N buffer doping	1×10^{17} cm^{-3}	1×10^{17} cm^{-3}
P^+ collector doping	1×10^{19} cm^{-3}	1×10^{19} cm^{-3}
p-SiC doping	—	1×10^{19} cm^{-3}
p polysilicon doping	—	1×10^{17} cm^{-3}
P^+ source region width	1 µm	1 µm
N^+ source region width	0.45 µm	0.45 µm
p-SiC width	—	1.1 µm
p polysilicon width	—	1.1 µm

2.2. Proposed Fabrication Procedure

Since the relevant flow test work has not been performed, Figure 2 shows the feasibility manufacturing process of H-IGBT. A p-poly layer, a p-SiC layer, a N^- drift layer, a CSL layer, a p-body layer, and so on are sequentially grown on the N^+ substrate [19,20], as shown in Figure 2a. Then, dry etching [21,22] is used to form gate trench regions, as shown in Figure 2b. The P^+ shield, P^+ source region, and N^+ source regions are formed by ion implantation [23,24], as shown in Figure 2c. The gate oxide layer is thermally grown in dry O_2 [25–27] and the trench regions are filled with polysilicon [28], as shown in Figure 2d. The substrate is removed and backside p-polySi and p-SiC epitaxial layers are dry-etched, as shown in Figure 2e. Epitaxial growing n buffer and P^+ collector in the etched portion are shown in Figure 2f, respectively. The reason why the gate oxide layer is grown under dry oxygen conditions is to avoid the problem of the oxide layer at the bottom of the trench being too thin under thermal oxidation conditions. Finally, all electrodes are metalized, including emitter, gate, and collector.

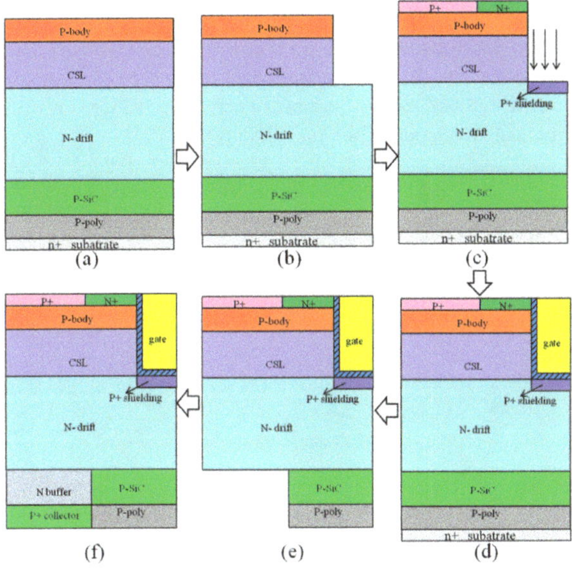

Figure 2. H-IGBT feasibility manufacturing process. (**a**) Epitaxial layers. (**b**) Forming a trench region by dry etching. (**c**) Source region and shield layer formed by ion implantation. (**d**) Forming a gate by growing an oxide layer and filling the polysilicon. (**e**) Forming a normal PN junction region by dry etching. (**f**) Forming a normal PN junction portion by epitaxy.

In the manufacturing process of the device, it should be noted that in order to obtain a high quality gate oxide layer, a dry oxygen process is selected. In actual production, it is often necessary to comprehensively consider the effects of film formation quality and production efficiency using a dry oxygen-wet oxygen-dry oxygen process; meanwhile, because SiC materials are special, their hardness is relatively large, so the etching is very difficult. SiO_2 can be used as a mask and etched by an inductively coupled plasma (ICP) etching method containing SF_6. The specific etching scheme can use a combination of $SF_6/O_2/A_r$ gases, with a flow rate of 4.2/8.4/28 sccm, a pressure and temperature of 0.4 Pa and 80 °C, respectively, an ICP power of 500 W and a bias power fixed at 15 W [29]. Attention should be paid to the formation of micro-grooves throughout the ICP etching process, which can cause an electric field concentration effect that in turn reduces the breakdown voltage of the device. After the etching, the surface of the trench will inevitably appear rough, which can be improved by the subsequent high temperature annealing process.

3. Simulation Results and Discussion

The material parameters and simulation models used in the simulation process are based on previous studies. Since these parameters have been widely used in the simulation of 4H-SiC IGBT, and the simulation results have been proved by experiments, these parameters and models have also been applied to this simulation. The models used in the simulation mainly include an energy band narrowing model (BGN), a parallel electric-field-dependent model (FLDMOB), a Fermi model, a concentration-dependent mobility model (CONMOB), and recombination models (Schockley-read-hall, AUGER) [30,31].

3.1. Forward Characteristics

Figure 3 shows the forward I-V characteristic curves of C-IGBT and H-IGBT and the hole concentration distribution through the drift region. It can be seen from the figure that the on-state characteristics of H-IGBT are slightly lower than C-IGBT. When V_{ge} = 20 V and I_{ce} = 100 A/cm^2, the on-state voltage drops of the two structures are 11.7 and 10.8 V, respectively. After data analysis, it can be concluded that the conduction voltage drop of the improved structure is increased by 8.3% compared to the conventional structure. From the heterojunction band diagram, the p-polySi/p-SiC junction contributes to hole injection, but the number of holes injected into the drift region is controlled by the upper PN junction, at which point the bias voltage of the PN junction above the heterojunction is small. As such, the total hole injection efficiency is lower than that of the left half of the PN junction and the forward characteristics of the C-IGBT are superior to those of the H-IGBT. The hole distribution of the drift region in the figure further confirms the above explanation.

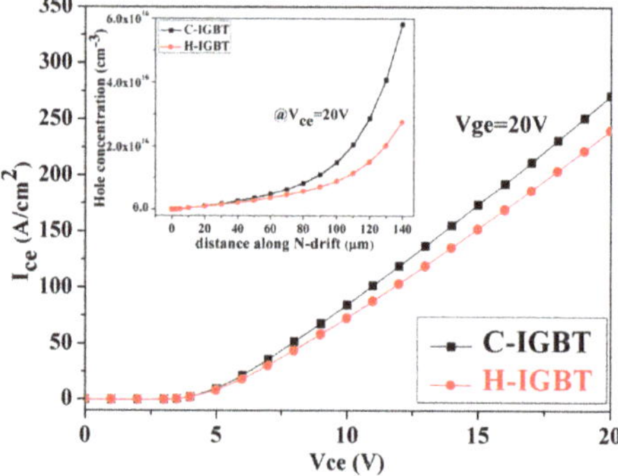

Figure 3. Forward I-V characteristics of C-IGBT and H-IGBT and the concentration distribution of holes in the entire drift region (inset) at x = 1.1μm.

3.2. Breakdown Characteristics

The two-dimensional electric field distribution when the device reaches its avalanche breakdown voltage is shown in Figure 4. At this time, the collector voltages of C-IGBT and H-IGBT are 15 and 17 kV, respectively. As can be seen from the figure, when the device reaches its breakdown voltage, the maximum internal electric fields of the conventional structure and proposed structure are 2.96 and 2.98 MV/cm, respectively. The breakdown voltage of the H-IGBT is larger than that of the C-IGBT because its hole injection efficiency is low, which is equivalent to lowering the doping concentration of the drift region. Therefore, when the device is blocked in the forward direction, the electric field strength of the entire drift region is raised, and the breakdown voltage of the H-IGBT is increased.

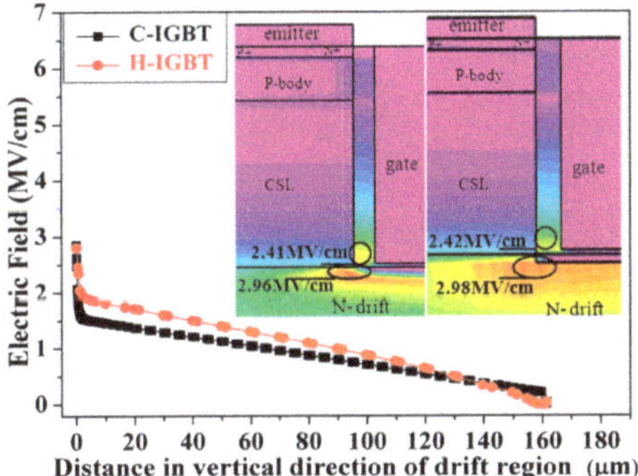

Figure 4. The electric field distribution of the C-IGBT and H-IGBT at the time of breakdown and the electric field cut of the entire drift region at x = 1.45 μm.

3.3. Turn-Off Characteristics

Since the device stores a large number of minority carriers in the drift region during the forward conduction process, the on-resistance is very small, but this is very disadvantageous for the turn-off process of the device. When the device is turned off, the carriers stored in the drift regions form a large tail current that extends the turn-off time of the device and greatly increases the power loss of the turn-off. In this simulation, we used the test circuit shown in Figure 5 to compare the shutdown performance of C-IGBT and H-IGBT. The clamped inductive load was modeled by a constant current source (2.1×10^{-6}) and the bus voltage was set to 60% of the breakdown voltage. A gate voltage of 5 kHz, a 50% duty cycle, and a voltage change from 20 to –5 V were used to control the turn-on and turn-off of the device.

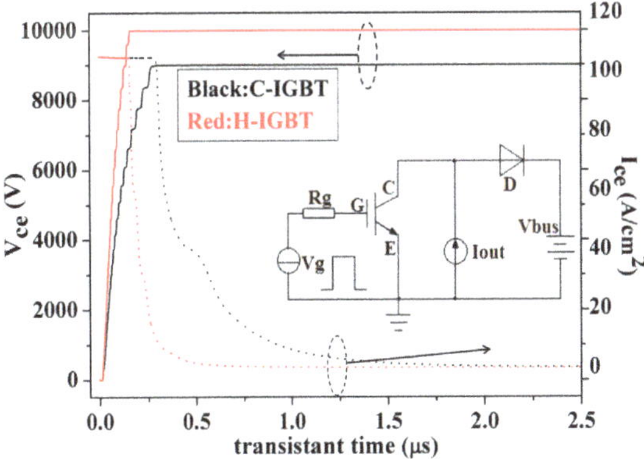

Figure 5. Turn-off characteristic curve and test circuit of C-IGBT and H-IGBT.

The turn-off characteristic curves of the conventional structure and the proposed structure are shown in Figure 5. It can be seen from the figure that the turn-off speed of the proposed structure is significantly better than that of the conventional structure, which is mainly due to the introduction of the p-polySi/p-SiC heterojunction on one side of the collector. After numerical calculation, it can be concluded that the turn-off losses of the traditional structure and the proposed structure were 7.7 and 3.2 mJ, respectively, and the turn-off loss was improved by about 58.4%. The reason why H-IGBT can have such excellent shutdown performance is mainly due to the introduction of a collector heterojunction. Through the adjustment of the doping concentration, the heterojunction can accelerate the extraction of electrons during the process of turning off the device, thereby reducing the turn-off loss. The above explanation can be further proven by the following analysis.

Figure 6 shows the electron concentration and carrier recombination rates near the collector of the conventional structure and the proposed structure during device turn-offs. It can be seen from the figure that the electron concentration and the carrier recombination rate of the conventional structure were higher than those of the proposed structure, indicating that the carriers generally disappeared by recombination when the conventional structure was turned off. The electron concentration and carrier recombination rate of the proposed structure were lower than those of the conventional structure, but the final turn-off loss was lower than that of the conventional structure, further indicating that the proposed structure had other bleed paths in addition to the composite bleed.

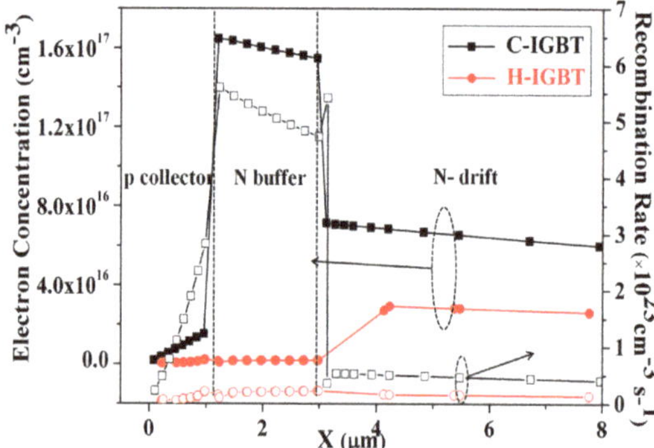

Figure 6. Electron concentration and carrier recombination rate on the collector side during voltage rise.

The electric field stubs and the p-polySi/p-SiC heterojunction energy band diagram near the collector during device turn-offs is shown in Figure 7. As can be seen from the figure, when the proposed structure was turned off, a much larger electric field spike was introduced to the collector side than that of the conventional structure, and this larger electric field could drive more electrons from the drift region to the collector, thus accelerating the turn-off of the device. In addition, it can be seen from the energy band diagram of the heterojunction that the heterojunction portion was more favorable for electron bleed than the ordinary PN junction portion. Ultimately, the combined effect of the two bleeds mechanisms greatly reduced the turn-off losses of the proposed structure.

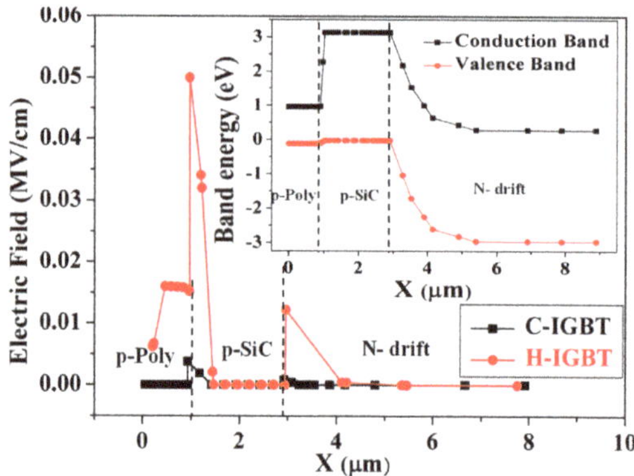

Figure 7. Electric field cut line and energy band diagram near the collector in the voltage rising phase.

Figure 8 shows a compromise between the on-state voltage and turn-off loss of the device for different drift region carrier lifetimes. It can be seen from the figure that as the carrier lifetime in the drift region decreases, the turn-on voltage drop gradually increases and the turn-off loss gradually decreases. The main reason for this result is that the carrier lifetime injected into the drift region decreases as the carrier lifetime decreases, which weakens the positive conductance modulation effect and increases the on-state voltage drop. When the device is turned off, the carrier bleed time is reduced since less carriers are stored in the drift region, and the device turn-off loss is reduced.

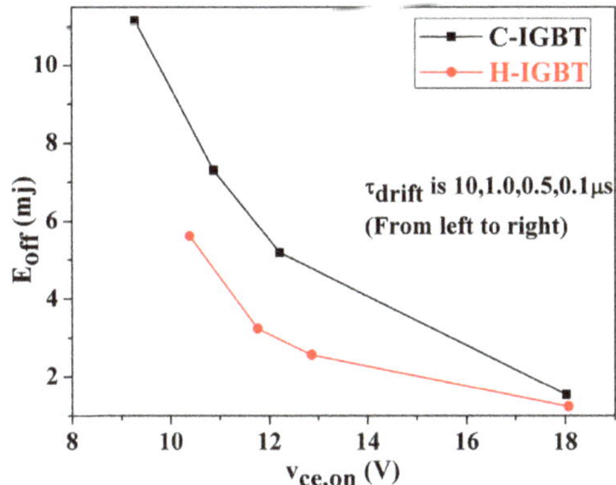

Figure 8. Tradeoff curve of conduction voltage drop and turn-off loss.

4. Conclusions

We presented a H-IGBT with partial p-polySi/p-SiC heterojunction at the backside of the device, and compared the electrical characteristics of H-IGBT and C-IGBT with ATLAS simulation software. The simulation results showed that, under the appropriate doping concentration, the heterojunction part has little effect on the forward conduction characteristics, and at the same it can greatly improve the turn-off speed of the device. In the case of forward blocking, the electric field in the entire drift region of the device is raised due to the introduction of the heterojunction, so the breakdown voltage of the device is improved. Finally, compared with C-IGBT, the turn-off loss of H-IGBT is reduced by 58.4%, the breakdown voltage is increased by 13.3%, and the on-state voltage drop is increased by 8.3%.

Author Contributions: Conceptualization, Y.W.; Software, X.W.; Investigation, F.-w.S., H.-k.M; writing—original draft preparation, H.-k.M.; writing—review and editing, Y.W.; supervision, Y.W.

Funding: This work was supported in part by the National Natural Science Foundation of China (No. 61774052) and in part by the Science and Technology on Analog Integrated Circuit Laboratory (No. 6142802180507).

Conflicts of Interest: The authors declare no conflict of interest.

References

1. Soto, C.; García-Rosales, C.; Echeberria, J.; Platacis, E.; Shisko, A.; Muktepavela, F.; Hernández, T.; Huertac, MM. Development, Characterization, and Testing of a SiC-Based Material for Flow Channel Inserts in High-Temperature DCLL Blankets. *IEEE Trans. Plasma Sci.* **2018**, *46*, 1561–1569. [CrossRef]
2. Chowdhury, S.; Chow, T.P. Performance tradeoffs for ultra-high voltage (15 kV to 25 kV) 4H-SiC n-channel and p-channel IGBTs. In Proceedings of the International Symposium on Power Semiconductor Devices and ICs (ISPSD), Prague, Czech Republic, 12–16 June 2016; pp. 75–78.
3. Terashima, T. Superior performance of SiC power devices and its limitation by self-heating. In Proceedings of the 2016 IEEE International Electron Devices Meeting (IEDM), San Francisco, CA, USA, 3–7 December 2016; pp. 10–15.

4. Regoutz, A.; Pobegen, G.; Aichinger, T. Interface chemistry and electrical characteristics of 4H-SiC/SiO_2 after nitridation in varying atmospheres. *J. Mater. Chem. C* **2018**, *6*, 12079–12085. [CrossRef]
5. Kunshan, Y.U.; Lijun, X.; Rui, J. Recent Development and Application Prospects of IGBT in Flexible HVDC Power System. *Autom. Electr. Power Syst.* **2016**, *40*, 139–143.
6. Deng, G.Q.; Luo, X.R.; Zhou, K.; He, Q.Y.; Ruan, X.L.; Liu, Q.; Sun, T.; Zhang, Bo. A snapback-free RC-IGBT with Alternating N/P buffers. In Proceedings of the 29th International Symposium on Power Semiconductor Devices and ICs (ISPSD), Sapporo, Japan, 28 May–1 June 2017; pp. 127–130.
7. Usman, M.; Nawaz, M. Device design assessment of 4H–SiC n-IGBT–A simulation study. *Solid-State Electron.* **2014**, *92*, 5–11. [CrossRef]
8. Das, A.; Islam, M.N.; Haq, M.M.; Khan, M.Z.R. Minority carrier lifetime profile into switching analysis of IGBT through parabolic approximation. In Proceedings of the 5th International Conference on Informatics, Electronics and Vision (ICIEV), Dhaka, Bangladesh, 13–14 May 2016; pp. 235–240.
9. Zhang, Q.C.; Das, M.; Sumakeris, J. 12-kV p-channel IGBTs with low on-resistance in 4H-SiC. *IEEE Electron Device Lett.* **2008**, *29*, 1027–1029. [CrossRef]
10. Deguchi, T.; Mizushima, T.; Fujisawa, H.; Takenaka, K.; Yonezawa, Y.; Fukuda, K.; Okumura, H.; Arai, M.; Tanaka, A.; Ogata, S.; et al. Static and dynamic performance evaluation of > 13 kV SiC p-channel IGBTs at high temperatures. In Proceedings of the IEEE 26th International Symposium on Power Semiconductor Devices & ICs (ISPSD), Waikoloa, HI, USA, 15–19 June 2014; pp. 261–264.
11. Chen, J.; Meng, H.; Jiang, F.X.C.; Lin, X.L. A snapback-free shorted-anode insulated gate bipolar transistor with an N-path structure. *Superlattices Microstruct.* **2015**, *78*, 201–209. [CrossRef]
12. Chen, X.; Cheng, J.; Teng, G.; Guo, H. Novel trench gate field stop IGBT with trench shorted anode. *J. Semicond.* **2016**, *37*, 054008. [CrossRef]
13. Shen, Z.; Zhang, F.; Tian, L.; Yan, G.; Wen, Z.; Zhao, W.; Wang, L.; Liu, X.; Sun, G.; Zeng, Y. Optimized P-emitter doping for switching-off loss of superjunction 4H-SiC IGBTs. In Proceedings of the 13th China International Forum on Solid State Lighting: International Forum on Wide Bandgap Semiconductors China (SSLChina: IFWS); Beijing, China, 15–17 November 2016, pp. 1–5.
14. Jiang, M.; Yin, X.; Shuai, Z.; Wang, J.; Shen, Z.J. An Insulated-Gate Bipolar Transistor With a Collector Trench Electron Extraction Channel. *IEEE Electron Device Lett.* **2015**, *36*, 935–937. [CrossRef]
15. Sung, W.; Wang, J.; Huang, A.Q.; Baliga, B.J. Design and investigation of frequency capability of 15kV 4H-SiC IGBT. In Proceedings of the 21st International Symposium on Power Semiconductor Devices & ICs, Barcelona, Spain, 14–18 June 2009; pp. 271–274.
16. Vechalapu, K.; Negi, A.; Bhattacharya, S. Comparative performance evaluation of series connected 15 kV SiC IGBT devices and 15 kV SiC MOSFET devices for MV power conversion systems. In Proceedings of the 2016 IEEE Energy Conversion Congress and Exposition (ECCE), Milwaukee, WI, USA, 18–22 September 2016; pp. 1–8.
17. Zhang, Q.; Wang, J.; Jonas, C.; Callanan, R.; Sumakeris, J.J.; Ryu, S.H.; Das, M.; Agarwal, A.; Palmour, J.; Huang, A.Q. Design and Characterization of High-Voltage 4H-SiC p-IGBTs. *IEEE Trans. Electron Device.* **2008**, *55*, 1912–1919. [CrossRef]
18. Ryu, S.H.; Capell, C.; Cheng, L.; Jonas, C.; Gupta, A.; Donofrio, M.; Clayton, J.; O'Loughlin, M.; Burk, A.; Grider, D.; et al. High performance, ultra high voltage 4H-SiC IGBTs. In Proceedings of the 2012 IEEE Energy Conversion Congress and Exposition (ECCE), Raleigh, NC, USA, 15–20 September 2012; pp. 3603–3608.
19. La Via, F.; Galvagno, G.; Foti, G.; Mauceri, M.; Leone, S.; Pistone, G.; Abbondanza, G.; Veneroni, A.; Masi, M.; Valente, GL.; et al. 4H-SiC Epitaxial Growth with Chlorine Addition. *Chem. Vap. Depos.* **2006**, *12*, 509–515. [CrossRef]
20. Kim, H.; Lee, H.; Kim, YS.; Lee, S.; Kang, H.; Heo, J.; Kim, H.J. On-axis Si-face 4H-SiC Epitaxial Growth with Enhanced Polytype Stability by Controlling Micro-steps during H_2 Etching Process. *CrystEngComm* **2017**, *19*, 2359–2366. [CrossRef]
21. Luna, L.E.; Tadjer, M.J.; Anderson, T.J.; Imhoff, E.A.; Hobart, K.D.; Kub, F.J. Deep reactive ion etching of 4H-SiC via cyclic SF_6/O_2 segments. *J. Micromech. Microeng.* **2017**, *27*, 095004. [CrossRef]
22. Hirai, H.; Kita, K. Difference of near-interface strain in SiO_2 between thermal oxides grown on 4H-SiC by dry-O_2 oxidation and H_2O oxidation characterized by infrared spectroscopy. *Appl. Phys. Lett.* **2017**, *110*, 956. [CrossRef]

23. Pastuović, Ž.; Siegele, R.; Capan, I.; Brodar, T.; Sato, S.I.; Ohshima, T. Deep level defects in 4H-SiC introduced by ion implantation: the role of single ion regime. *J. Phys. Condens. Matter* **2017**, *29*, 475701. [CrossRef] [PubMed]
24. Chen, Y. Study on mechanism and key techniques of ion implantation surface modification method for ultra-precision machining. Ph.D. Thesis, The University of Queensland, Brisbane, Australia, 2015.
25. Hatakeyama, T.; Suzuki, T.; Ichinoseki, K.; Matsuhata, H.; Fukuda, K.; Shinohe, T.; Arai, K. Impact of Oxidation Conditions and Surface Defects on the Reliability of Large-Area Gate Oxide on the C-Face of 4H-SiC. *Mater. Sci. Forum* **2010**, *645–648*, 799–804. [CrossRef]
26. Cabello, M.; Soler, V.; Montserrat, J.; Rebollo, J.; Rafi, J.M.; Godignon, P. Impact of boron diffusion on oxynitrided gate oxides in 4H-SiC metal-oxide-semiconductor field-effect transistors. *Appl. Phys. Lett.* **2017**, *111*, 321–337. [CrossRef]
27. Zhong, S.; Liao, N.; Luo, C.; Que, L.; Kou, L.; Wu, M. Effects of Gate Dielectric Process on Inter-layer Dielectric of Polysilicon of Charged-coupled Devices. *Semicond. Optoelectron.* **2017**, *38*, 345–348.
28. Lizotte, S.C. Process for filling deep trenches with polysilicon and oxide. U.S. Patent 6229194B1, 8 May 2001.
29. Kawada, Y. Method for Manufacturing Silicon Carbide Semiconductor Device. U.S. Patent 7510977B2, 31 March 2009.
30. Karner, H.W.; Kernstock, C.; Stanojević, Z.; Baumgartner, O.; Schanovsky, F.; Karner, M.; Helms, D.; Eilers, R.; Metzdorf, M. TCAD-based characterization of logic cells: Power, performance, area, and variability. In Proceedings of the 2017 International Symposium on VLSI Technology, Systems and Application (VLSI-TSA), Hsinchu, Taiwan, 24–27 April 2017; pp. 1–2.
31. Silvaco, I. *ATLAS User's Manual: Device Simulation Software*; Silvaco Int.: Santa Clara, CA, USA, 2018.

Article

Electrical Performance and Bias-Stress Stability of Amorphous InGaZnO Thin-Film Transistors with Buried-Channel Layers

Ying Zhang, Haiting Xie and Chengyuan Dong *

Department of Electronic Engineering, Shanghai Jiao Tong University, Shanghai 200240, China; Zhangying0036@sjtu.edu.cn (Y.Z.); haitingx@163.com (H.X.)

* Correspondence: cydong@sjtu.edu.cn; Tel.: +86-021-3420-8271

Received: 17 October 2019; Accepted: 11 November 2019; Published: 14 November 2019

Abstract: To improve the electrical performance and bias-stress stability of amorphous InGaZnO thin-film transistors (a-IGZO TFTs), we fabricated and characterized buried-channel devices with multiple-stacked channel layers, i.e., a nitrogen-doped a-IGZO film (front-channel layer), a conventional a-IGZO film (buried-channel layer), and a nitrogen-doped a-IGZO film (back-channel layer). The larger field-effect mobility (5.8 $cm^2V^{-1}s^{-1}$), the smaller subthreshold swing value (0.8 V/dec, and the better stability (smaller threshold voltage shifts during bias-stress and light illumination tests) were obtained for the buried-channel device relative to the conventional a-IGZO TFT. The specially designed channel-layer structure resulted in multiple conduction channels and hence large field-effect mobility. The in situ nitrogen-doping caused reductions in both the front-channel interface trap density and the density of deep states in the bulk channel layers, leading to a small subthreshold swing value. The better stability properties may be related to both the reduced trap states by nitrogen-doping and the passivation effect of the nitrogen-doped a-IGZO films at the device back channels.

Keywords: amorphous InGaZnO; thin-film transistor; nitrogen-doping; buried-channel; stability

---

1. Introduction

Amorphous silicon thin-film transistors (a-Si TFTs) are the mainstream of active-matrix devices for flat panel displays (FPDs); additionally, polycrystalline (p-Si) TFTs are used to address some high-standard FPD products. In recent years, amorphous InGaZnO (a-IGZO) has been extensively studied as a potential material for the channel layers of TFT devices. In fact, a-IGZO TFTs are considered to replace silicon TFTs owing to their high mobility, low-temperature deposition, good large-area uniformity, and simple processing methods [1–3].

However, the electrical performance and stability properties of a-IGZO TFTs still need further improvements for their applications in FPDs and other fields [4,5]. Recently, some researchers—including our group—reported that nitrogen-doping (N-doping) effectively improved the electrical properties (e.g., subthreshold swing (SS) and bias-stress stability) of a-IGZO TFTs by decreasing the number of deep states and oxygen vacancies (Vo) in the device channel layers and reducing the channel/dielectric interface trap density with N atoms incorporated into the a-IGZO film [6–8]. However, the field-effect mobility (μ_{FE}) of the nitrogen-doped a-IGZO (a-IGZO:N) TFT devices also decreases due to the suppression of the oxygen vacancy (Vo) level in their channel layers, the main source of free electrons in oxide semiconductors [9,10]. Therefore, N-doping is an effective method for improving the electrical properties of a-IGZO TFTs, but its shortcomings still need to be overcome.

In this study, buried-channel a-IGZO:N TFTs, i.e., devices with multiple-stacked channel layers, were fabricated and evaluated. These devices showed better performance and stability than the conventional devices. Specifically, the channel layer of the buried-channel device consisted of an a-IGZO:N layer (front-channel layer), an a-IGZO layer (buried-channel layer), and an a-IGZO:N layer (back-channel layer). We believe this device structure is a feasible approach to improving the electrical properties of a-IGZO TFTs.

2. Materials and Methods

Inverted staggered a-IGZO TFTs were prepared on p+ heavily doped silicon wafers with 100 nm thick thermal oxide (SiO_2). The silicon wafers and thermal SiO_2 were used as the gate electrodes and gate insulators of the TFT devices, respectively. For comparison, three types of TFT devices with different channel layers were fabricated, as shown in Figure 1. A 30 nm thick single channel layer was formed with conventional a-IGZO for Device A and with a-IGZO:N for Device B, respectively. Then, Device C with the buried-channel layer structure (10 nm thick a-IGZO:N (front-channel layer) + 10 nm thick a-IGZO (buried-channel layer) + 10 nm thick a-IGZO:N (back-channel layer)) was designed and prepared.

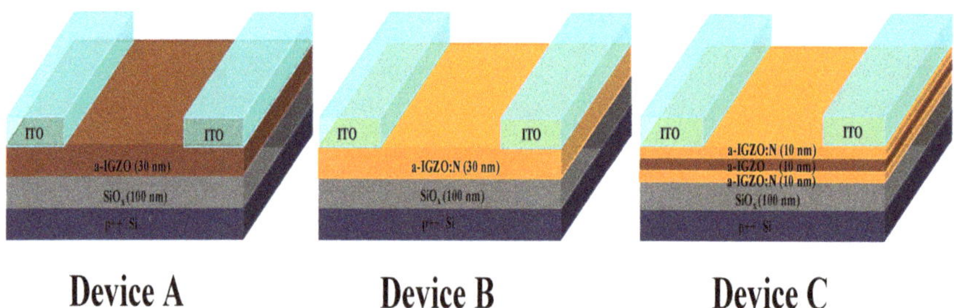

Figure 1. Schematic cross sections of the amorphous InGaZnO thin-film transistors (a-IGZO TFTs) prepared in this study.

The channel layers and the source/drain (S/D) electrodes were deposited with the RF-magnetron sputtering technique. The sputtering chamber was evacuated to a base pressure ($<3 \times 10^{-6}$ Torr) before the film depositions. The gas pressure was fixed at 3×10^{-3} Torr during the sputtering process. The channel layers were prepared at room temperature (RT) using an IGZO target (In_2O_3:Ga_2O_3:ZnO = 1:1:1 mol%); the RF power and the Ar flow rate were 60 W and 10 sccm, respectively. As for the depositions of the a-IGZO:N films, the nitrogen gas (N_2) was fed into the sputtering chamber at the flow rate of 1.2 sccm, and the Ar flow rate was fixed at 10 sccm. Then, the 100 nm thick indium-tin-oxide (ITO) layers were prepared as S/D electrodes in the same sputtering chamber. For simplicity, no passivation layers were prepared in this study. Both the channel layers and S/D electrodes were patterned by the shadow masks during sputtering; the channel width (W) and length (L) of the a-IGZO TFTs were fixed at 1000 and 250 μm, respectively. Finally, the TFT devices were annealed in N_2 atmosphere at 380 °C for 1 h.

The electrical measurements for the a-IGZO TFTs were performed at RT using an electrical analyzer (Keithley 4200, Keithley, Cleveland, OH, USA). For the transfer curve tests, the drain-source voltage (V_{DS}) was fixed at 10 V, and the gate-source voltage (V_{GS}) ranged from −20 to 40 V.

3. Results and Discussion

Figure 2 and Table 1 show the transfer curves and the corresponding extracted electrical parameters of the TFT devices, respectively. Here the μ_{FE} was obtained graphically from the square root of drain current ($I_{DS}^{1/2}$) versus the gate voltage (V_{GS}) in the saturation region using the intercept and maximum slope [11]. The threshold voltage (V_{TH}) was extracted from the gate voltage value where $I_{DS}/(W/L)$ = 1 nA. The subthreshold swing (SS) was defined as the half value of the difference between the gate voltages corresponding to the drain currents of 10^{-10} A and 10^{-8} A. The on–off current ratio (I_{ON}/I_{OFF})

was obtained from the ratio of the maximum and minimum drain current values within the V_{GS} range of −20 to ~40 V.

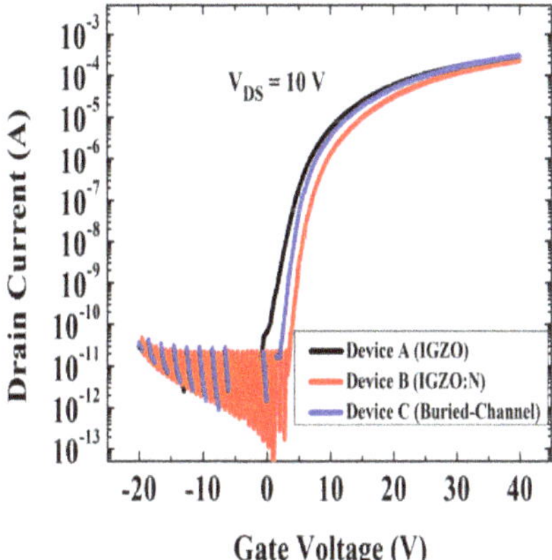

Figure 2. Transfer curves of the prepared TFTs (Devices A, B, and C).

Table 1. Extracted electrical parameters of the amorphous InGaZnO thin film transistors (a-IGZO TFTs).

Device	A	B	C
μ_{FE} ($cm^2V^{-1}s^{-1}$)	5.8	4.7	5.8
SS (V/dec)	1.3	0.8	0.8
V_{TH} (V)	2.0	5.0	3.5
I_{ON}/I_{OFF}	2.2×10^8	3.4×10^8	5.6×10^8

As switching devices in FPDs, TFTs are expected to have high field-effect mobility and low subthreshold swing, which can lead to better switching speed and smaller power consumption. As shown in Figure 2 and Table 1, the a-IGZO:N TFT (Device B) showed an improved SS value (0.8 V/dec) but degraded μ_{FE} (4.7 $cm^2V^{-1}s^{-1}$) compared with those of the conventional IGZO TFT (Device A, SS = 1.3 V/dec and μ_{FE} = 5.8 $cm^2V^{-1}s^{-1}$). In fact, this result was consistent with the other reported a-IGZO:N TFTs [7,9]. Besides, from Figure 2 one may clearly observe that the N-doping process could cause a positive V_{TH} shift for a-IGZO TFTs. As shown in Table 1, the V_{TH} increased from 2.0 V (Device A) to 5.0 V (Device B) with the nitrogen doped into the channel layers of the a-IGZO TFT. This result also agreed well with the findings of a recent study of a-IGZO:N TFTs [12]. Most importantly, the buried-channel TFT (Device C) showed the best electrical performance in this study. As shown in Table 1, Device C exhibited the smallest SS value (0.8 V/dec) as well as the largest μ_{FE} (5.8 $cm^2V^{-1}s^{-1}$) among all three devices. In addition, the V_{TH} of the buried-channel device (3.5 V) lay between those of the conventional a-IGZO TFT (Device A) and the a-IGZO:N TFT (Device B). It should be noted here that all the tested devices exhibited reasonably good I_{ON}/I_{OFF} values (>10^8), as shown in Figure 2 and Table 1.

Figure 3 shows the transfer curves of the a-IGZO TFTs under positive bias stress (PBS) tests. For measurements, the gate electrodes were applied by +30 V with the drain and source grounded; after a period, the transfer curves were instantly measured (V_{DS} = 10 V). As shown in Figure 3, the transfer curves of the three devices shifted positively to different extents as the stressing time elapsed. As shown in Figure 3a, the positive V_{TH} shift of the conventional a-IGZO TFT (Device A) was as large as +4.5 V under 2500 s of PBS testing. In contrast, the a-IGZO:N TFT (Device B) showed a much smaller V_{TH} shift (+3 V) than that of the conventional a-IGZO TFT, as shown in Figure 3b. As for the buried-channel device (Device C), the maximum V_{TH} shift value (+3.5 V, as shown in Figure 3c) lay between those of Devices A and B. In fact, the PBS stability of Device C was quite close to that of

Device B, as shown in Figure 3. In other words, during PBS tests, the buried-channel TFT (Device C) showed more stable properties than the conventional a-IGZO TFT (Device A).

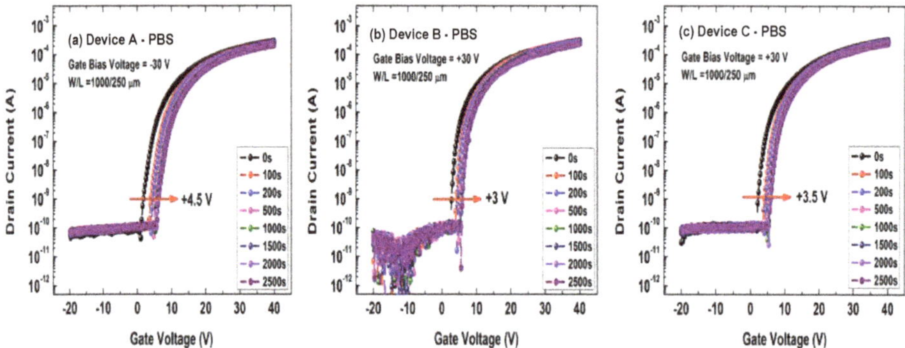

Figure 3. Positive bias stress (PBS) testing results of the a-IGZO TFTs: (**a**) the conventional a-IGZO TFT (Device A), (**b**) the a-IGZO:N TFT (Device B), and (**c**) the buried-channel TFT (Device C). The stressing conditions were V_{GS} = +30 V and V_{DS} = 0 V.

Figure 4 shows the transfer curves of the a-IGZO TFT devices under the negative bias stress (NBS) tests, where the stressing conditions were as follows: V_{GS} = −30 V and V_{DS} = 0 V. The NBS measurement operation was the same as that of PBS. One may observe from Figure 4 that the transfer curves of the three devices exhibited different negative shifts during NBS tests. As shown in Figure 4a,b, the maximum V_{TH} shift of the conventional a-IGZO TFT (−3.3 V) was much larger than that of the a-IGZO:N TFT (−1 V). Importantly, Device C showed a similar V_{TH} shift (−1.7 V after 2500 s of NBS testing) to that of Device B, implying that the buried-channel device (Device C) also exhibited better stability than the conventional a-IGZO TFT (Device A) during NBS tests.

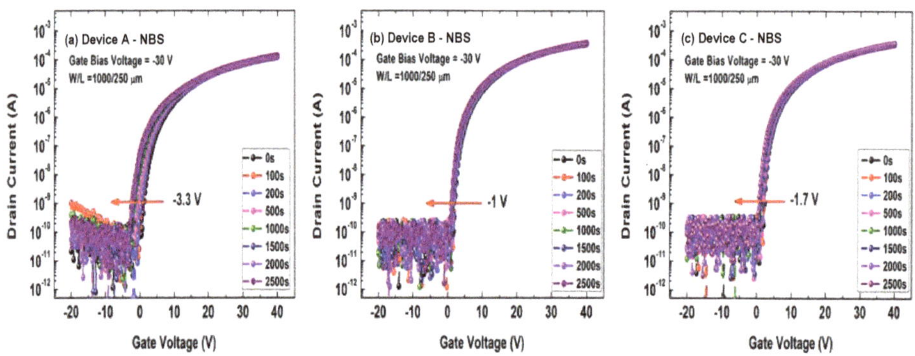

Figure 4. Negative bias stress (NBS) testing results of the a-IGZO TFTs: (**a**) the conventional a-IGZO TFT (Device A), (**b**) the a-IGZO:N TFT (Device B), and (**c**) the buried-channel TFT (Device C). The stressing conditions were V_{GS} = −30 V and V_{DS} = 0 V.

Figure 5 shows the transfer curves of the a-IGZO TFT devices under negative bias stress with ultraviolet (UV) light illumination (NBIS) tests. The stressing voltage was V_{GS} = −20 V; the wavelength and power intensity of the UV light were 380 nm and 0.1 mW/cm^2, respectively. Both the light illumination and gate stressing were applied for a period, and then the transfer curves were immediately measured (V_{DS} = 10 V). According to Figure 5, one may notice that all three devices exhibited serious negative shifts during NBIS tests. As shown in Figure 5a, the maximum V_{TH} shift of the conventional a-IGZO TFT (Device A) was up to −7 V, which was the worst among all the tested samples. However, this negative V_{TH} shift could be effectively reduced to −3 V using the buried-channel structure (as shown in Figure 5c) or to −2 V by adopting the N-doping technology (as shown in Figure 5b). Apparently, the NBIS stability of Device C was more similar to that of Device B than that of Device A. In other words, the experimental results proved that the buried-channel device (Device C) could lead to a significant improvement in NBIS stability compared with the conventional a-IGZO TFT (Device A).

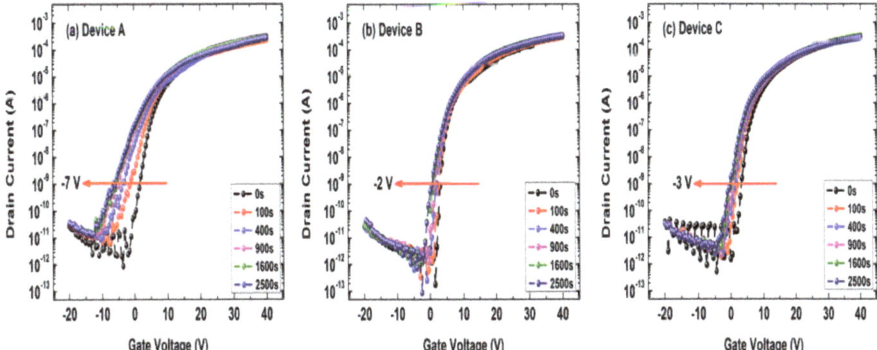

Figure 5. Negative bias stress with ultraviolet (UV) light illumination (NBIS) testing results of the a-IGZO TFTs: (**a**) the conventional a-IGZO TFT (Device A), (**b**) the a-IGZO:N TFT (Device B), and (**c**) the buried-channel TFT (Device C). The stressing voltage was V_{GS} = −20 V; the wavelength and power intensity of the UV light used in this test were 380 nm and 0.1 mW/cm^2, respectively.

It has been reported by our group and other researchers that the field-effect mobilities of oxide semiconductor TFTs were evidently improved by employing the double-stacked channel layers (DSCL) with a high defect-density channel layer and a low defect-density channel layer [13–15]. In this study, we further designed and prepared the buried-channel a-IGZO:N TFTs using the multiple-stacked channel layers composed of a 10 nm thick a-IGZO:N layer (front-channel layer), a 10 nm thick a-IGZO layer (buried-channel layer), and a 10 nm thick a-IGZO:N layer (back-channel layer). This design exhibited the best electrical performance (e.g., the smallest SS value, the largest μ_{FE}, and the optimum V_{TH}), as shown in Table 1. Furthermore, the buried-channel device exhibited similar stable properties to those of the a-IGZO:N TFT during PBS, NBS, and NBIS tests, which were much better than those of the conventional a-IGZO TFT. The related physical mechanisms could be ascertained by analyzing the channel-layer structure of these devices.

Figure 6 shows the energy band diagrams for Devices A, B, and C. As reported previously [7,9], N-doping caused a mobility reduction due to the significant suppression of the Vo level in the bulk channel layer. Former studies [15–17] indicated that N-doping reduced the Vo and defect density in the a-IGZO films. Hence, there was more Vo acting as the origin of free carriers in the conventional a-IGZO layer compared with that in the a-IGZO:N film. Therefore, one could assume that fewer free electrons took part in conductivity for Device B compared with the case for Device A (as shown in Figure 6a,b). This resulted in a lower mobility for Device B.

But, why did the buried-channel device not show degraded mobility as the a-IGZO:N TFT did? We attributed this fact to the particular energy band structure of the buried-channel device, as shown in Figure 6c. The electrons tended to inject from the conventional a-IGZO layer towards both the bottom a-IGZO:N layer and the top a-IGZO:N layer owing to the electron concentration difference between the conventional a-IGZO layer and the a-IGZO:N layer. Due to the positive polarity of the a-IGZO layer, the electrons accumulated in both the bottom a-IGZO:N/a-IGZO interface and the top a-IGZO/a-IGZO:N interface. During the operation of the buried-channel TFT device, these accumulated electrons also took part in conductivity, forming a first sub-channel at the bottom a-IGZO:N/a-IGZO interface and a second sub-channel at the top a-IGZO/a-IGZO:N interface. In other words, the conduction current in the buried-channel TFT was not only dominated by the main channel (channel layer/dielectric layer interface), but also the two additional sub-channels (bottom a-IGZO:N/a-IGZO interface and top a-IGZO/a-IGZO:N interface). The total carrier concentration from these three conduction channels in the buried-channel device might be comparable only with that from the front channel in the conventional a-IGZO TFT, as shown in Figure 6a,c, leading to the μ_{FE} values of the buried-channel device (Device C) that were nearly equal to those of the conventional a-IGZO TFT (Device A). In fact, these qualitative analyses might also be confirmed by numerical simulations, which are commonly used in explaining and discussing electrical properties of semiconductor devices [18,19]. The related study is still ongoing.

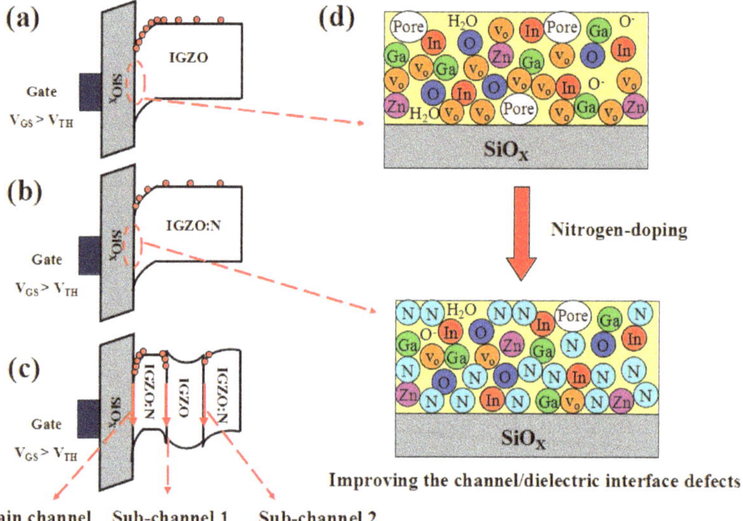

Figure 6. Energy band diagrams for (**a**) the conventional a-IGZO TFT (Device A), (**b**) the a-IGZO:N TFT (Device B), and (**c**) the buried-channel a-IGZO TFT (Device C). (**d**) The schematic illustration of the improvement effect for the N-doping process on the trap states in the bulk-layers and interfaces.

As mentioned before, N-doping effectively decreased the interface trap density as well as the deep defect density in the channel layer by incorporating the N atoms into the a-IGZO film (as shown in Figure 6d). Since the SS values of TFTs are closely related to the channel/dielectric trap density and the number of deep states in the channel layers [7,9,20], it is quite reasonable that the a-IGZO:N TFT had smaller SS values than the conventional a-IGZO TFT. As shown in Figure 6b,c, both the front-channel interface and the nearby channel layer of the buried-channel device were the same as those of the a-IGZO:N TFT, which naturally led to the same SS values for both cases.

The evident stability improvements (during PBS, NBS, and NBIS tests) of the a-IGZO:N TFT with respect to the conventional a-IGZO TFTs were assumed to result from the following reasons. First, the incorporation of N atoms into the a-IGZO film led to the suppression of trap states in the bulk channel layer as well as those in the front-channel interface (as shown in Figure 6d). Second, the nitrogen atoms doped at the back channel might act as a passivation layer, which could effectively protect the device from the influence of ambient gas (such as O_2, moisture, etc.). Third, the back-channel a-IGZO:N layer might shield the channel from UV light influence to a certain extent. As for the buried-channel device, the aforementioned stability-improving effects were also valid. However, the existence of the buried-channel layer (a-IGZO) more or less increased the trap states in the bulk channel layers compared with the a-IGZO:N TFT (Device B). Therefore, the stability properties (PBS,

NBS, and NBIS) of the buried-channel device (Device C) were a little worse than those of the a-IGZO:N TFT (Device B), but still much better than those of the conventional a-IGZO TFT (Device A).

So far, we can make some overall comments on the buried-channel TFT devices. Compared with the conventional a-IGZO TFT, the buried-channel device exhibited better electrical performance (smaller SS value and non-degraded μ_{FE}) and more stable properties (much smaller V_{TH} shifts during PBS, NBS, and NBIS tests). Compared with the a-IGZO:N TFT, the buried-channel device showed a little worse stability under PBS, NBS, and NBIS tests, but better electrical properties (larger μ_{FE} and the same SS value). Notably, the buried-channel structure could be easily achieved by one-pump-down depositions in the sputtering machines. Therefore, we believe the use of buried-channel devices is a feasible approach for improving the electrical performance and stability properties of amorphous oxide thin-film transistors.

4. Conclusions

Buried-channel TFTs with multiple-stacked channel layers including an a-IGZO:N film (front-channel layer), an a-IGZO film (buried-channel layer), and an a-IGZO:N film (back-channel layer) were fabricated and measured in this work. Compared with those of the conventional a-IGZO TFT, the better electrical performance (e.g., smaller SS value, larger μ_{FE}, and optimum V_{TH}) as well as the improved stability properties (e.g., smaller V_{TH} shifts during PBS, NBS, and NBIS tests) were obtained for the buried-channel device. The a-IGZO:N film used as the front-channel layer in the buried-channel structure could reduce both the interface trap density and the bulk-layer deep state density, leading to a smaller SS value (0.8 V/dec) in comparison with the conventional a-IGZO TFT device (SS = 1.3 V/dec). The non-degraded μ_{FE} (5.8 $cm^2V^{-1}s^{-1}$) for the buried-channel TFT might be attributed to its three conduction channels, including the main channel at the interface between the channel layer and the dielectric layer, the first sub-channel formed at the bottom a-IGZO:N/a-IGZO interface, and the second sub-channel formed at the top a-IGZO/a-IGZO:N interface. In addition, the improved bias-stress and NBIS stability of the buried-channel TFT compared with those of the conventional a-IGZO TFT might be mainly due to the suppression of the trap states in the bulk channel layers, the reduction in the defects at the front-channel interface, and the passivation effect created by using the a-IGZO:N film as the back-channel layer.

Author Contributions: Y.Z. and H.X. fabricated and measured all the a-IGZO TFTs. Y.Z. and C.D. designed the experiments and contributed to the theoretical explanations. The manuscript was written and revised by all the authors.

Funding: This work was supported by Natural Science Foundation of China (Grant No. 61474075).

Conflicts of Interest: The authors declare no conflict of interest.

References

1. Nomura, K.; Ohta, H.; Takagi, A.; Kamiya, T.; Hirano, M.; Hosono, H. Room-temperature fabrication of transparent flexible thin-film transistors using amorphous oxide semiconductors. *Nature* **2004**, *432*, 488–492. [CrossRef] [PubMed]
2. Zhou, Y.; Dong, C. Influence of passivation layers on positive gate bias-stress stability of amorphous InGaZnO thin-film transistors. *Micromachines* **2018**, *9*, 603. [CrossRef] [PubMed]
3. Ide, K.; Ishikawa, K.; Tang, H.; Katase, T.; Hiramatsu, H.; Kumomi, H.; Hosono, H.; Kamiya, T. Effects of base pressure on growth and optoelectronic properties of amorphous In-Ga-Zn-O: Ultralow optimum oxygen supply and bandgap widening. *Phys. Status Solidi A* **2019**, *216*. [CrossRef]
4. Choi, S.; Kim, J.Y.; Kang, H.; Ko, D.; Rhee, J.; Choi, S.J.; Kim, D.M.; Kim, D.H. Effect of oxygen content on current stress-induced instability in bottom-gate amorphous InGaZnO thin-film transistors. *Materials* **2019**, *12*, 3149. [CrossRef] [PubMed]
5. Dong, C.; Xu, J.; Zhou, Y.; Zhang, Y.; Xie, H. Light illumination stability of amorphous InGaZnO thin film transistors in oxygen and moisture ambience. *Solid State Electron.* **2019**, *153*, 74–78. [CrossRef]

6. Liu, P.T.; Chou, Y.T.; Teng, L.F.; Li, F.H.; Shieh, H.P. Nitrogenated amorphous InGaZnO thin film transistor. *Appl. Phys. Lett.* **2011**, *98*. [CrossRef]
7. Raja, J.; Jang, K.; Balaji, N.; Yi, J. Suppression of temperature instability in InGaZnO thin-film transistors by in situ nitrogen doping. *Semicond. Sci. Technol.* **2013**, *28*, 115010. [CrossRef]
8. Xie, H.; Zhou, Y.; Zhang, Y.; Dong, C. Chemical bonds in nitrogen-doped amorphous InGaZnO thin film transistors. *Results Phys.* **2018**, *11*, 1080–1086. [CrossRef]
9. Raja, J.; Jang, K.; Balaji, N.; Choi, W.; Trinh, T.T.; Yi, J. Negative gate-bias temperature stability of N-doped InGaZnO active-layer thin-film transistors. *Appl. Phys. Lett.* **2013**, *102*. [CrossRef]
10. Ryu, B.; Noh, H.K.; Choi, E.A.; Chang, K.J. O-vacancy as the origin of negative bias illumination stress instability in amorphous In–Ga–Zn–O thin film transistors. *Appl. Phys. Lett.* **2010**, *97*, 022108. [CrossRef]
11. Aikawa, S.; Darmawan, P.; Yanagisawa, K.; Nabatame, T.; Abe, Y.; Tsukagoshi, K. Thin-film transistors fabricated by low-temperature process based on Ga- and Zn-free amorphous oxide semiconductor. *Appl. Phys. Lett.* **2013**, *102*, 102101. [CrossRef]
12. Huang, X.; Wu, C.; Lu, H.; Ren, F.; Chen, D.; Zhang, R.; Zheng, Y. Enhanced bias stress stability of a-InGaZnO thin film transistors by inserting an ultra-thin interfacial InGaZnO:N layer. *Appl. Phys. Lett.* **2013**, *102*, 193505. [CrossRef]
13. Park, J.C.; Lee, H.N. Improvement of the performance and stability of oxide semiconductor thin-film transistors using double-stacked active layers. *IEEE Electron Dev. Lett.* **2012**, *33*, 818–820. [CrossRef]
14. Zhan, R.; Dong, C.; Yang, B.R.; Shieh, H.P. Modulation of interface and bulk states in amorphous InGaZnO thin film transistors with double stacked channel layers. *Jpn. J. Appl. Phys.* **2013**, *52*, 090205. [CrossRef]
15. Xie, H.; Wu, Q.; Xu, L.; Zhang, L.; Liu, G.; Dong, C. Nitrogen-doped amorphous oxide semiconductor thin film transistors with double-stacked channel layers. *Appl. Surf. Sci.* **2016**, *387*, 237–243. [CrossRef]
16. Nguyen, T.T.; Renault, O.; Aventurier, B.; Rodriguez, G.; Barnes, J.P.; Templier, F. Analysis of IGZO thin-film transistors by XPS and relation with electrical characteristics. *J. Disp. Technol.* **2013**, *9*, 770–774. [CrossRef]
17. Raja, J.; Jang, K.; Balaji, N.; Hussain, S.Q.; Velumani, S.; Chatterjee, S.; Kim, T.; Yi, J. Aging effects on the stability of nitrogen-doped and un-doped InGaZnO thin-film transistors. *Mater. Sci. Semicond. Process.* **2015**, *37*, 129–134. [CrossRef]
18. Bencherif, H.; Dehimi, L.; Pezzimenti, F.; Corte, F.G.D. Temperature and SiO_2/4H-SiC interface trap effects on the electrical characteristics of low breakdown voltage MOSFETs. *Appl. Phys. A* **2019**, *125*, 294. [CrossRef]
19. Megherbi, M.L.; Pezzimenti, F.; Dehimi, L.; Saadoune, M.A.; Corte, F.G.D. Analysis of trapping effects on the forward current-voltage characteristics of al-implanted 4H-SiC p-i-n Diodes. *IEEE Trans. Electron Dev.* **2018**, *65*, 3371–3378. [CrossRef]
20. Nomura, K.; Kamiya, T.; Hosono, H. Highly stable amorphous In-Ga-Zn-O thin-film transistors produced by eliminating deep subgap defects. *Appl. Phys. Lett.* **2011**, *99*, 053505. [CrossRef]

Article

Measurement of Heat Dissipation and Thermal-Stability of Power Modules on DBC Substrates with Various Ceramics by SiC Micro-Heater Chip System and Ag Sinter Joining

Dongjin Kim [1,2], Yasuyuki Yamamoto [3], Shijo Nagao [2,*], Naoki Wakasugi [4], Chuantong Chen [2,*] and Katsuaki Suganuma [2]

1 Department of Adaptive Machine Systems, Graduate School of Engineering, Osaka University, Osaka 565-0871, Japan; djkim@eco.sanken.osaka-u.ac.jp
2 The Institute of Scientific and Industrial Research, Osaka University, Osaka 567-0047, Japan; suganuma@sanken.osaka-u.ac.jp
3 Specialty products development, Tokuyama Co., Yamaguchi 746-0006, Japan; yasu-yamamoto@tokuyama.co.jp
4 Division of system development, Yamato Scientific Co., Ltd., Tokyo 135-0047, Japan; naoki.wakasugi@yamato-net.co.jp
* Correspondence: shijo.nagao@sanken.osaka-u.ac.jp (S.N.); chenchuantong@sanken.osaka-u.ac.jp (C.C.); Tel.: +81-6-6879-8521 (S.N. & C.C.)

Received: 4 October 2019; Accepted: 28 October 2019; Published: 31 October 2019

Abstract: This study introduced the SiC micro-heater chip as a novel thermal evaluation device for next-generation power modules and to evaluate the heat resistant performance of direct bonded copper (DBC) substrate with aluminum nitride (AlN-DBC), aluminum oxide (DBC-Al_2O_3) and silicon nitride (Si_3N_4-DBC) ceramics middle layer. The SiC micro-heater chips were structurally sound bonded on the two types of DBC substrates by Ag sinter paste and Au wire was used to interconnect the SiC and DBC substrate. The SiC micro-heater chip power modules were fixed on a water-cooling plate by a thermal interface material (TIM), a steady-state thermal resistance measurement and a power cycling test were successfully conducted. As a result, the thermal resistance of the SiC micro-heater chip power modules on the DBC-Al_2O_3 substrate at power over 200 W was about twice higher than DBC-Si_3N_4 and also higher than DBC-AlN. In addition, during the power cycle test, DBC-Al_2O_3 was stopped after 1000 cycles due to Pt heater pattern line was partially broken induced by the excessive rise in thermal resistance, but DBC-Si_3N_4 and DBC-AlN specimens were subjected to more than 20,000 cycles and not noticeable physical failure was found in both of the SiC chip and DBC substrates by a x-ray observation. The results indicated that AlN-DBC can be as an optimization substrate for the best heat dissipation/durability in wide band-gap (WBG) power devices. Our results provide an important index for industries demanding higher power and temperature power electronics.

Keywords: power cycle test; SiC micro-heater chip; direct bonded copper (DBC) substrate; Ag sinter paste; wide band-gap (WBG); thermal resistance

1. Introduction

Silicon carbide (SiC) and gallium nitride (GaN), are wide band-gap (WBG) semiconductors, are strongly recognized as the best materials for the power electronics applications demanding higher power and higher temperature [1,2]. Since these WBG semiconductors have superior properties, kind of a wide band gap (>3 eV), a high critical electric field (>3 MV/cm) and a high saturation velocity

(>2 × 10^7 cm/s), SiC and GaN can enable to overcome the ultimate performances reached by silicon (Si) based devices, in terms of power conversion efficiency [3]. In addition, WBG semiconductor devices can be operating much higher temperatures (>250 °C) than Si based devices (<150 °C) [4–8], this means massive, complex and heavy cooling systems can be eliminated from power conversion systems. Inverters and converters automotive components can be change to smaller by the simply heat dissipation design of in the high temperature environments [9]. In general, the structure of the power module has multi-layer structures of a semiconductor chip and a heat dissipation/insulation plate. All the layers have different material properties like a coefficient of thermal expansion (CTE), which causes thermos-mechanical stress during repetitive operating [10,11].

Besides, to withstand higher power and higher thermal density, how to insulate electricity and dissipate heat is one of the key issues of WBG power conversion systems. Therefore, there have been some issues, for instance how to interconnect these multi-layers to implement in high temperature and have a thermal-stable reliability. In addition, power electronic substrate plays an important role in dissipating heat to prevent power electronic module failure [12–18]. Heat dissipation/insulation substrate, which is a direct bonded copper (DBC) and a direct bonded aluminum (DBA), are existed between power semiconductor device and heat sink. Heat is transferred from surface of the power semiconductor device to the heat sink. The DBC and DBA were metallized both side of ceramic plate, to improve the thermal conductivity of ceramic substrates and form circuits [19], and were considered as the most promising substrates for power modules due to their excellent thermal conductivity, low thermal resistance, and high insulation voltage.

However, the sandwich structure of the DBC and DBA substrate will cause thermo-mechanical stresses due to the CTE mismatch. It has been reported that heat resistance of the substrate depending on the type of ceramics and metals [20–22]. In this regard, our group carried out to evaluate thermo-mechanical stability of various ceramic applied active metal brazed (AMB) DBC substrates with the Ni plating layer during a harsh thermal shock cycling test at the temperature range of from −50 °C to 250 °C [23]. The DBC substrate with silicon nitride (Si_3N_4-DBC) ceramics middle layer indicate no serious damage within 1,000 cycles of thermal shock cycles, while those of an aluminum nitride (AlN) and alumina (Al_2O_3) caused by Cu layer severe delamination after the same cycles [23].

The results indicated that the failures may impress that the AlN and Al_2O_3 have critical disadvantage of higher CTE value and lower toughness than Si_3N_4. Since this previous study just investigated the DBC substrate itself, the failures must be induced by the mismatch between Cu and ceramic, meaning that the stress occurred from the DBC. Such thermal cycles give only homogeneous temperature change in the substrate specimens as shown in Figure 1a. However, the realistic thermal distributions in a ceramic substrate must have a large temperature gradient to transfer the generated heat from surface of the device chips to the cooling system as shown in Figure 1b. The stress induced by the CTE mismatch during the thermal conduction from chips to cooling system was more complexed than single DBC substrate. Therefore, the evaluation of such a practical temperature distribution is urgently needed for designing power module. In this context, power cycling test is the strongly useful evaluation methods to evaluate the reliability of a device packaging used in a condition similar to those in actual operations. In this process, the heat transfer performance of such a substrate mainly depends on the thermal properties of employed ceramic materials, e.g., Al_2O_3, AlN, and Si_3N_4. Power modules was required with a comprehensive combination of conductive Cu and insulation ceramics, involving the layer thicknesses and circuit patterns to achieve both thermal management and the least power loss at the same time. Usually, to measure the thermal resistance of devices after a power cycling test, it needs an equipment of power cycling system and also a T3ster system which means the Thermal Transient Tester system. It is a technology to monitor the thermal transport through the devices. However, both of the power cycling system and the T3ster system are very expensive and need to take up a lot of space. A simple, fast and miniaturized thermal measurement system is actually necessary.

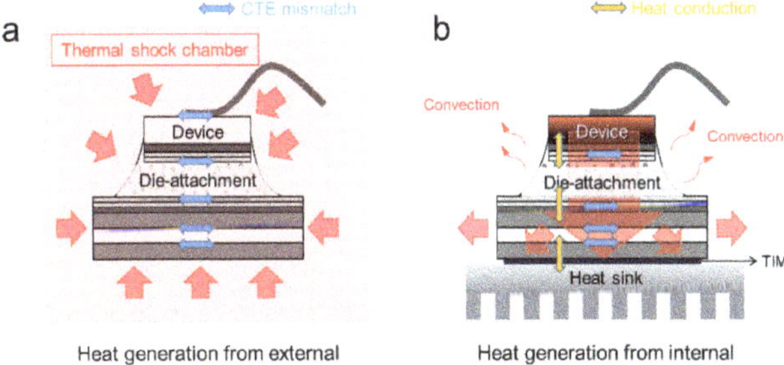

Figure 1. Comparison of the test environments and its heat transfer path: (**a**) thermal shock or aging condition as non-driving condition and (**b**) actual power cycling test as driving condition.

In this study, SiC micro-heater chip was designed and used to bond with the various type of ceramic DBC substrates such as AlN and Al_2O_3, and Si_3N_4. The Ag sinter joining was used as a die to attach material because which can withstand the high temperature applications and reported in many previous studies. Steady-state thermal resistance from the SiC micro-heater chip to cooling system with different type DBC were measured. In addition, the power cycling test also was performed to investigated the high temperature reliability of the SiC micro-heater chip die attach structure on each type DBC substrate. The failure after power cycling was analyzed by a micro-focus 3D computed tomography (CT) X-Ray system. This method can significantly distinguish with traditional thermal cycling test because thermal properties of material can be considered during repetitive thermal environments. Our method of power cycling test with SiC heater chip may provide in important evaluation index of future packaging systems with WBG semiconductors targeted for high-temperature/power density power modules.

2. Materials and Methods

2.1. SiC Micro-Heater Chip and Direct Bonded Copper (DBC) Substrate

Geometrically 3D designed SiC micro-heater chip was fabricated with n-doped 4H-SiC by using a lithography technology (see Figure 2a), also it consists of a heater, a probe and an electrode, the material of heater and probe are platinum (Pt). The DBC substrate with five lands of the top side was introduced, and the SiC heater chip was die attached onto the center land and its geometric dimension is as displayed in Figure 2b. At the backside of SiC micro-heater chip and the top side of DBC substrates were metallized in the order of Ti and Ag layer by a sputtering process.

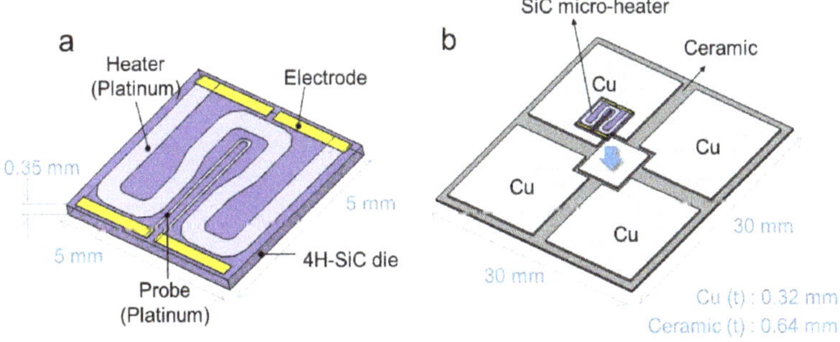

Figure 2. SiC heater die-attached structure: (**a**) SiC micro-heater chip structure (**b**) SiC heater chip on the substrate.

2.2. Die Attach Material

Two types of Ag particles were mixed and used to make Ag paste. The first type comprised flake-shaped Ag particles with an average lateral diameter of 8 µm and a thickness of 260 nm, as shown in Figure 3a (AgC-239, Fukuda Metal Foil and Powder Co., Ltd., Kyoto, Japan). The second type comprised spherical-shape particles with an average diameter of 400 nm, as shown in Figure 3b (FHD, MitsuiMining and Smelting Co., Ltd., Tokyo, Japan). The two kinds of Ag particles were mixed as the Ag filler with a weight ratio of 1:1 [24]. These hybrid particles were then stirred magnetically for 10 min and vibrated ultrasonically for 30 min with viscosity of 100–200 Pa·s solvent (alkylene glycols and polyols) using a hybrid mixer (HM-500, Keyence Corporation, Osaka, Japan) to fabricate Ag paste to achieve a uniform mixture. The amount of the solvent was maintained at approximately 10 wt % in the paste to keep a suitable viscosity of 150–250 centi poise (cPs) at room temperature. The Ag paste was then printed onto the center land of the DBC substrates which were purchased from Mitsubishi Materials Corporation. For the bonding process, a step-by-step profile with low pressure (0.4 MPa), including 120 °C and 180 °C for 5 min as pre-heating, and then 250 °C for 60 min, was applied to prevent pores formation caused by evaporation of organic solvent (see Figure 3b). Finally, the sintered cross section is formed as shown in Figure 3c. For power supply, 10EA Au wires of 45 µm diameter (TANAKA DENSHI KOGYO, Yoshida, Japan) on the four electrode were each interconnected by an ultrasonic wire bonding machine (Kulicke & Soffa Industries Inc, 4500 digital series, Singapore).

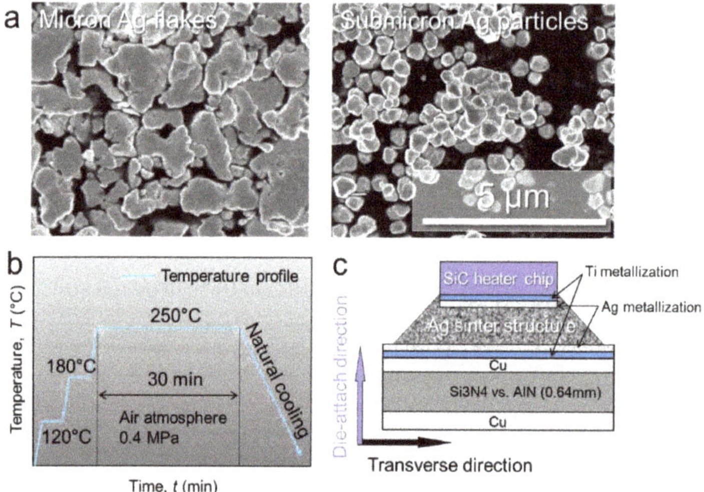

Figure 3. (**a**) Ag particles, (**b**) sintering condition and (**c**) schematic diagram of the assembled image.

2.3. Test Machine and Method

Figure 4 shows the machine of power cycling test and thermal resistance measurement and its component. The machine consists of four parts, power supplier, controller, monitor and water-cooling system. Over 200 W of power is supplied from the power source to the electrode, and when heat is generated from the top of the heater chip, heat is transferred to the heat sink (25 °C) through the die attach and the DBC substrate. At this time, SiC heater/DBC die-attached specimen is mounted on the water cooling plate by diamond thermal grease (thermal interface material, TIM). To prevent delamination between specimen and cooling plate during power cycling, the DBC substrate was fixed to the water-cooling plate with a force of 20 N. Then, ΔT is calculated by the temperature of the top surface of the SiC heater chip (T_{chip}) and the temperature of the thermocouple attached to the water-cooling plate ($T_{thermocouple}$), the equation is as follows:

$$\Delta T = T_{chip} - T_{thermocouple} \tag{1}$$

Here, the thermal resistance (R_{th}) can be calculated by dividing this temperature difference (ΔT) by the input power (Q) as follows:

$$R_{th} = \Delta T / Q \tag{2}$$

where, Q (W) is Joule heating by the electrical power of SiC micro-heater chip and T is the temperature (K).

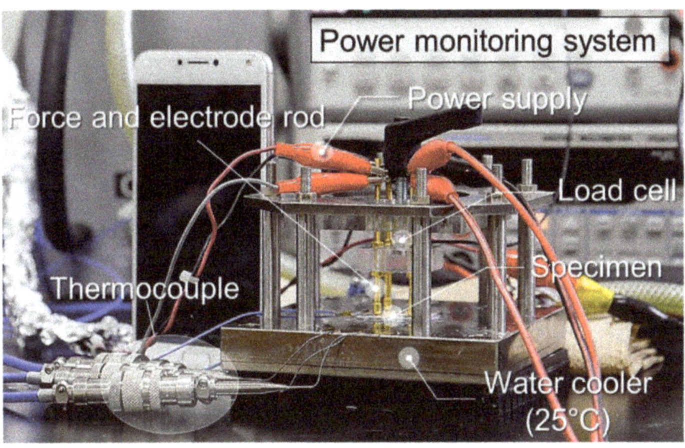

Figure 4. Power cycling test system with SiC micro-heater chip.

3. Results and Discussion

3.1. Steady-State Thermal Resistance

For thermal resistances measurement, a specimen was packaged by the method introduced in the previous section (see Figure 3). After the die bonding, the specimen interconnected without noticeable defects such as oxidation, and its actual manufactured and interconnected SiC heater die attached specimen are shown in Figure 5a. Furthermore, the electrode and temperature sensor of SiC heater chip was Au wire bonded with the DBC substrate after die-attach bonding. The Au wire bonded with the DBC substrate was given at upper left corner in Figure 5a. In addition, an enlarged image of the packaged SiC micro-heater chip circuit is shown in Figure 5b. When the power is supplied, the heat transfer path is shown in Figure 6a, where each layer has a thermal resistance. Here, since all materials except ceramic are the same, input power dependent thermal resistance performance can be directly compared according to ceramic type. The results of thermal resistance performance are shown in Figure 6b. The thermal resistance of the SiC micro-heater chip power modules on the DBC-Al_2O_3 substrate at power over 200 W was about twice higher than DBC-Si_3N_4 and also higher than DBC-AlN. The thermal resistance of DBC-AlN was lower than that of DBC-Si_3N_4. In addition, thermal resistance of the SiC micro-heater chip power modules on the DBC-Al_2O_3 substrate tended to increase thermal resistance significantly with increasing power but there is almost not so much change for the DBC-AlN and DBC-Si_3N_4 with the increased power. The reason of the lowest thermal resistance of AlN is that the thermal conductivity of AlN (285 $W{\cdot}m^{-1}{\cdot}K^{-1}$, for single crystals) is significantly higher than Al_2O_3 (25 $W{\cdot}m^{-1}{\cdot}K^{-1}$) and Si_3N_4 (60–90 $W{\cdot}m^{-1}{\cdot}K^{-1}$) [21,25]. Therefore, AlN-DBC can be considered to have the best heat dissipation performance of an excellent power module when applied properly in the area where heat dissipation is intensively required.

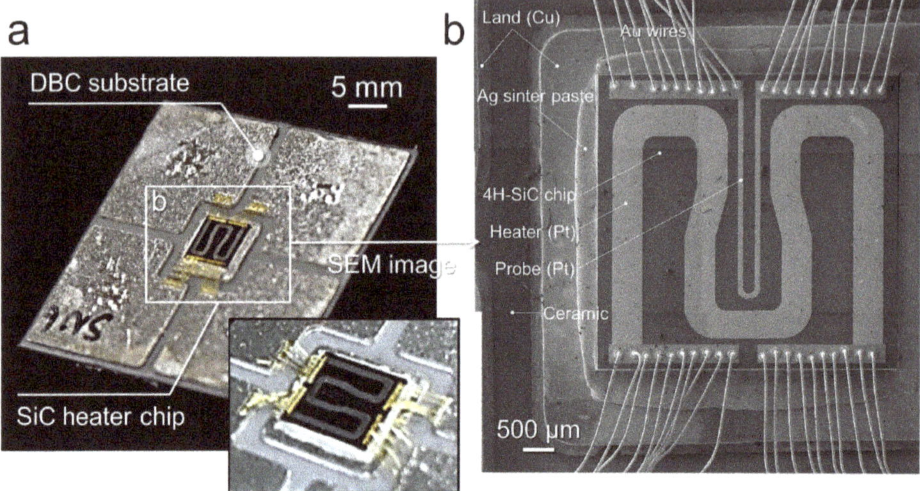

Figure 5. Test specimen description: (**a**) interconnected SiC heater chip on direct bonded copper (DBC) substrate and (**b**) its SEM image of detail components.

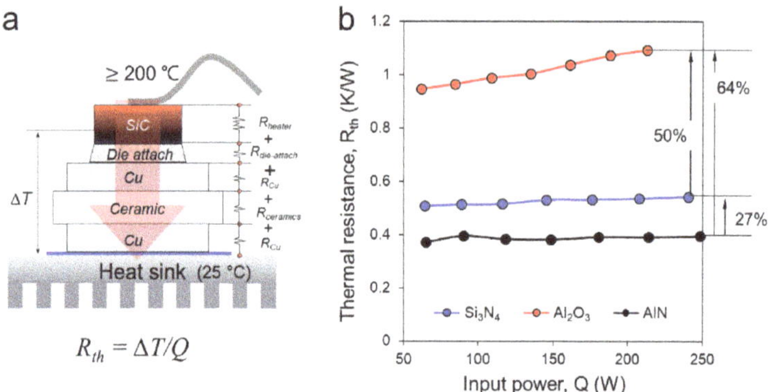

Figure 6. (**a**) Schematic diagram of heat path and its thermal resistance measurement during power cycling (**b**) results of various ceramics of thermal resistance.

3.2. Power Cycling Test

For the power cycling test, the temperature of the cooling plate was constantly controlled to be 50 °C. The water flow rate was 5.4 L/min. The constant current for heating were tested, and constant current is necessary to suppress the surge spike when heating was started. The typical temperature profile of the power cycling is plotted in Figure 7. The input electric current in the test was 2.1 A. The condition of power cycling test was applied 2 s ON and 5 s OFF at the input power of 200 W.

The results indicated that Al_2O_3 was stopped during the power cycle test due to excessive rise in thermal resistance before 1000 cycles, Si_3N_4 and AlN specimens were each subjected to more than 20,000 cycles of power cycle tests. In the case of Al_2O_3 stopping, die-attach and ceramic component were not any degradation but the Pt heater pattern line was partially broken as shown in Figure 8a due to the poor heat dissipation performance of Al_2O_3. The Pt heater pattern line accumulated a lot of heat, leading to the temperature increase significantly, leading to the film of pattern line partially melted as shown in Figure 8b.

Figure 9a shows temperature swing behavior of the Si_3N_4-DBC substrate, the heater temperature at the beginning was about 240 °C, and when it reached 20,000 cycles, the temperature was about 260 °C. In the case of AlN-DBC substrate, the starting temperature was about 220 °C, and the temperature after 20,000 cycles was about 280 °C (see Figure 9b). The temperature ranges of Si_3N_4 and AlN are not significantly different, but the temperature change tendency was remarkable. Unlike Si_3N_4 substrate, which has a small change in temperature, AlN was largely divided into two stages. The stage 1 is from the starting point to 4000 cycles, it reached up to saturation temperature. The stage 2 is from 4000 cycles to 20,000 cycles, at a relatively stable temperature, the temperature slowly reached its peak point.

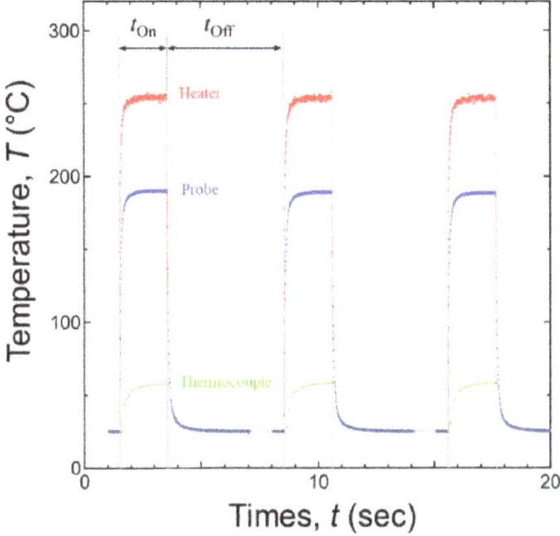

Figure 7. Temperature swing during power cycling tests.

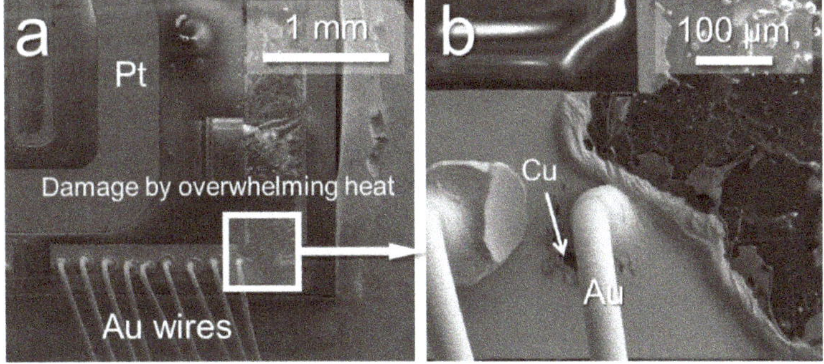

Figure 8. Pt heater pattern line was partly broken (**a**) and its magnified view (**b**).

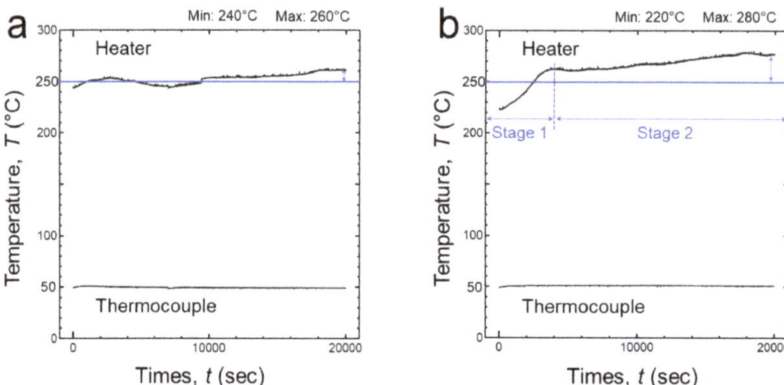

Figure 9. Temperature swing behavior during power cycling tests: (**a**) Si_3N_4 and (**b**) AlN.

To investigate a factor of different temperature behavior, a nondestructive test was performed by a micro-focus 3D computed tomography (CT) X-Ray system (XVA-160N, Uni-Hite System Corporation, Yamato City, Japan) on the specimens after power cycling tests. Figure 10 displays nondestructive test results (X-ray) after power cycling tests after power cycling tests (20,000 cycles). Figure 10a,b show the SiC micro-heater chip power modules on the DBC-Si_3N_4 substrate and its high magnification image, respectively. Figure 10b,c the SiC micro-heater chip power modules on the DBC-Si_3N_4 substrate of AlN and its high magnification image Figure 10d, respectively. There was no relationship between the different temperature behavior of the two substrates and the substantial damage to the die attach structure. Therefore, no delamination of the ceramic or degradation of the Ag sinter joint occurred during power cycling tests, and it was clearly found that the temperature rise did not cause physical failure. Despite the excellent thermal conductivity of AlN, the cause of temperature rise shown in Figure 9b can be regarded as an issue of the TIM. The mechanism of air entrainment into the TIM are shown in Figure 11. In the initial state, the DBC substrate and the water cooling plate were closely adhered by TIM (see Figure 11a), after experiencing expansion and shrinkage, there should be some gaps generation between the TIM and substrate, the adhesion strength changed to lower with a little gap generation as displayed in Figure 11b,c. Air entrainment into the TIM also can lead to the temperature swing increase.

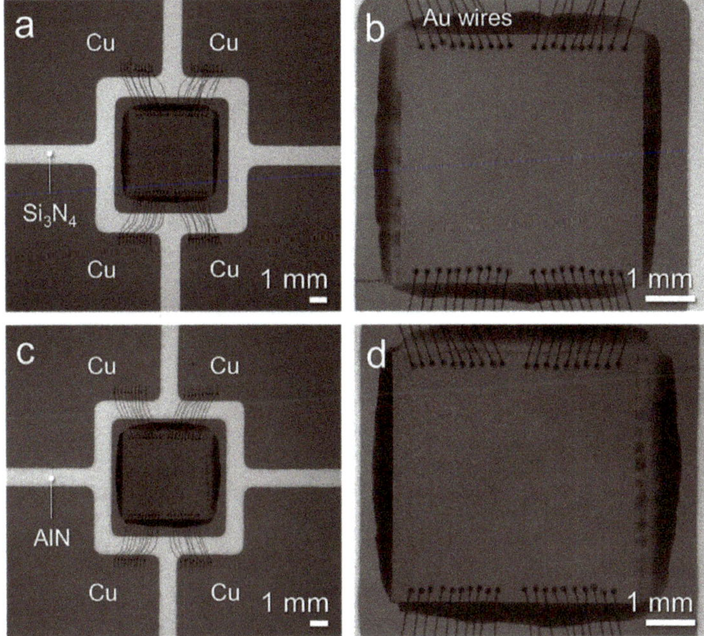

Figure 10. Nondestructive test results (X-ray) after power cycling tests (20,000 cycles): (**a**) specimen of Si_3N_4 and (**b**) its high magnification image, (**c**) specimen of AlN and (**d**) its high magnification image.

Thus, to avoid this phenomenon, it can be prevented by using a soft TIM or warpage suppression mechanism that does not harden under repeated substrate warpage behavior. The results of the power cycling test using the SiC micro-heater chip show the heat resistance performance of the DBC substrate, are significantly different from the traditional thermal shock test. Consequently, the AlN-DBC substrates may have the best heat dissipation and durability, perhaps by optimizing TIM performance for WBG power conversion systems.

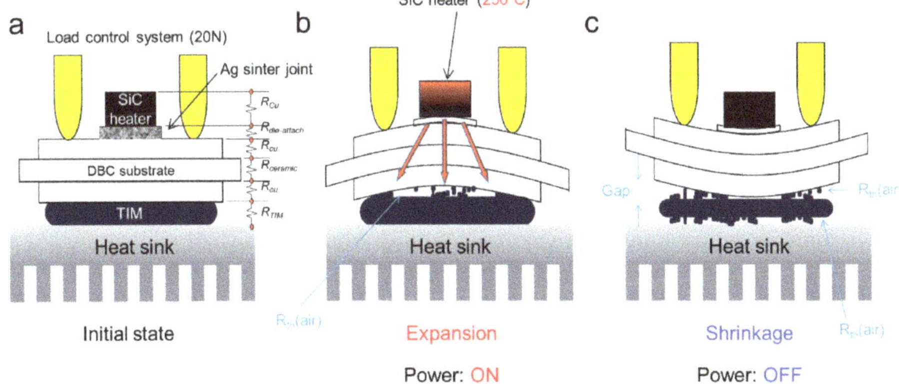

Figure 11. Schematic diagram of AlN temperature rise mechanism by mechanical behavior during power cycling test: (**a**) initial state (**b**) expansion (**c**) shrinkage.

4. Conclusions

In this study, a power cycling test system based on a SiC micro-heater chip was developed to simulate the active SiC devices where the temperature of the heater chip can be recorded through the power cycling test in real-time. The new approach is useful for the evaluation of high-power devices with a similar condition to the real packaging system. The thermal resistance of of the SiC micro-heater chip power modules on DBC-AlN was lower than that of DBC-Al_2O_3 and DBC-Si_3N_4. In addition, thermal resistance of the SiC micro-heater chip power modules on the DBC-Al_2O_3 substrate tended to increase thermal resistance significantly with increasing power but there is almost not so much change for the DBC-AlN and DBC-Si_3N_4 with the same increased power. Furthermore, no delamination of the ceramic or degradation of the Ag sinter joint occurred during power cycling tests for both DBC-AlN and DBC-Si_3N_4, and it was clearly found that the temperature rise did not cause physical failure. The results exhibit that both DBC-AlN and DBC-Si_3N_4 are excellent candidates for high power modules with WBG semiconductor devices, suitable for industries demanding high energy efficiency, when the packaging like the die-attach and the cooling system is well optimized.

Author Contributions: Conceptualization, S.N. and K.S.; methodology, Y.Y.; software, S.N.; validation, C.C.; formal analysis, N.W., D.K.; investigation, D.K.; writing—original draft preparation, D.K.; writing—review and editing, C.C. and S.N.; visualization, D.K., C.C. and S.N.; supervision, K.S.; project administration, S.N. and K.S.; funding acquisition, S.N. and K.S.

Funding: The present study is supported by the METI project: Standardization for Energy Conservation as International Standardization of Thermal Property Measurements for Next Generation Power Electronics. The authors are grateful to all the members of the standardization project. The present study is supported by Yamato Scientific. Co. Ltd.

Conflicts of Interest: The authors declare no conflict of interest.

References

1. Gao, Y.; Zhang, H.; Li, W.; Jiu, J.; Nagao, S.; Sugahara, T.; Suganuma, K. Die bonding performance using bimodal Cu particle paste under different sintering atmospheres. *J. Electron. Mater.* **2017**, *46*, 4575–4581. [CrossRef]
2. Wong, K.Y.; Chen, W.; Liu, X.; Zhou, C.; Chen, K.J. GaN smart power IC technology. *Phys. Status Solidi Basic Res.* **2010**, *247*, 1732–1734. [CrossRef]
3. Roccaforte, F.; Giannazzo, F.; Iucolano, F.; Eriksson, J.; Weng, M.H.; Raineri, V. Surface and interface issues in wide band gap semiconductor electronics. *Appl. Surf. Sci.* **2010**, *256*, 5727–5735. [CrossRef]
4. Kim, D.; Chen, C.; Pei, C.; Zhang, Z.; Nagao, S.; Suetake, A.; Sugahara, T.; Suganuma, K. Thermal shock reliability of a GaN die-attach module on DBA substrate with Ti/Ag metallization by using micron/submicron Ag sinter paste. *Jpn. J. Appl. Phys.* **2019**, *58*, SBBD15. [CrossRef]
5. Noh, S.; Choe, C.; Chen, C.; Suganuma, K. Heat-resistant die-attach with cold-rolled Ag sheet. *Appl. Phys. Express* **2018**, *11*, 016501. [CrossRef]
6. Kunimune, T.; Kuramoto, M.; Ogawa, S.; Sugahara, T.; Nagao, S.; Suganuma, K. Ultra thermal stability of LED die-attach achieved by pressureless Ag stress-migration bonding at low temperature. *Acta Mater.* **2015**, *89*, 133–140. [CrossRef]
7. Millan, J.; Godignon, P.; Perpina, X.; Perez-Tomas, A.; Rebollo, J. A survey of wide bandgap power semiconductor devices. *IEEE Trans. Power Electron.* **2014**, *29*, 2155–2163. [CrossRef]
8. Noh, S.; Choe, C.; Chen, C.; Zhang, H.; Suganuma, K. Printed wire interconnection using Ag sinter paste for wide band gap power semiconductors. *J. Mater. Sci. Mater. Electron.* **2018**, *29*, 15223–15232. [CrossRef]
9. Pérez-Tomás, A.; Fontserè, A.; Placidi, M.; Jennings, M.R.; Gammon, P.M. Modelling the metal–semiconductor band structure in implanted ohmic contacts to GaN and SiC. *Model. Simul. Mater. Sci. Eng.* **2013**, *21*, 035004. [CrossRef]
10. Hung, T.Y.; Chiang, S.Y.; Huang, C.J.; Lee, C.C.; Chiang, K.N. Thermal-mechanical behavior of the bonding wire for a power module subjected to the power cycling test. *Microelectron. Reliab.* **2011**, *51*, 1819–1823. [CrossRef]

11. Chen, C.; Suganuma, K.; Iwashige, T.; Sugiura, K.; Tsuruta, K. High-temperature reliability of sintered microporous Ag on electroplated Ag, Au, and sputtered Ag metallization substrates. *J. Mater. Sci. Mater. Electron.* **2018**, *29*, 1785–1797. [CrossRef]
12. Park, S.; Nagao, S.; Kato, Y.; Ishino, H.; Sugiura, K.; Tsuruta, K.; Suganuma, K. High-temperature die attachment using Sn-plated Zn solder for power electronics. *IEEE Trans. Compon. Packag. Manuf. Technol.* **2015**, *5*, 902–909. [CrossRef]
13. Yu, F.; Cui, J.; Zhou, Z.; Fang, K.; Johnson, R.W.; Hamilton, M.C. Reliability of Ag sintering for power semiconductor die attach in high-temperature applications. *IEEE Trans. Power Electron.* **2017**, *32*, 7083–7095. [CrossRef]
14. Chen, C.; Suganuma, K. Microstructure and mechanical properties of sintered Ag particles with flake and spherical shape from nano to micro size. *Mater. Des.* **2019**, *162*, 311–321. [CrossRef]
15. Zhang, Z.; Chen, C.; Yang, Y.; Zhang, H.; Kim, D.; Sugahara, T.; Nagao, S.; Suganuma, K. Low-temperature and pressureless sinter joining of Cu with micron/submicron Ag particle paste in air. *J. Alloys Compd.* **2019**, *780*, 435–442. [CrossRef]
16. Park, S.; Nagao, S.; Sugahara, T.; Suganuma, K. Mechanical stabilities of ultrasonic Al ribbon bonding on electroless nickel immersion gold finished Cu substrates. *Jpn. J. Appl. Phys.* **2014**, *53*, 2–9. [CrossRef]
17. Park, S.; Nagao, S.; Sugahara, T.; Suganuma, K. Heel crack propagation mechanism of cold-rolled Cu/Al clad ribbon bonding in harsh environment. *J. Mater. Sci. Mater. Electron.* **2015**, *26*, 7277–7289. [CrossRef]
18. Jiu, J.; Zhang, H.; Nagao, S.; Sugahara, T.; Kagami, N.; Suzuki, Y.; Akai, Y.; Suganuma, K. Die-attaching silver paste based on a novel solvent for high-power semiconductor devices. *J. Mater. Sci.* **2016**, *51*, 3422–3430. [CrossRef]
19. Lindemann, A.; Strauch, G. Properties of direct aluminium bonded substrates for power semiconductor components. *PESC Rec. IEEE Annu. Power Electron. Spec. Conf.* **2004**, *6*, 4171–4177.
20. Xu, L.; Wang, M.; Zhou, Y.; Qian, Z.; Liu, S. An optimal structural design to improve the reliability of Al_2O_3-DBC substrates under thermal cycling. *Microelectron. Reliab.* **2016**, *56*, 101–108. [CrossRef]
21. Fei, M.; Fu, R.; Agathopoulos, S.; Fang, J.; Wang, C.; Zhu, H. A preparation method for Al/AlN ceramics substrates by using a CuO interlayer. *Mater. Des.* **2017**, *130*, 373–380. [CrossRef]
22. Lin, C.Y.; Tuan, W.H. Direct bonding of aluminum to alumina for thermal dissipation purposes. *Int. J. Appl. Ceram. Technol.* **2016**, *13*, 170–176. [CrossRef]
23. Choe, C.; Chen, C.; Noh, S. Thermal shock performance of DBA/AMB substrates plated by Ni and Ni—P layers for high-temperature applications of power device modules. *Materials* **2018**, *11*, 2394. [CrossRef] [PubMed]
24. Chen, C.; Nagao, S.; Suganuma, K.; Jiu, J.; Sugahara, T.; Zhang, H.; Iwashige, T.; Sugiura, K.; Tsurutab, K. Macroscale and microscale fracture toughness of microporous sintered Ag for applications in power electronic devices. *Acta Mater.* **2017**, *129*, 41–51. [CrossRef]
25. Ning, X.S.; Lin, Y.; Xu, W.; Peng, R.; Zhou, H.; Chen, K. Development of a directly bonded aluminum/alumina power electronic substrate. *Mater. Sci. Eng. B Solid State Mater. Adv. Technol.* **2003**, *99*, 479–482. [CrossRef]

Article

Proton Irradiation Effects on the Time-Dependent Dielectric Breakdown Characteristics of Normally-Off AlGaN/GaN Gate-Recessed Metal-Insulator-Semiconductor Heterostructure Field Effect Transistors

Dongmin Keum and Hyungtak Kim *

School of Electronic and Electrical Engineering, Hongik University, Seoul 04066, Korea; rmaehdalf@hanmail.net
* Correspondence: hkim@hongik.ac.kr; Tel.: +82-2-320-3013

Received: 14 September 2019; Accepted: 23 October 2019; Published: 26 October 2019

Abstract: In this work, we investigated the time-dependent dielectric breakdown (TDDB) characteristics of normally-off AlGaN/GaN gate-recessed metal–insulator–semiconductor (MIS) heterostructure field effect transistors (HFETs) submitted to proton irradiation. TDDB characteristics of normally-off AlGaN/GaN gate-recessed MISHFETs exhibited a gate voltage (V_{GS}) dependence as expected and showed negligible degradation even after proton irradiation. However, a capture emission time (CET) map and cathodoluminescence (CL) measurements revealed that the MIS structure was degraded with increasing trap states. A technology computer aided design (TCAD) simulation indicated the decrease of the vertical field beneath the gate due to the increase of the trap concentration. Negligible degradation of TDDB can be attributed to this mitigation of the vertical field by proton irradiation.

Keywords: AlGaN/GaN; proton irradiation; time-dependent dielectric breakdown (TDDB); reliability; normally off

---

1. Introduction

Semiconductor technology used in satellites or exploration robots in harsh environments is mainly based on silicon semiconductor technology, but uses modules for heat dissipation, the hermetic structure, and the shielding structure for extreme environments, such as high temperature and radiation; however, these modules are generally heavy and complicated parts. This burden can be relieved if robust semiconductor materials can be employed in the electronics used for harsh environments. Among attractive candidates, AlGaN/GaN heterostructure field effect transistors (HFETs) are attracting much intention as power switching devices for harsh environmental applications thanks to GaN's superior radiation resistance [1]. Recently, studies on the radiation characteristics of GaN-based transistors have been widely conducted [2–4]. Especially, studies on the irradiation effects of protons occupying the majority of low earth orbits (LEO) were carried out [5–7]. In general, AlGaN/GaN HFETs irradiated with protons exhibit a positive shift in the threshold voltage (V_{th}) and a reduction in the drain current (I_{DS}), which can be attributed to the displacement damage near the two-dimensional electron gas (2-DEG) [8,9]. Exceptionally, the improvement of carrier concentration also has been reported at a relatively low dose [10]. In this paper, a sufficiently high dose (5×10^{14} cm^{-2}) was used to deteriorate the irradiated devices and the irradiated devices followed the generally reported results [11].

The main advantage of an AlGaN/GaN heterostructure is a natural formation of a 2-DEG channel without intentional doping [12], which leads to high mobility with a high sheet charge density [13].

Therefore, AlGaN/GaN HFETs inherently operate as normally-on devices. However, for the circuit configuration and stable operation, it is essential to implement a normally-off operation [14]. A gate-recessed metal–insulator–semiconductor (MIS) structure was employed to realize a normally-off operation and has exhibited a stable V_{th} over 1 V and a low gate leakage [15,16]. Gate reliability has been one of the critical issues of AlGaN/GaN HFETs and can be aggravated in gate-recessed MIS structures due to the processes of AlGaN barrier etching and insulator deposition [17–19]. In this study, we fabricated normally-off AlGaN/GaN MISHFETs by using a gate-recessed MIS structure and investigated the effects of proton irradiation on the gate reliability of normally-off AlGaN/GaN gate-recessed MISHFETs.

2. Experimental

AlGaN/GaN-on-Si wafer was provided by Enkris, Suzhou, China. As presented in the data sheet from Enkris, the epitaxial structure, grown using metal organic chemical vapor deposition (MOCVD) (AIXTRON, Herzogenrath, Germany), consisted of a 10-nm in situ SiN_x layer, a 4-nm undoped GaN capping layer, a 23-nm undoped $Al_{0.23}Ga_{0.77}N$ barrier, and 5-μm undoped-GaN buffer layer on a GaN-on-Si substrate. A cross-sectional diagram of the fabricated MISHFET is shown in Figure 1.

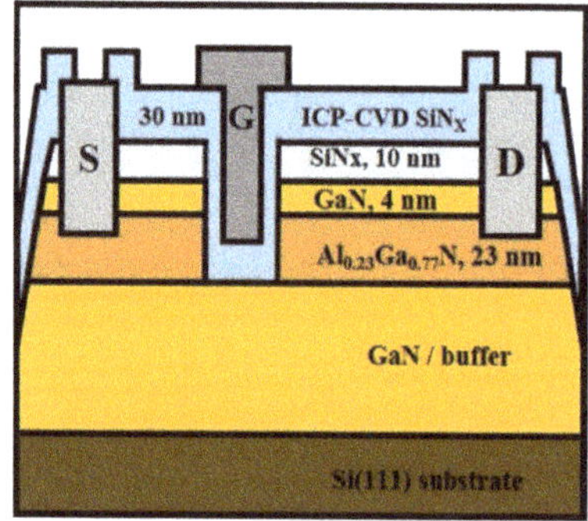

Figure 1. The cross-sectional view of the fabricated device. (S = Source, G = Gate, D = Drain, ICP-CVD = Inductively coupled plasma-chemical vapor deposition).

The fabrication process was as follows. Ohmic contacts were formed using an e-beam evaporated Ti/Al/Ni/Au (20/120/25/50 nm) metal stack and alloyed via rapid thermal annealing at 830 °C for 30 s. After the ohmic process, the mesa isolation and gate recess followed. The AlGaN barrier was fully recessed using inductively coupled plasma–reactive ion etching (ICP-RIE) (BMR Technology Corporation, Placentia, CA, USA) with a power of 5 W and a Cl_2/BCl_3 ambient atmosphere. A 30-nm SiN_X layer was deposited using ICP chemical vapor deposition (ICP-CVD) (BMR Technology Corporation, Placentia, CA, USA) using SiH_4/NH_3 gas at 350 °C, and a Ni/Au (40/350 nm) metal stack was evaporated for the gate contact. The gate length/width, gate-to-drain distance, and gate-to-source distance were 2/100 μm, 15 μm, and 3 μm, respectively.

TDDB characteristics were measured using the gate voltages of 13, 13.5, and 14 V at the temperature of 150 °C. Proton irradiation was carried out using a MC-50 cyclotron (Scanditronix, Vislanda, Sweden) at the Korea Institute of Radiological and Medical Sciences (KIRAMS) with an energy of 5 MeV, and a total fluence of 5×10^{14} cm^{-2} was chosen to deteriorate the irradiated devices. Proton irradiation was performed at room temperature. Electrical characteristics and cathodoluminescence (CL) were measured using a Agilent 4155A semiconductor parameter analyzer (Agilent Technologies, Santa Clara, CA, USA) and a JXA-8530F (JEOL Ltd., Tokyo, Japan), respectively.

3. Results and Discussion

Figure 2a shows a representative result of the time-zero breakdown (TZB) and TDDB characteristics with V_{GS} = 13 V at 150 °C. For TDDB measurements, we used the constant voltage method. In the constant voltage method, the gate voltages close to TZB were applied and the gate current (I_{GS}) was measured periodically. The gate current typically decreased before the time-dependent breakdown. I_{GS} increased after certain period of time and this time was defined as the time to breakdown (t_{BD}) [20]. In order to investigate the TDDB characteristics of normally-off AlGaN/GaN gate-recessed MISHFETs, we carried out constant voltage stress tests with gate voltages of 13, 13.5, and 14 V. Figure 2b shows the relationship between the TDDB and V_{GS} before and after the proton irradiation. TDDB characteristics showed an almost negligible change, although the devices were irradiated with protons.

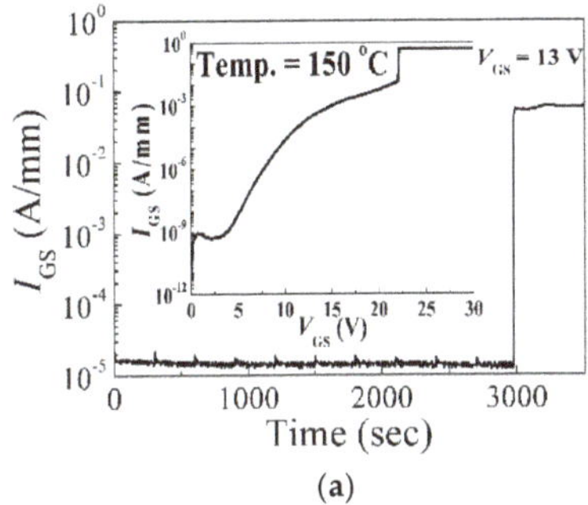

(a)

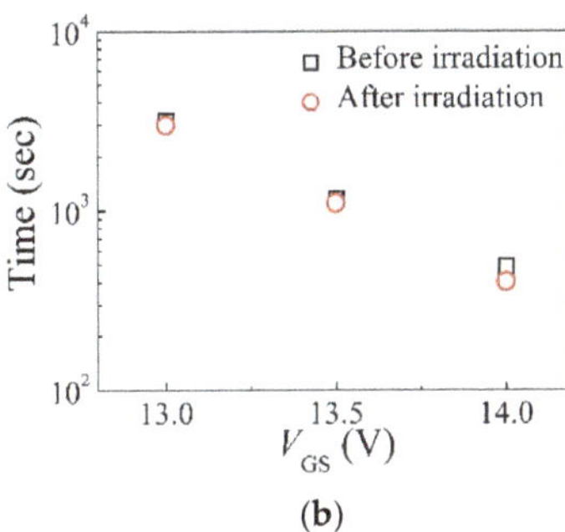

(b)

Figure 2. (**a**) The results of time-dependent dielectric breakdown (TDDB) characteristics carried out on one representative device (inset) time-zero breakdown (TZB) characteristics. (**b**) The V_{GS} dependence of TDDB characteristics at 150 °C with V_{GS} = 13, 13.5, and 14 V.

In order to investigate a broad distribution of overall traps through the gate region, which are widely believed to be applicable to GaN HFETs, capture emission time (CET) maps [21–23] were extracted using a stress–recovery sequence before and after proton irradiation. A CET map can be constructed from the shift of the I–V or Capacitance -voltage (C–V) characteristics. Every defect has a capture time (τ_C) and emission time (τ_e) during stress and recovery. Empty defects are charged and charged defects will emit its electron after the stress and recovery times of τ_C and τ_e, respectively. This behavior of defects can be described using the τ-axis and is called a CET map. A typical measurement procedure of CET maps is as follows. A positive voltage is assigned to the gate to trap electrons in the interface between the insulator and GaN, and then the time-dependent recovery characteristics are observed. During the stress sequence, traps with a capture time constant smaller than the stress time are occupied. During the recovery sequence, traps with an emission time constant smaller than the recovery time are released. According to this scheme, CET maps can extract traps with a specific capture time and recovery time by repeating the time-based stress-recovery experiments. From the V_{th} variation value obtained through the stress–recovery experiments, the following Equation (1) is used to obtain the overall interface trap level (N_{it}) of the gate region:

$$N_{it} = \frac{\epsilon_0 \epsilon_d}{q} \frac{\Delta V_{th}}{t_d} \quad (1)$$

where ϵ_0 and ϵ_d are the dielectric constants of air and insulator, respectively, and t_d is the thickness of the insulator. N_{it} was derived from $Q = C{\cdot}\Delta V$. The bias stress instability was induced by the 8 V of V_{GS}. V_{th} was extracted at the point where the drain current of 100 μA/mm flowed into the transfer curve with V_{DS} = 1 V (linear region). The V_{th} variation through the typical stress and recovery experiments for this analysis are summarized in Figure 3. ΔV_{th} increased with stress time in the stress plot and decreased with recovery time in the recovery plot for several extraction points. The squares of CET maps represent the behavior of N_{it} with each capture time and recovery time. CET maps obtained from the V_{th} variation and Equation (1) before and after proton irradiation are described in Figure 4. The overall darkening of the squares after proton irradiation qualitatively indicates the increase of trap states under the gate region. Further investigation needs to be carried out to understand the more pronounced increase in the traps within a certain time window. Figure 5 shows the change of ΔV_{th} and N_{it} as the stress time increased up to 1000 s with a $V_{\text{GS}} = 8$ V before and after the proton irradiation. N_{it} was increased from 5.6×10^{11} cm^{-2} to 1.6×10^{12} cm^{-2} at the stress time of 100 s with the V_{GS} of 8 V. The values of N_{it} obtained from this experiment are comparable with and even lower than those in the literature with/without a gate-recessed structure [5,11,24–26]. ΔV_{th} and N_{it} increased after proton irradiation. Table 1 summarizes the correlation of the interface traps states (N_{it} or D_{it}) and proton irradiation on GaN-based transistors. D_{it} has the same meaning with N_{it} at the specific energy of trap levels.

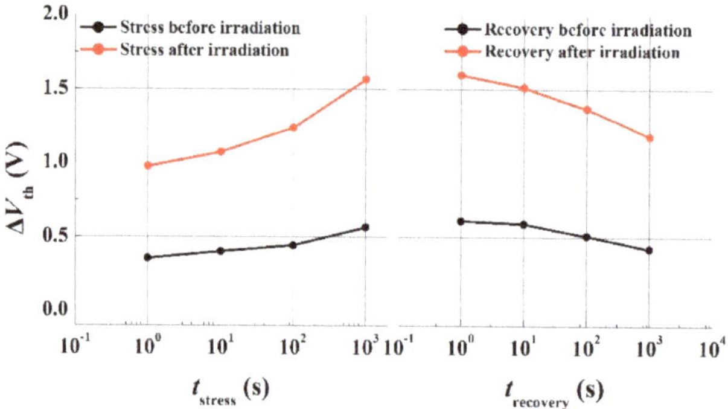

Figure 3. Stress and recovery behavior of normally-off AlGaN/GaN gate-recessed metal–insulator–semiconductor heterostructure field effect transistors (MISHFETs) up to 1000 s with a $V_{\text{GS}} = 8$ V before and after proton irradiation.

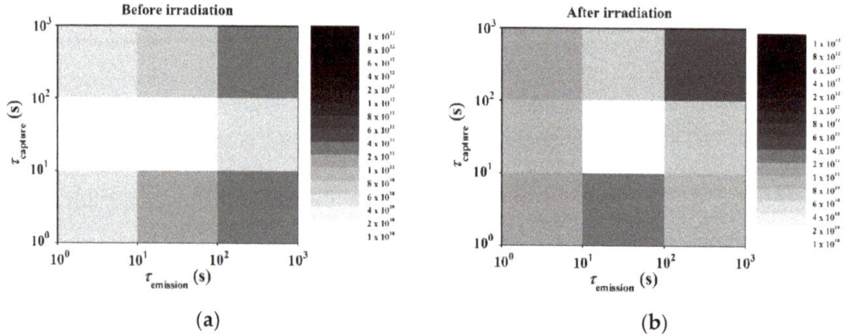

Figure 4. Capture emission time (CET) maps in normally-off AlGaN/GaN gate-recessed MISHFETs: (**a**) before proton irradiation and (**b**) after proton irradiation.

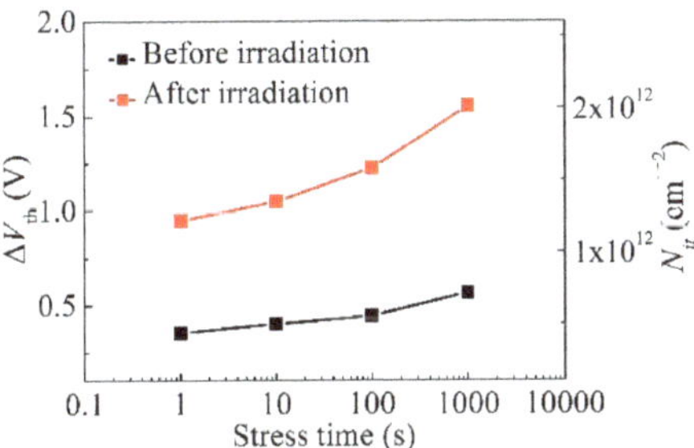

Figure 5. ΔV_{th} and N_{it} drift over stress time up to 1000 s measured using a V_{GS} of 8 V before and after proton irradiation.

Table 1. The correlation between the interface traps states and the proton irradiation on GaN-based transistors.

Operation Mode	N_{it} or D_{it} (cm^{-2} or cm^{-2}·eV^{-1})	Irradiation Dose (cm^{-2})	Irradiation Energy (MeV)	Reference
Normally Off	1.2×10^{12}–2×10^{12}	5×10^{14}	5	This work
Normally Off	1.1×10^{12}–6×10^{13}	5×10^{14}	5	[5]
Normally Off	1.3×10^{13}–2.6×10^{13}	5×10^{14}	5	[11]
Normally On	1.8×10^{12}–1.8×10^{13}	5×10^{14}	3	[25]
Normally On	1.4×10^{13}	10^{15}	5	[26]

The CL spectra of normally-off AlGaN/GaN gate-recessed MISHFETs shown in Figure 6a was measured to understand the proton irradiation effects on the optical properties before and after irradiation. The typical measurement procedure of CL is as follows. First, the electron beam is irradiated onto the target semiconductor. Then, the interaction of the electron beam with the target semiconductor results in the promotion of electrons from the valence band to the conduction band. When the promoted electron and hole recombine, the exposed semiconductor provides information about its optical property. This optical property can be collected using a retractable parabolic mirror. CL was measured in the access region between the gate and drain, as shown in Figure 6b. White circles indicate the points irradiated by the electron beam. The decay of CL intensity was decreased by 31.7% after the proton irradiation. This suggests that the trap states generated by the proton irradiation reduced the recombination of electron–hole pairs generated by the electron beam.

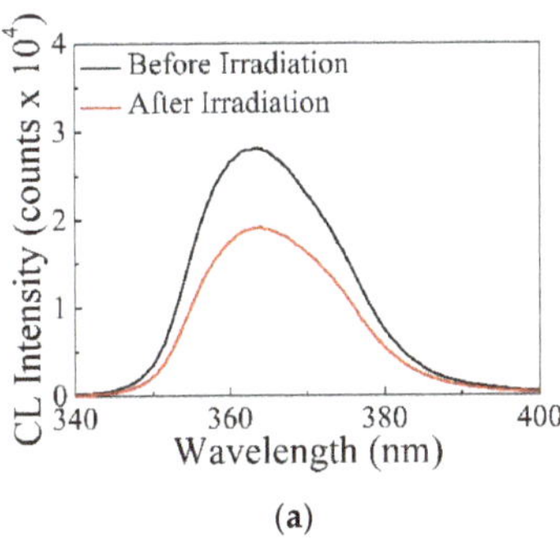

(a)

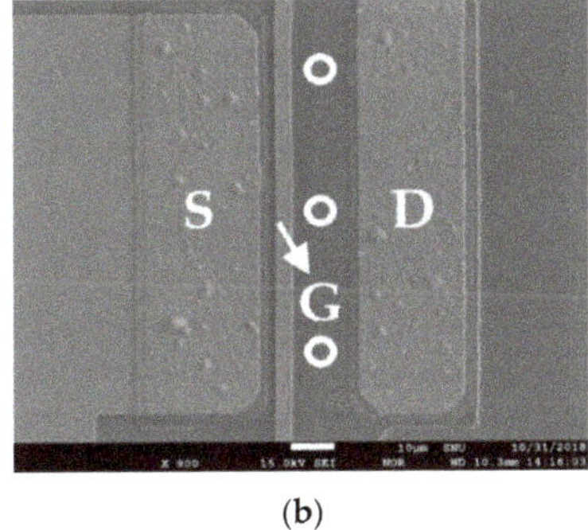

(b)

Figure 6. (**a**) Cathodoluminescence (CL) spectra of normally-off AlGaN/GaN gate-recessed MISHFETs before and after proton irradiation. (**b**) SEM image of the analyzed device.

The main degradation mechanism of TDDB characteristics is attributed to the breakdown of the gate dielectric around the gate overhang [27]. Therefore, we carried out a TCAD simulation using Silvaco ATLAS (Silvaco Atlas, Santa Clara, CA, USA) to profile the vertical electric field distribution of the normally-off AlGaN/GaN gate-recessed MISHFETs. Figure 7 shows the simulated transfer curves compared with the measured ones (a) and the vertical electric field within the gate dielectric under the gate for V_{GS} = 14 V (b) before and after the proton irradiation. Proton irradiation was applied to the simulation by employing negatively charged traps in accordance with Patrick et al. [28]. Proton irradiation can generate Ga and N vacancies in the irradiated devices via collisions. Ga vacancies can act as acceptor-like traps [29], and N vacancies can act as both acceptor- and donor-like traps [30]. These two vacancies can also compensate each other, but the quantitative analysis is still unclear. However, proton irradiation results in a V_{th} shift in the positive direction, which can infer that the acceptor-like traps (negatively charged traps) are dominant in the irradiated device. The volume density of the traps was calculated using stopping and range of ions in matter (SRIM) and its value was reported to be about the order of 10^{17} cm^{-3} [28,31,32]. Within GaN, there are pre-existing traps with various activation energies from shallow to deep level states. Proton irradiation increases the concentration of pre-existing traps and new trap states with different activation energies [33]. Therefore, the activation energies of the trap level applied to the TCAD simulation were distributed uniformly through the bandgap of GaN. As negatively charged traps were applied, the vertical electric field of the gate dielectric under the gate was significantly decreased by 83%. CET maps and CL spectra verified the deterioration of the irradiated devices, but it was also confirmed through the TCAD simulation that the trap states induced via proton irradiation reduced the vertical electric field of the dielectric under the gate region. It is presumed that TDDB characteristics negligibly changed, even after the proton irradiation, due to the offset of these two opposite effects. C–V measurements can provide useful information for understanding the defect level and should be analyzed in our future work.

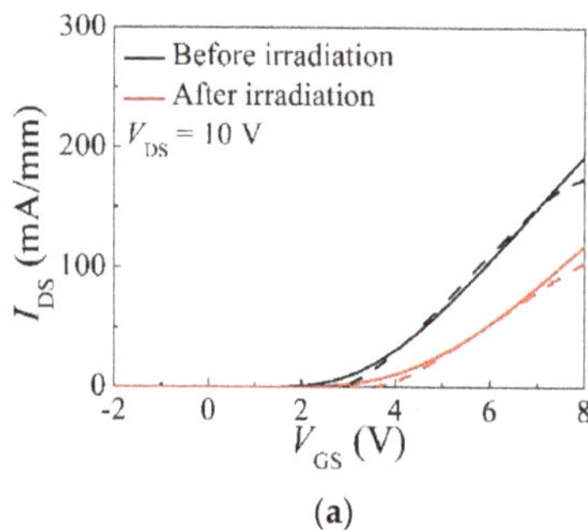

(a)

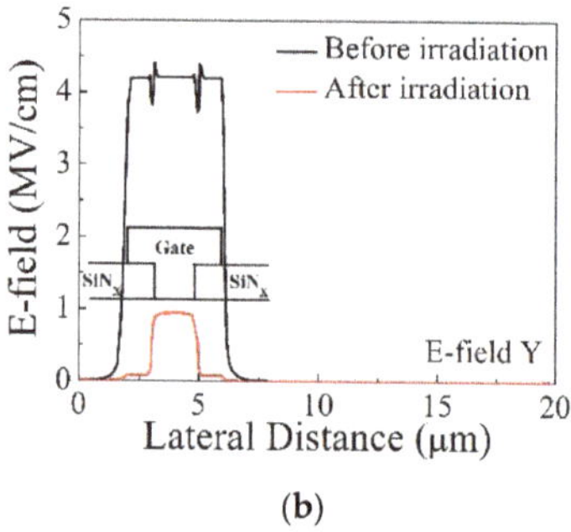

(b)

Figure 7. (**a**) The simulated transfer curves (dashed lines) compared with the measured ones (solid lines). (**b**) The vertical electric field distribution of the gate dielectric with V_{GS} = 14 V.

4. Conclusions

TDDB characteristics of normally-off AlGaN/GaN gate-recessed MISHFETs were investigated before and after proton irradiation. After proton irradiation, the irradiated devices exhibited the same V_{GS} dependence and a negligible change. Although the interface and trap states were deteriorated by proton irradiation, it was observed using a TCAD simulation that the vertical electric field under the gate was significantly reduced as the trap concentration increased. The field reduction via proton irradiation seemed to be linked to unchanged TDDB characteristics despite the deterioration of interface and trap states. Further investigation is needed to figure out the definite origin of the unchanged TDDB characteristics of normally-off AlGaN/GaN gate-recessed MISHFETs.

Author Contributions: Data curation, D.K.; Formal analysis, D.K.; writing—original draft, D.K.; project administration, H.K; supervision, H.K, writing—review and editing, H.K.

Funding: This work was supported by Korea Electric Power Corporation (R18XA02) and National Research Foundation of Korea Government (NRF-2019R1H1A2078240).

Acknowledgments: The authors express their sincere thanks to the staff of the MC-50 Cyclotron Laboratory (KIRAMS) for the excellent operation and their support during the experiment.

Conflicts of Interest: The authors declare no conflict of interest.

References

1. Hazdra, P.; Popelka, S. Radiation resistance of wide-bandgap semiconductor power transistors. *Phys. Status Solidi (a)* **2017**, *214*, 1600447. [CrossRef]
2. Mararo, J.; Nicolas, G.; Nhut, D.M.; Forestier, S.; Rochette, S.; Vendier, O.; Langrez, D.; Cazaux, J.; Feudale, M. GaN for space application: Almost ready for flight. *Int. J. Microw. Wirel. Technol.* **2010**, *2*, 121–133. [CrossRef]
3. Pearton, S.J.; Ren, F.; Patrick, E.; Law, M.E.; Polylakov, A.Y. Review-ionizing radiation damage effects on GaN devices. *ECS J. Solid State Sci. Technol.* **2016**, *5*, Q35–Q60. [CrossRef]
4. Chen, J.; Puzyrev, Y.S.; Jiang, R.; Zhang, E.X.; McCurdy, M.W.; Fleetwood, D.M.; Schrimpf, R.D.; Pantelides, S.T.; Arehart, A.R.; Ringel, S.A.; et al. Effects of applied bias and high field stress on the radiation response of GaN/AlGaN HEMTs. *IEEE Trans. Nucl. Sci.* **2015**, *62*, 2423–2430. [CrossRef]
5. Keum, D.M.; Cha, H.Y.; Kim, H. Proton bombardment effects on normally-off AlGaN/GaN-on-Si recessed MISHeterostructure FETs. *IEEE Trans. Nucl. Sci.* **2015**, *62*, 3362–3368. [CrossRef]
6. Lv, L.; Ma, X.; Zhang, J.; Bi, Z.; Liu, L.; Shan, H.; Hao, Y. Proton irradiation effects on AlGaN/AlN/GaN heterojunctions. *IEEE Trans. Nucl. Sci.* **2015**, *62*, 300–305. [CrossRef]
7. Kim, D.S.; Lee, J.H.; Yeo, S.; Lee, J.H. Proton irradiation effects on AlGaN/GaN HEMTs with different isolation methods. *IEEE Trans. Nucl. Sci.* **2018**, *65*, 579–582. [CrossRef]
8. Anderson, T.J.; Koehler, A.D.; Specht, P.; Weaver, B.D.; Greenlee, J.D.; Tadjer, M.J.; Hite, J.K.; Mastro, M.A.; Porter, M.; Wade, M.; et al. Failure mechanisms in AlGaN/GaN HEMTs irradiated with 2MeV protons. *ECS Trans.* **2015**, *66*, 15–20. [CrossRef]
9. Weaver, B.D.; Martin, P.A.; Boos, J.B.; Cress, C.D. Displacement damage effects in AlGaN/GaN high electron mobility transistors. *IEEE Trans. Nucl. Sci.* **2012**, *59*, 3077–3080. [CrossRef]
10. Lee, I.H.; Lee, C.; Choi, B.K.; Yun, Y.; Chang, Y.J.; Jang, S.Y. Proton-induced conductivity enhancement in AlGaN/GaN HEMT devices. *J. Korean. Phys. Soc.* **2018**, *72*, 920–924. [CrossRef]
11. Keum, D.; Kim, H. Proton-irradiation effects on charge trapping-related instability of normally-off AlGaN/GaN recessed MISHFETs. *J. Semicond. Technol. Sci.* **2019**, *19*, 214–219. [CrossRef]
12. Ambacher, O.; Smart, J.; Shealy, J.R.; Weimann, N.G.; Chu, K.; Murphy, M.; Schaff, W.J.; Eastman, L.F.L.; Dimitrov, R.; Wittmer, L.; et al. Two-dimensional electron gases induced by spontaneous and piezoelectric polarization charges in N- and Ga-face AlGaN/GaN heterostructures. *J. Appl. Phys.* **1999**, *85*, 3222–3233. [CrossRef]
13. Smorchkova, I.P.; Elsass, C.R.; Ibbetson, J.P.; Vetury, R.; Heying, B.; Fini, P.; Haus, E.; DenBaars, S.P.; Speck, J.S.; Mishra, U.K. Polarization-induced charge and electron mobility in AlGaN/GaN heterostructures grown by plasma-assisted molecular-beam epitaxy. *J. Appl. Phys.* **1999**, *86*, 4520–4526. [CrossRef]
14. Su, M.; Chen, C.; Rajan, S. Prospects for the application of GaN power devices in hybrid electric vehicle drive systems. *Semicond. Sci. Technol.* **2013**, *28*, 074012. [CrossRef]
15. Choi, W.; Seok, O.; Ryu, H.; Cha, H.Y.; Seo, K.S. High-voltage and low -leakage-current gate recessed normally-off GaN MIS-HEMTs with dual gate insulator employing PEALD-SiN_X/RF-Sputtered-HfO_2. *IEEE Electron Device Lett.* **2014**, *35*, 175–177. [CrossRef]
16. Park, B.R.; Lee, J.G.; Choi, W.; Kim, H.; Seo, K.S.; Cha, H.Y. High-quality ICPCVD SiO_2 for normally off AlGaN/GaN-on-Si recessed MOSHFETs. *IEEE Electron Device Lett.* **2013**, *34*, 354–356. [CrossRef]
17. Wu, T.L.; Marcon, D.; Jaeger, B.D.; Hove, M.V.; Bakeroot, B.; Stoffels, S.; Groeseneken, G.; Decoutere, S. Time dependent dielectric breakdown (TDDB) evaluation of PE-ALD SiN gate dielectrics on AlGaN/GaN recessed gate D-mode MIS-HEMTs and E-mode MIS-FETs. In Proceedings of the International Reliability Physics Symposium (IPRS), Monterey, CA, USA, 19–23 April 2015. [CrossRef]
18. Kim, H.S.; Eom, S.K.; Seo, K.S.; Kim, H.; Cha, H.Y. Time-dependent dielectric breakdown of recessed AlGaN/GaN-on-Si MOS-HFETs with PECVD SiO_2 gate oxide. *Vacuum* **2018**, *155*, 428–433. [CrossRef]

19. Wuerfl, J.; Bahat-Treidel, E.; Brunner, F.; Cho, E.; Hilt, O.; Ivo, P.; Knauer, A.; Kurpas, P.; Lossy, R.; Schulz, M.; et al. Reliability issues of GaN based high voltage power devices. *Microelectron. Reliab.* **2011**, *51*, 1710–1716. [CrossRef]
20. Schroder, D.K. Reliability and failure analysis. In *Semiconductor Material and Device Characterization*, 3rd ed.; John Wiley & Sons, Inc.: Hoboken, NJ, USA, 2015; pp. 689–740.
21. Lagger, P.; Ostermaier, C.; Pobegen, G.; Pogany, D. Towards understanding the origin of threshold voltage instability of AlGaN/GaN MISHEMTs. In Proceedings of the International Electron Devices Meeting (IEDM), San Francisco, CA, USA, 10–13 December 2012. [CrossRef]
22. Reisinger, H.; Grasser, T.; Gustin, W.; Schluandnder, C. The statistical analysis of individual defects constituting NBTI and its implications for modeling DC- and AC-stress. In Proceedings of the International Reliability Physics Symposium (IPRS), Anaheim, CA, USA, 2–6 May 2010. [CrossRef]
23. Ostermaier, C.; Lagger, P.; Reiner, M.; Pogany, D. Review of bias-temperature instabilities at the III-N/dielectric interface. *Microelectron. Reliab.* **2018**, *82*, 62–83. [CrossRef]
24. Wang, Y.; Wang, M.; Xie, B.; Wen, C.P.; Wang, J.; Hao, Y.; Wu, W.; Chen, K.J.; Shen, B. High-Performance normally-off Al_2O_3/GaN MOSFET using a wet etching-based gate recess technique. *IEEE Electron Device Lett.* **2013**, *34*, 1370–1372. [CrossRef]
25. Zheng, X.F.; Dong, S.S.; Ji, P.; Wang, C.; He, Y.L.; Lv, L.; Ma, X.H.; Hao, Y. Characterization of bulk traps and interface states in AlGaN/GaN heterostructure under proton irradiation. *Appl. Phys. Lett.* **2018**, *112*, 233504. [CrossRef]
26. Kim, B.J.; Ahn, S.; Ren, F.; Pearton, S.J.; Yang, G.; Kim, J. Effects of proton irradiation and thermal annealing on off-state step-stressed AlGaN/GaN high electron mobility transistors. *J. Vac. Sci. Technol. B* **2016**, *34*, 041231. [CrossRef]
27. Alamo, J.A.; Guo, A.; Warnock, S. Gate dielectric reliability and instability in GaN metal-insulator-semiconductor high-electron-mobility transistors for power electronics. *J. Mater. Res.* **2017**, *32*, 3458–3468. [CrossRef]
28. Patrick, E.; Law, M.E.; Liu, L.; Cuervo, C.V.; Xi, Y.; Ren, F.; Pearton, S.J. Modeling proton irradiation in AlGaN/GaN HEMTs: Understanding the increase of critical voltage. *IEEE Trans. Nucl. Sci.* **2013**, *60*, 4103–4108. [CrossRef]
29. Look, D.C. Defect-related donors, acceptors, and traps in GaN. *Phys. Status Solidi B* **2001**, *228*, 293–302. [CrossRef]
30. Ganchenkova, M.G.; Nieminen, R.M. Nitrogen vacancies as major point defects in gallium nitride. *Phys. Rev. Lett.* **2006**, *96*, 196402. [CrossRef]
31. Ziegler, J.F.; Ziegler, M.D.; Biersack, J.P. SRIM—the stopping and range of ions and matter (2010). *Nucl. Instrum. Methods Phys. Res. B* **2010**, *268*, 1818–1823. [CrossRef]
32. Lv, L.; Ma, J.G.; Cao, Y.R.; Zhang, J.C.; Zhang, W.; Li, L.; Xu, S.R.; Ma, X.H.; Ren, X.T.; Hao, Y. Study of proton irradiation effects on AlGaN/GaN high electron mobility transistors. *Microelectron. Reliab.* **2011**, *51*, 2168–2172. [CrossRef]
33. Zhang, Z.; Arehart, A.R.; Cinkilic, E.; Chen, J.; Zhang, E.X.; Fleetwood, D.M.; Schrimpf, R.D.; McSkimming, B.; Speck, J.S.; Ringel, S.A. Impact of proton irradiation on deep level states in n-GaN. *Appl. Phys. Lett.* **2013**, *103*, 042102. [CrossRef]

 micromachines

Article

InGaN/GaN Distributed Feedback Laser Diodes with Surface Gratings and Sidewall Gratings

Zejia Deng [1,2], Junze Li [1,2,*], Mingle Liao [1,2], Wuze Xie [1,2] and Siyuan Luo [1,2]

1 Microsystems and Terahertz Research Center, China Academy of Engineering Physics, Chengdu 610200, China; dengzejia@mtrc.ac.cn (Z.D.); liaomingle@mtrc.ac.cn (M.L.); xiewuze@mtrc.ac.cn (W.X.); luosiyuan@mtrc.ac.cn (S.L.)

2 Institute of Electronic Engineering, China Academy of Engineering Physics, Mianyang 621999, China

* Correspondence: lijunze@mtrc.ac.cn

Received: 15 September 2019; Accepted: 12 October 2019; Published: 14 October 2019

Abstract: A variety of potential applications such as visible light communications require laser sources with a narrow linewidth and a single wavelength emission in the blue light region. The gallium nitride (GaN)-based distributed feedback laser diode (DFB-LD) is a promising light source that meets these requirements. Here, we present GaN DFB-LDs that share growth and fabrication processes and have surface gratings and sidewall gratings on the same epitaxial substrate, which makes LDs with different structures comparable. By electrical pulse pumping, single-peak emissions at 398.5 and 399.95 nm with a full width at half maximum (FWHM) of 0.32 and 0.23 nm were achieved, respectively. The surface and sidewall gratings were fabricated alongside the p-contact metal stripe by electrical beam lithography and inductively coupled plasma etching. DFB LDs with 2.5 μm ridge width exhibit a smaller FWHM than those with 5 and 10 μm ridge widths, indicating that the narrow ridge width is favorable for the narrowing of the line width of the DFB LD. The slope efficiency of the DFB LD with sidewall gratings is higher than that of surface grating DFB LDs with the same ridge width and period of gratings. Our experiment may provide a reliable and simple approach for optimizing gratings and GaN DFB-LDs.

Keywords: GaN laser diode; distributed feedback (DFB); surface gratings; sidewall gratings

---

1. Introduction

Gallium nitride (GaN) laser diodes (LDs) are potentially used in displays, medical application, visible light communications (VLC), etc [1–4]. Some applications require a single peak and narrow-linewidth laser source. Distributed feedback lasers diodes (DFB-LDs) with these advantages have attracted extensive concern in both academia and the industry. In optical atomic clocks, an extremely narrow-linewidth blue laser is required to aim at atomic cooling transition [5]. In medical diagnostics, the DFB LD is a promising light source for fluorescence spectroscopy, where the emission wavelength can be precisely targeted [6].

GaN-based DFB LDs have been achieved by buried gratings [7–9], sidewall gratings, and surface gratings [10–13]. In the beginning, the first order DFB LD was achieved by establishing the buried gratings [14]. Subsequently, the third and 39th DFB LDs were fabricated by etching sidewall gratings [15–17]. Additionally, the single longitudinal mode emissions of the 10th DFB LDs with surface gratings were realized under optical pumping and electrical impulse driving [18,19]. In contrast to burying gratings, DFB LDs with sidewall and surface gratings do not require high-cost and hard crystal regrowth processing, preventing the device from being damaged in the regrowth process [20,21]. Surface gratings and sidewall gratings are preferred for DFB LDs due to simpler fabrication processes. Though there have been varieties of DFB LDs fabricated using surface gratings or sidewall gratings, to the best of our knowledge, all of them, regardless of grating type, are achieved on different epitaxial wafers; there are no reports on GaN DFB-LDs that share growth and fabrication processes and have surface gratings and sidewall gratings on the same epitaxial substrate. Therefore, it is inevitable that there are distinct differences in the optical and electrical properties of DFB LDs with different kinds of grating structures. It is difficult to systematically compare and analyze the performance of DFB LDs with surface and sidewall gratings, because they are fabricated on diverse epitaxial wafers. Hence, in order to further understand the performance differences of DFB LDs with different structures, it is necessary to fabricate DFB LDs with different types of gratings on the same epitaxial substrate, which can provide some experimental evidences for the optimal design of DFB LDs.

In this work, the finite-difference-time-domain (FDTD) tool was used to analyze the influence of the width of the ridge and the types of gratings on the properties of DFB LDs. Based on the designed structure of LDs, DFB-LDs with surface and sidewall gratings of different periods and ridge widths were fabricated on the same epitaxial wafer. Fabry–Pérot (F–P) LDs with the same ridge widths were also fabricated on the same epi-wafer for comparison. The well-proportioned gratings were defined alongside the p-contact metal stripe by electrical beam lithography (EBL) and inductively coupled plasma (ICP) etching. The morphology characteristics of all the LDs were observed by a scanning electron microscope (SEM) with an FEI Nova NanoSEM 450. The electrical and optical characteristics of the DFB LDs, including the linewidth, slope efficiency, and threshold current, were measured and analyzed.

2. Simulation and Fabrication

According to the coupled mode theory, the coupling coefficient can be achieved approximately by the following equation [22]:

$$\kappa = \frac{2(n_2 - n_1)}{\lambda_1}\frac{\sin(\pi\, m\, \gamma)}{m} \tag{1}$$

where m is the grating order, λ_1 is the wavelength, n_1 is the effective modal index in the broad area including the sections of the ridge and grating, n_2 is the effective modal index in the section of the ridge, and γ is the duty-cycle of gratings.

The effective modal index (n_{eff}) of the waveguide in a GaN LD can be calculated using the finite-difference-time-domain (FDTD), where the epitaxial structure and device structure of the simulation can be found in the parameters displayed in the following sections. Generally, there is a decreasing trend in modal index with ridge width at a certain etch depth. Typical optical field distributions are given in Figure 1, with ridge widths of 2.5 and 5 μm. It is obvious that a reduction of the waveguide width restriction the optical field more, resulting in a decreasing in mode number.

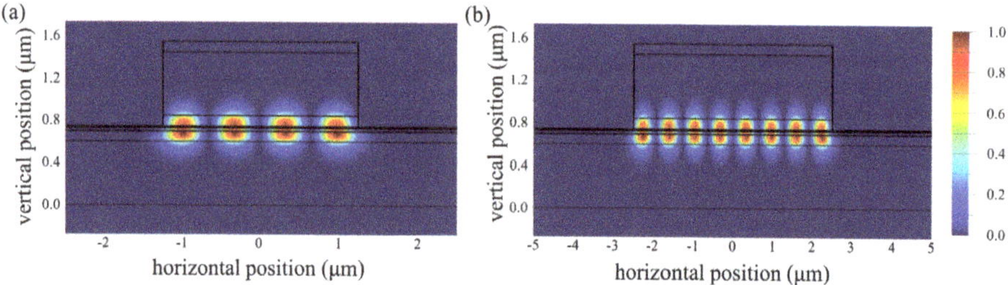

Figure 1. Optical field profiles for ridge widths of (**a**) 2.5 μm and (**b**) 5 μm.

The refractive modal index increases with ridge width and gradually approaches the maximum value when the ridge width is relatively large, which is shown in Figure 2a. It is worthy pointing out that coupling length kL has a great influence on the characteristics of the DFB LDs. Figure 2b shows the coupling length kL as a function of the ridge width when the etch depth was 500 nm and the cavity length was 600 μm. For the sidewall grating, the broad section including the ridge and grating had a width twice the size of the section of the ridge, whereas for the surface grating, the broad section possessed a fixed width of 80 μm. Obviously, *kL* showed a decreasing trend as the ridge width increased. It must be pointed out that this calculation was based on the fundamental mode; in the high order, mode *kL* had a relatively large value, leading to a moderate reflectivity. Since the surface gratings had a greater width (and thus refractive index) in the broad area for the same ridge width (compared with the sidewall gratings), they had a bigger refractive index difference $\Delta n = (n_2 - n_1)$. As demonstrated in Equation (1), the coupling length *kL* was proportional to refractive index difference Δn when the order and duty ratio of the gratings were defined. As such, the *kL* of the DFB LD with surface gratings was slightly higher than that of the one with sidewall gratings when the ridge widths of the two kinds of structures were identical.

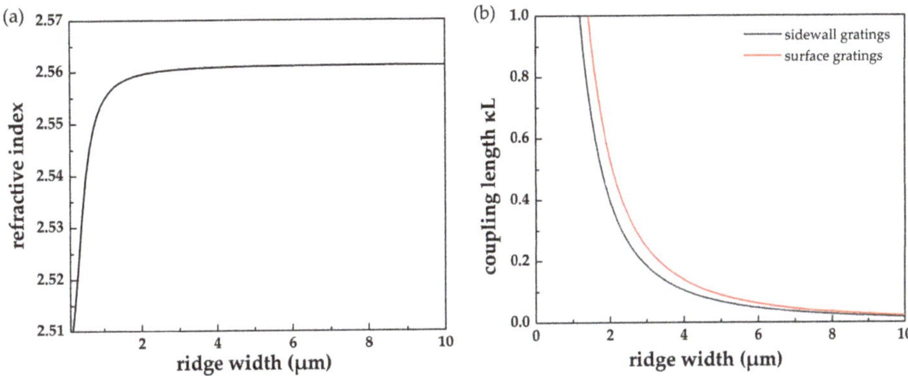

Figure 2. (**a**) Refractive index as a function of the ridge width. (**b**) Coupling length *kL* as a function of the ridge width of the sidewall and surface gratings. Etching depth: 500 nm.

According to the Bragg condition, $\Lambda = \lambda m/2n_{\text{eff}}$ and the characterization of the emission at around 403 nm of the F–P LDs fabricated on the epitaxial wafer, the periods of the 10th-order and 20th-order Bragg grating were calculated and found to be 824 and 1648 nm, respectively. Another parameter for the gratings is the duty cycle $\gamma = (\Lambda - o)/\Lambda$, where o denotes the width of the etched trench. A duty cycle of around 0.8 was adjusted for the gratings.

The designed complete structures of the DFB LDs with surface and sidewall gratings are depicted in Figure 3. It is worth noting that the surface gratings had a larger grating width than the sidewall grating and were much wider than the width of the ridge waveguide, while the width of the sidewall grating was closer to that of the ridge waveguide. The LDs were fabricated from the AlInGaN epi-structure composed of n-type and p-type epitaxial layers and InGaN/GaN multiple quantum wells (MQWs) designed for the emission of 403 nm, including a lower cladding layer of 850 nm $Al_{0.075}Ga_{0.925}N$:Si, a 60-nm-thick $In_{0.03}Ga_{0.97}N$:Si lower waveguide layer, a multiple quantum well with three 6.5- or 2-nm-thick undoped GaN quantum barriers and two 2.7 nm $In_{0.1}Ga_{0.9}N$ wells, an upper waveguide layer of doped 60 nm InGaN:Mg, a 20 nm $Al_{0.13}G_{0.87}N$:Mg electron blocking layer, a 450 nm $Al_{0.05}Ga_{0.95}N$:Mg upper cladding layer, and a 10 nm GaN:Mg subcontact layer. The F–P LDs and DFB LDs were fabricated on the same epi-wafer by the same manufacturing process except for the building of the surface and sidewall gratings of DFB LDs. The 2.5-, 5- and 10-μm-wide Ni/Au p-contact stripes with a thickness of 10/30 nm were formed on the wafer by electron beam evaporation and a rapid thermal process. The 150-nm-thick SiO_2 layer was deposited on the whole wafer by plasma-enhanced chemical vapor deposition (PECVD) as the hard mask of etching of gratings and ridge waveguides for DFB LDs. The 250-nm-thick positive polymethyl methacrylate (PMMA) resist was used to define the gratings with periods of 824 and 1648 nm, the patterns of the gratings were constructed on both sides of the p-contact stripes and perpendicular to them with this method of electron beam lithography (EBL), the model of the EBL was Raith i-line plus with a write resolution of 7 nm. The patterns of gratings and ridge waveguides were transferred to the epi-structure of the wafer through reactive ion etching based on CHF_3/SF_6 gas and the dry etching of inductively coupled plasma (ICP) using Cl_2/BCl_3 gas in succession. The 300 nm-thick SiO_2 insulating layer was formed on the wafer by PECVD, and the 2-, 4.5- and 8.5-μm-wide openings on the ridge were given shape by dry etching the SiO_2 layer using CHF_3/SF_6 mixed gas before the 50/250 nm Ti/Au metal pad was deposited. The back side of the wafer was thinned and polished to a thickness of 120 μm, and then an n-contact Ti/Pt/Au metal layer with a thickness of 50/50/250 nm was deposited on it. The processed sample was cut into bars with a cavity length of 600 μm, and both front and back facets were uncoated.

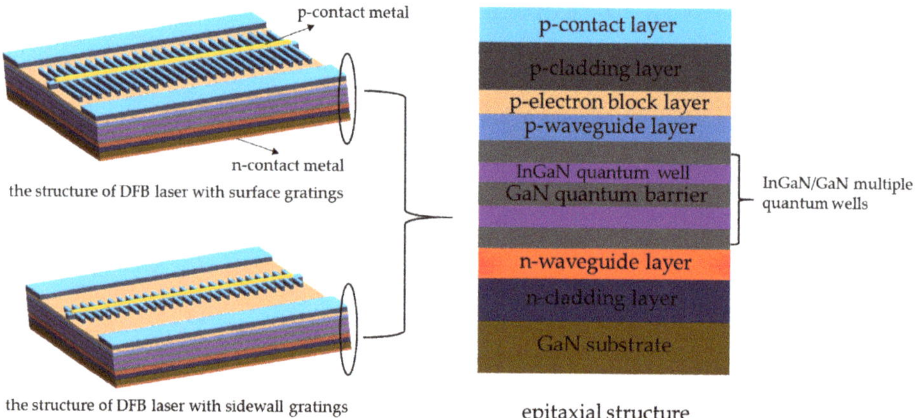

Figure 3. The structures of processed distributed feedback laser diodes (DFB-LDs) with surface gratings and sidewall gratings, as well as the diagrammatic drawing of the cross profile of the gallium nitride (GaN)-based epitaxial wafer.

3. Results and Discussions

3.1. The DFB LDs with the Surface Gratings

Table 1 compares the structure parameters of different manufactured LDs. Devices 1 and 4 are F–P LDs with ridge widths of 2.5 and 10 μm, respectively, that are compared with the fabricated DFB-LDs with surface gratings. Devices 2 and 3 have the same ridge width and width and duty ratio of surface gratings, but they have different grating periods of 824 and 1648 nm. The same parameters also apply to Devices 5 and 6 except for the ridge width. All of them were fabricated from the same epitaxial wafer and shared the same processing.

Table 1. The structure parameters of the fabricated distributed feedback laser diodes (DFB-LDs) of surface gratings and Fabry–Pérot (F–P) LDs.

Sample	Device 1	Device 2	Device 3	Device 4	Device 5	Device 6
Ridge width	2.5 μm	2.5 μm	2.5 μm	10 μm	10 μm	10 μm
Width of grating of each side	-	40 μm	40 μm	-	40 μm	40 μm
Period of gratings	z	824 nm	1648 nm	-	824 nm	1648 nm
Duty ratio of grating	-	80%	80%	-	80%	80%

The structures of the ridge and gratings of the DFB LDs with surface gratings were well formed, as shown in Figure 4. The high resolution SEM graph in Figure 4a shows a lateral view of the DFB LD with surface gratings, thus displaying the general structural features of the device. The actual parameters of the ridge and the grating of the processed DFB LD are shown in Figure 4b, which were close to the designed values shown in Table 1. Figure 4c presents the sectional feature of the surface gratings with a period of 1648 nm. It can be concluded that the gratings have the desired quality and a depth of 481 nm close to the head of the upper waveguide layer, which means that they basically meet our design requirements.

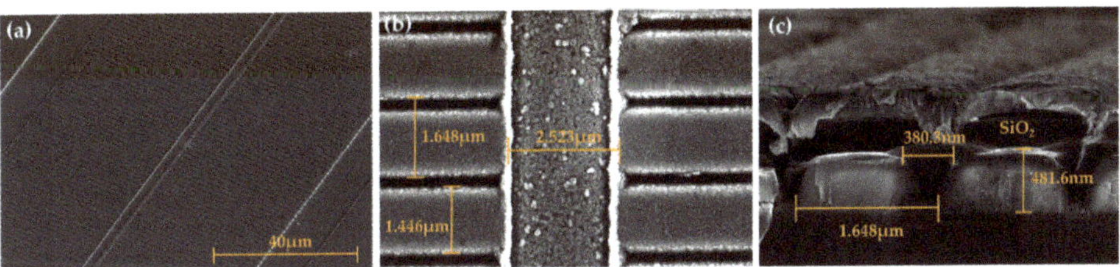

Figure 4. Scanning electron microscope (SEM) images of (**a**) an oblique view of the fabricated DFB laser diodes with surface gratings, (**b**) a top vision of the 20th order surface gratings alongside a metal stripe, and (**c**) a sectional structure of the surface gratings perpendicular to the facet of LDs.

The spectral measurements were conducted by a fiber spectrum analyzer (BIM6002, Brolight, Hangzhou, China) with a resolution of 0.16 nm under a pulsed operation of a 500 ns pulse length and 1kHz repetition frequency. Figure 5 shows the emission spectrum chart of the DFB LDs (Devices 1–6) with surface gratings and F–P LDs. All of them were operated at the impulsive condition of 1.2 times the threshold current by the fiber spectrometer. Since the emission peaks of those LDs had different intensities under certain test currents, the differences in the position of the bottom of the spectrums of LDs were caused by the normalization process of dividing by the different highest intensities. The emission peaks of Devices 1–6 were located at 402.13, 398.86, 398.50, 402.86, 400.50 and 400.86 nm, respectively. Obviously, the DFB LDs had a shorter lasing wavelength than the F–P LDs with the same ridge width because of the interaction between gratings and multiple quantum wells. The offset of the 2~3 nm emission wavelength between the DFB and the F–P laser also confirms the modulation effect of the grating. The full width at half maxima (FWHM) corresponding to DFB LDs for Devices 2, 3, 5, and 6 were 0.37, 0.32, 0.52, and 0.50 nm, respectively. We can conclude that the DFB LDs exhibited a narrower emission width and a single peak emission because of the modulation of the gratings, while the F–P LDs for Devices 1 and 4 had distinct multimode characteristics. Additionally, the FWHM of the emission of the DFB LD with the same ridge seems to have rarely been related to the period of the gratings, while the DFB LD with the 2.5 μm ridge width had a lower FWHM than that with the 10 μm ridge width. This can be partly explained by the fact that, as depicted in Figure 1, the simultaneous oscillation of lateral modes sharing the same longitudinal mode number eventually results in a wider spectral width. In addition, according to the equation $L = \lambda m/2n_{\text{eff}}$—where $n_{\text{eff}} \approx 2.5$ and $L = 600$ μm—the expected free spectral range (FSR) of the longitudinal modes of the F–P laser could be approximately 0.05 nm at around 400 nm. Therefore, the DFB LDs possibly showed multi-mode operation, but this phenomenon was not been confirmed because of the lack of the measurement condition of the high resolution spectrum and the lateral far field.

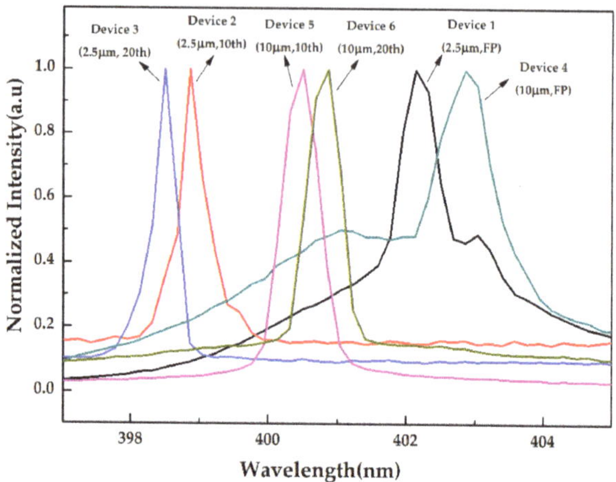

Figure 5. Emission spectrograms of fabricated laser diodes (Devices 1–6) driven by electrical impulses with a 500 ns pulse width and a 1 kHz repetition rate.

3.2. The DFB LDs with the Sidewall Gratings

In addition, the DFB LDs with sidewall gratings were fabricated on the same epitaxial wafer, and they shared the same processing steps with the surface grating DFB LDs. The structure parameters of the fabricated LDs are shown in Table 2. Devices 7 and 8 had the same ridge width of 2.5 μm, gratings width of 1.25 μm, and 80% duty ratio of gratings, but they had different grating periods of 824 and 1648 nm, respectively. The same parameters for DFB LDs with sidewall gratings except for the widths of the ridge and gratings were applied to Devices 10 and 11. Device 9 was the F–P LDs with a ridge width of 5 μm, which was in contrast with the DFB LDs for Devices 10 and 11.

Table 2. Structure parameters of the fabricated DFB LDs of sidewall gratings and F–P LDs.

Sample	Device 7	Device 8	Device 9	Device 10	Device 11
Ridge width	2.5 μm	2.5 μm	5 μm	5 μm	5 μm
Width of gratings of each side	1.25 μm	1.25 μm	-	2.5 μm	2.5 μm
Period of gratings	824 nm	1648 nm	-	824 nm	1648 nm
Duty ratio of gratings	80%	80%	-	80%	80%

Figure 6 shows that the DFB LD with 20th order sidewall gratings was well fabricated. Figure 6a is the top view of the DFB LD with sidewall gratings, and it depicts the structure of ridge, sidewall gratings, double etching grooves, and the p-contact metal on the ridge. Figure 6b,c shows the actual parameters of the ridge and gratings of the processed device, which are basically in conformity to the values in Table 2. Sidewall gratings with a period of 1648 nm, a duty ratio of approximately 80%, and a width of 1.25 μm of each side alongside the ridge waveguide were observed. Since the corresponding etchings processes were under the same experimental condition, the surface and sidewall gratings had a similar etching depth and sectional morphology, as shown in Figure 4. The notches of the gratings had a tilt angle as a result of the ICP etch process, so the average duty ratio of the fabricated gratings was greater than 80% [23,24]. Moreover, the simulation results showed that the influence of the relative large grating angle of 70° could be ignored [25], offering a high enough reflectivity for the DFB LDs.

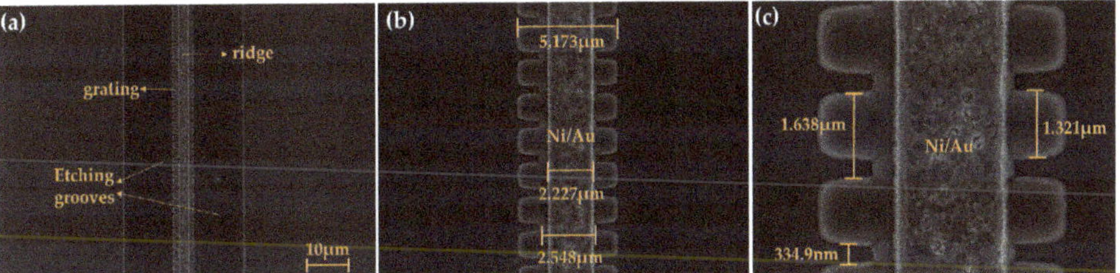

Figure 6. SEM images of a top view of the fabricated DFB laser diodes with 20th order sidewall gratings. (**a**) An overall view of the DFB LD with double etching grooves; (**b**) a low power and (**c**) a high power perspective of the ridge waveguide and the sidewall gratings along it.

The emission spectrum of the DFB LDs with sidewall gratings and the F–P LDs for Devices 7–11 can be seen in Figure 7, and the LDs were operated at a current around 20% above threshold current. The emission peaks of the Devices 7–11 were 399.59, 399.95, 403.76, 401.4 and 401.77 nm, respectively. The FWHMs of the DFB LDs for Devices 7, 8, 10, and 11 were 0.25, 0.23, 0.42, and 0.48 nm, respectively, which were slightly higher than the resolution of the fiber spectrometer. In the context of comparing the multimode morphology of F–P LDs with the ridge of 5 μm during lasing, the emission spectrum of the DFB LDs showed a single peak because of the existence of the gratings. Similar to the previous results, the DFB LDs of sidewall gratings had a lower lasing wavelength than the F–P LDs with the same ridge width. In addition, the DFB LDs with a 2.5 μm ridge width had a lower FWHM than those with the 5 μm ridge width, thus indicating that a narrow ridge width is favorable for the narrowing of the linewidth of the DFB LD. According to the spectrum characteristics, we could speculate that the effect of grating periods on the line width of LDs is not very evident.

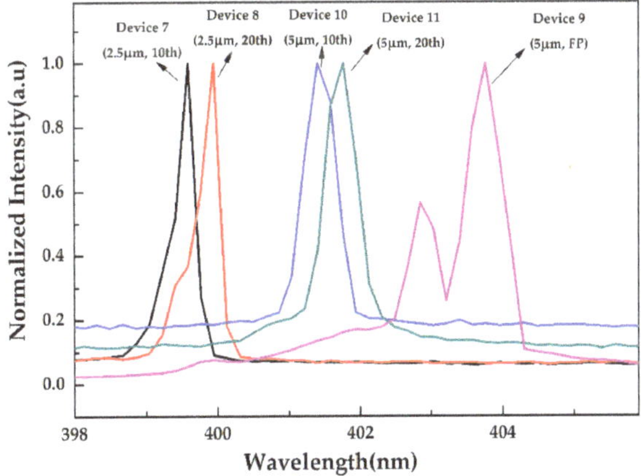

Figure 7. Emission spectrograms of processed laser diodes for Devices 7–11 driven by electrical impulses with a 500 ns pulse width and a 1 kHz repetition rate.

3.3. The Comparison of Properties of the DFB LDs with Surface and Sidewall Gratings

Meanwhile, based on the good linewidth characteristics of DFB LDs with a ridge width of 2.5 μm, the characteristics of DFB LDs with sidewall and surface gratings and the F–P LD were compared in the case that their ridge widths were all 2.5 μm. The sidewall gratings need lower costs for writing a smaller area of patterns than surface gratings which use EBL. The light output power and voltage versus current (L–I–V) characteristics of the LDs were tested under the pulsed driving condition with a pulse width of 1 μs and a pulse repetition frequency of 10 kHz at room temperature to avoid generating excess heat. Figure 8 shows L–I–V characteristics of the DFB LDs and the F–P LDs with the same ridge width of 2.5 μm. The values of the threshold currents of Devices 1, 2, 3, 7, and 8 were between 450 and 560 mA, and their slope efficiencies were approximately 0.24, 0.12, 0.17, 0.13, and 0.19 W/A, respectively. The deviations of the threshold current and slopes efficiency of the LDs with the same structures were less than 8% and 13%, respectively. However, all of the F–P and DFB LDs tested showed comparatively high threshold currents and lower slope efficiencies, which was mainly deemed to owe to the performance of the epitaxial wafer. More intuitive and detailed results are shown in Table 3.

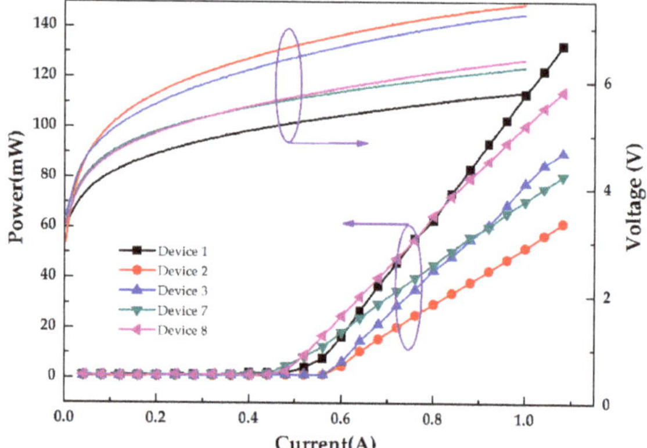

Figure 8. Light output power–current–voltage characteristics of Devices 1, 2, 3, 7, and 8 operated in pulsed mode with a 1 μs pulse width and a 10 kHz repetition rate.

Table 3. Electrical properties of the fabricated DFB LDs and F–P LDs.

Sample	Device 1	Device 2	Device 3	Device 7	Device 8
Type of gratings	-	Surface	Surface	Sidewall	Sidewall
Order of gratings	-	10th	20th	10th	20th
Threshold current	(518 ± 12) mA	(539 ± 4) mA	(550 ± 10) mA	(458 ± 7) mA	(485 ± 19) mA
The slopes efficiency	(235 ± 15) mW/A	(116 ± 3) mW/A	(172 ± 11) mW/A	(129 ± 7) mW/A	(189 ± 6) mW/A
FWHM	1.96 nm	0.37 nm	0.32 nm	0.25 nm	0.23 nm

From the data in Figure 8 and Table 3, it can be seen that DFB LDs with the same ridge width and a lower grating period exhibited lower slope efficiency because of the higher value of kL compared to those with high order gratings [26]. In addition, the slope efficiency of DFB LDs with sidewall gratings was slightly higher than that of the DFB LDs with surface gratings but smaller than F–P LDs because of the scatter losses and diffraction losses from the gratings. Besides, the FWHM of DFB LDs with sidewall gratings was slightly lower than that of DFB LDs with surface gratings, and this may have been due to the fact that the sidewall grating DFB LD had a narrower effective width in the whole structure compared to the surface grating of the DFB LD with the same ridge width. Given its smaller grating area, the sidewall grating DFB LD had fewer transverse modes, thus resulting in a smaller FWHM compared to the surface grating DFB LD. Additionally, the linewidth of the DFB LDs with a single peak emission showed a drastic reduction compared to that of the F–P LDs with multiple lasing peaks.

4. Conclusions

In conclusion, the GaN-based DFB LDs that integrated with the surface and sidewall gratings were investigated by the FDTD method. The surface and sidewall gratings alongside the p-contact metal stripe on the ridge waveguide were fabricated by EBL and ICP etching on the same epitaxial wafer. The DFB LD with the 20th order, 80% duty-cycle surface gratings showed a single wavelength emission at 398.86 nm with an FWHM of 0.32 nm under the electrical pulsed driving, and the DFB LD with the 20th order, 80% duty-cycle sidewall gratings obtained a peak emission at 399.95 nm with an FWHM of 0.23 nm. Additionally, both of the DFB LDs showed a narrower linewidth compared to that of the F–P LDs. Moreover, the FWHM of the DFB LDs with the ridge width of 2.5 μm was obviously lower than that of the DFB LDs using the same type of gratings with ridge widths of 5 or 10 μm, which indicates that the narrow ridge width was favorable for the narrowing of the linewidth. Furthermore, the sidewall grating DFB LDs possessed a slightly higher slope efficiency than that of the surface grating DFB LDs with the same ridge width and period of gratings. Given that, the DFB LD with sidewall grating required a lower fabrication cost and achieved better device performance compared to the surface grating DFB LDs, which makes it a better choice for these applications. In addition, in order to further improve the performance of the DFB LDs, the optimization of the structure of the gratings and the conduction of the cavity surface coating process are required to reduce threshold current and improve slope efficiency. Additionally, the side-mode suppression ratio and the high resolution spectral measurement are also required in future work.

Author Contributions: Conceptualization—Z.D. and J.L.; data curation—Z.D. and M.L.; methodology—Z.D. and M.L.; project administration—J.L.; supervision—J.L.; validation—W.X. and S.L.; writing original draft—Z.D.

Funding: This work was funded by Science Challenge Project (No. TZ2016003-2), National Natural Science Foundation of China (No. 61804140), and National Key R&D Program of China (No. 2017YFB0403103).

Acknowledgments: The authors would like to thank the Nanofabrication facility in Suzhou Institute of Nano-tech and Nano-bionics (CAS) for equipment access.

Conflicts of Interest: The authors declare no conflict of interest.

References

1. Chi, Y.C.; Hsieh, D.H.; Tsai, C.T.; Chen, H.Y.; Kuo, H.C.; Lin, G.R. 450-nm GaN laser diode enables high-speed visible light communication with 9-Gbps QAM-OFDM. *Opt. Express* **2015**, *23*, 13051–13059. [CrossRef] [PubMed]
2. Watson, S.; Tan, M.; Najda, S.P.; Perlin, P.; Leszczynski, M.; Targowski, G.; Grzanka, S.; Kelly, A.E. Visible light communications using a directly modulated 422 nm GaN laser diode. *Opt. Lett.* **2013**, *38*, 3792–3794. [CrossRef] [PubMed]
3. Zeller, W.; Naehle, L.; Fuchs, P.; Gerschuetz, F.; Hildebrandt, L.; Koeth, J. DFB Lasers Between 760 nm and 16 μm for Sensing Applications. *Sensors* **2010**, *10*, 2492–2510. [CrossRef] [PubMed]
4. Balci, M.H.; Chen, F.; Cunbul, A.B.; Svensen, Ø.; Akram, M.N.; Chen, X. Comparative study of blue laser diode driven cerium-doped single crystal phosphors in application of high-power lighting and display technologies. *Opt. Rev.* **2018**, *25*, 166–174. [CrossRef]
5. Shimada, Y.; Chida, Y.; Ohtsubo, N.; Aoki, T.; Takeuchi, M.; Kuga, T.; Torii, Y. A simplified 461-nm laser system using blue laser diodes and a hollow cathode lamp for laser cooling of Sr. *Rev. Sci. Instrum.* **2013**, *84*, 063101. [CrossRef]
6. Najda, S.P.; Perlin, P.; Leszczyński, M.; Slight, T.J.; Meredith, W.; Schemmann, M.; Moseley, H.; Woods, J.A.; Valentine, R.; Kalra, S.; et al. A multi-wavelength (uv to visible) laser system for early detection of oral cancer. In *Imaging, Manipulation, and Analysis of Biomolecules, Cells, and Tissues XIII*; International Society for Optics and Photonics: San Francisco, CA, USA, March 2015; p. 932809.
7. Luo, Y.; Nakano, Y.; Tada, K.; Inoue, T.; Hosomatsu, H.; Iwaoka, H. Fabrication and characteristics of gain-coupled distributed feedback semiconductor lasers with a corrugated active layer. *IEEE J. Quantum Electron.* **1991**, *27*, 1724–1731. [CrossRef]
8. Hofstetter, D.; Romano, L.T.; Paoli, T.L.; Bour, D.P.; Kneissl, M. Realization of a complex-coupled InGaN/GaN-based optically pumped multiple-quantum-well distributed-feedback laser. *Appl. Phys. Lett.* **2000**, *76*, 2337–2339. [CrossRef]
9. Masui, S.; Tsukayama, K.; Yanamoto, T.; Kozaki, T.; Nagahama, S.I.; Mukai, T. First-order AlInGaN 405 nm distributed feedback laser diodes by current injection. *Jpn. J. Appl. Phys.* **2006**, *45*, L749. [CrossRef]

10. Schweizer, H.; Gräbeldinger, H.; Dumitru, V.; Jetter, M.; Bader, S.; Brüderl, G.; Weimar, A.; &Härle, V. Laterally coupled InGaN/GaN DFB laser diodes. *Phys. Status Solidi A* **2002**, *192*, 301–307. [CrossRef]
11. Hofmann, R.; Gauggel, H.P.; Griesinger, U.A.; Gräbeldinger, H.; Adler, F.; Ernst, P.; Bolay, H.; Harle, V.; Schoz, F.; Schweizer, H.; et al. Realization of optically pumped second-order GaInN-distributed-feedback lasers. *Appl. Phys. Lett.* **1996**, *69*, 2068–2070. [CrossRef]
12. Zhang, H.; Cohen, D.A.; Chan, P.; Wong, M.S.; Mehari, S.; Becerra, D.L.; Nakamura, S.; DenBaars, S.P. Continuous-wave operation of a semipolar InGaN distributed-feedback blue laser diode with a first-order indium tin oxide surface grating. *Opt. Lett.* **2019**, *44*, 3106–3109. [CrossRef] [PubMed]
13. Holguín-Lerma, J.A.; Ng, T.K.; Ooi, B.S. Narrow-line InGaN/GaN green laser diode with high-order distributed-feedback surface grating. *Appl. Phys. Express* **2019**, *12*, 042007. [CrossRef]
14. Masui, S.; Tsukayama, K.; Yanamoto, T.; Kozaki, T.; Nagahama, S.I.; Mukai, T. CW operation of the first-order AlInGaN 405 nm distributed feedback laser diodes. *Jpn. J. Appl. Phys.* **2006**, *45*, L1223. [CrossRef]
15. Slight, T.J.; Odedina, O.; Meredith, W.; Docherty, K.E.; Kelly, A.E. InGaN/GaN distributed feedback laser diodes with deeply etched sidewall gratings. *IEEE Photonics Technol. Lett.* **2016**, *28*, 2886–2888. [CrossRef]
16. Slight, T.J.; Yadav, A.; Odedina, O.; Meredith, W.; Docherty, K.E.; Rafailov, E.; Kelly, A.E. InGaN/GaN laser diodes with high order notched gratings. *IEEE Photonics Technol. Lett.* **2017**, *29*, 2020–2022. [CrossRef]
17. Slight, T.J.; Stanczyk, S.; Watson, S.; Yadav, A.; Grzanka, S.; Rafailov, E.; Perlin, P.; Najda, S.P.; Leszczynski, M.; Gwyn, S.; et al. Continuous-wave operation of (Al, In) GaN distributed-feedback laser diodes with high-order notched gratings. *Appl. Phys. Express* **2018**, *11*, 112701. [CrossRef]
18. Kang, J.H.; Martens, M.; Wenzel, H.; Hoffmann, V.; John, W.; Einfeldt, S.; Wernicke, T.; Kneissl, M. Optically pumped DFB lasers based on GaN using 10th-order laterally coupled surface gratings. *IEEE Photonics Technol. Lett.* **2016**, *29*, 138–141. [CrossRef]
19. Kang, J.H.; Wenzel, H.; Hoffmann, V.; Freier, E.; Sulmoni, L.; Unger, R.S.; Einfeldt, S.; Wernicke, T.; Kneissl, M. DFB laser diodes based on GaN using 10th order laterally coupled surface gratings. *IEEE Photonics Technol. Lett.* **2017**, *30*, 231–234. [CrossRef]
20. Li, J.Z.; Huang, F.; Yang, H.J.; Deng, Z.J.; Liao, M.L. The MOCVD overgrowth studies of III-Nitride on Bragg grating for distributed feedback lasers. In *14th National Conference on Laser Technology and Optoelectronics (LTO 2019)*; International Society for Optics and Photonics: Shanghai, China, May 2019; p. 111701X.
21. Hofmann, R.; Wagner, V.; Neuner, M.; Off, J.; Scholz, F.; Schweizer, H. Optically pumped GaInN/GaN-DFB lasers: Overgrown lasers and vertical modes. *Mater. Sci. Eng. B* **1999**, *59*, 386–389. [CrossRef]
22. Cao, Y.L.; Hu, X.N.; Luo, X.S.; Song, J.F.; Cheng, Y.; Li, C.M.; Liu, C.Y.; Wang, H.; Liow, T.Y.; Lo, G.Q.; et al. Hybrid III–V/silicon laser with laterally coupled Bragg grating. *Opt. Express* **2015**, *23*, 8800–8808. [CrossRef]
23. Wenzel, H.; Fricke, J.; Decker, J.; Crump, P.; Erbert, G. High-power distributed feedback lasers with surface gratings: Theory and experiment. *IEEE J. Quantum. Elect.* **2015**, *21*, 352–358. [CrossRef]
24. Zolotarev, V.V.; Leshko, A.Y.; Pikhtin, N.A.; Slipchenko, S.O.; Sokolova, Z.N.; Lubyanskiy, Y.V.V.; Voronkova, N.V.; Tarasov, I.Y.S. Integrated high-order surface diffraction gratings for diode lasers. *Quantum. Electron.* **2015**, *45*, 1091. [CrossRef]
25. Bao, S.; Song, Q.; Xie, C. The influence of grating shape formation fluctuation on DFB laser diode threshold condition. *Opt. Rev.* **2018**, *25*, 330–335. [CrossRef]
26. Crump, P.; Schultz, C.M.; Wenzel, H.; Erbert, G.; Tränkle, G. Efficiency-optimized monolithic frequency stabilization of high-power diode lasers. *J. Phys. D Appl. Phys.* **2012**, *46*, 013001. [CrossRef]

Article

Improved Output Power of GaN-based VCSEL with Band-Engineered Electron Blocking Layer

Huiwen Luo [1,2], Junze Li [1,2,*] and Mo Li [1,2,*]

1 Institute of Electronic Engineering, China Academy of Engineering Physics, Mianyang 621999, China; luohuiwen@mtrc.ac.cn

2 Microsystem and Terahertz Research Center, China Academy of Engineering Physics, Chengdu 610200, China

* Correspondence: lijunze@mtrc.ac.cn (J.L.); limo@mtrc.ac.cn (M.L.); Tel.: +86-0286-572-6054 (J.L.); +86-0286-572-6011 (M.L.)

Received: 15 September 2019; Accepted: 10 October 2019; Published: 12 October 2019

Abstract: The vertical-cavity surface-emitting laser (VCSEL) has unique advantages over the conventional edge-emitting laser and has recently attracted a lot of attention. However, the output power of GaN-based VCSEL is still low due to the large electron leakage caused by the built-in polarization at the heterointerface within the device. In this paper, in order to improve the output power, a new structure of p-type composition-graded $Al_xGa_{1-x}N$ electron blocking layer (EBL) is proposed in the VCSEL, by replacing the last quantum barrier (LQB) and EBL in the conventional structure. The simulation results show that the proposed EBL in the VCSEL suppresses the leaking electrons remarkably and contributes to a 70.6% increase of the output power, compared with the conventional GaN-based VCSEL.

Keywords: GaN-based vertical-cavity surface-emitting laser (VCSEL); composition-graded $Al_xGa_{1-x}N$ electron blocking layer (EBL); electron leakage

1. Introduction

Vertical-cavity surface-emitting lasers (VCSELs) exhibit several advantages over edge-emitting lasers (LDs), including high-speed direct modulation, circular mode profile, low threshold current, etc. [1–3]. Recently, although GaAs-based VCSELs that emit red or infrared light have been commercialized and applied to various products, it is still hoped that VCSELs will not only cover the red area, but also the blue and green spectrum region, thus constituting the ternary color of light for illumination, display and communication of the next generation [4]. As wide-bandgap materials, the bandgap width of gallium nitride (GaN) and its alloy materials are continuously adjustable from 0.7 eV to 6.2 eV, covering near-red, green, blue, and ultraviolet. They have become the main material for manufacturing short-wavelength light-emitting diodes (LEDs) and LDs. Recently, the GaN-based materials have been applied to VCSELs and achieved important progress [5]. Universities and research institutes such as National Chiao-Tung University [6], Nichia Corporation [7], University of California, Santa Barbara [8], Sony Corporation [1], Xiamen University [9] and Meijo University [10] conducted good work on the designing and manufacturing of blue GaN-based VCSELs. In 2018, Stanley Electric Co., Ltd. demonstrated a GaN-based VCSEL with an output power of 7.6 mW by reducing both the internal loss and the reflectivity of the front cavity mirror [11]. This structure achieved the highest output power of a GaN-based VCSEL ever produced to the best of our knowledge. However, it is still relatively low as a result of ignoring other mechanisms. For example, for InP and other semiconductor-based lasers, lower threshold and higher output power can be achieved by designing the micro-cavities, which may also have good effects in GaN-based VCSELs [12–14].

Beyond designing the micro-cavities, in order to obtain higher output power, methods which maximize the radiative recombination rate have been adopted by minimizing the electron leakage out of the active region. Among those methods, $Al_xGa_{1-x}N$ electron blocking layer (EBL) is usually placed between the active region and p-GaN layer, since it provides a higher energy barrier in the conduction band between quantum wells (QWs) and the p-layers. The higher energy barrier in the conduction band contributes to the confinement of the electrons in the active region, so stronger output power can be achieved [15]. However, the very commonly used $Al_xGa_{1-x}N$ EBL sometimes cannot efficiently reduce electron leakage due to the polarization charge in that layer, which decreases the barrier height for electron transport [16,17]. As such, the blocking effects of $Al_xGa_{1-x}N$ EBL on electron overflow are limited [18]. In addition, there are big differences in the polarization value and the energy barrier height in the conduction band between $Al_xGa_{1-x}N$ EBL and $In_xGa_{1-x}N$ last quantum barrier (LQB), which results in a large built-in polarization field and significant free electron accumulation at the heterointerface between the two layers. This free electron accumulation outside of the active region severely degrades the internal quantum efficiency. To solve this problem, Zhang et al. proposed a design which replaces the $In_xGa_{1-x}N$ LQB with a tapered $Al_xGa_{1-x}N$ LQB in GaN-based LD [19], and Lin et al. proposed another structure which replaces the $Al_xGa_{1-x}N$ EBL with a composition-graded $Al_xGa_{1-x}N$ EBL in the VCSEL [20].

In this work, our concept is to improve the output power by optimizing the carrier transport. We proposed an improved GaN-based VCSEL structure called GVCSEL to reduce the electron leakage. In GVCSEL, the LQB and EBL in the conventional GaN-based VCSEL are replaced with a new layer which consists of a composition-graded p-$Al_xGa_{1-x}N$ layer and a p-$Al_xGa_{1-x}N$ layer. The physical and optical properties of the GVCSEL are investigated numerically with the Photonic Integrated Circuit Simulator in 3D (PICS3D) software (Crosslight Corporation, Vancouver, BC, Canada). The results show that the GVCSEL effectively confines the electrons in the active region and has a more uniform carrier distribution. This contributes to a better radiative recombination and helps achieve a 70.6% increase of the output power compared with the conventional GaN-based VCSEL.

2. Device Structure and Simulation Parameters

In this work, the conventional GaN-based VCSEL was used for comparison, as shown in Figure 1a [21], which consists of 11 pairs of Ta_2O_5/SiO_2 as the bottom and top distributed Bragg reflectors (DBRs), respectively. Then, there is a 5.3 μm-thick n-type GaN layer (n-doping = 2.5×10^{18} cm^{-3}) and five periods of $In_{0.1}Ga_{0.9}N/In_{0.035}Ga_{0.965}N$ multi-quantum wells (MWQs). The thicknesses of the well and the barrier are 4 nm and 8 nm, respectively, the n-doping of the barrier is 1×10^{18} cm^{-3}, whilst the LQB is undoped. Next there is a 20 nm-thick $Al_{0.21}Ga_{0.79}N$ (p-doping = 5×10^{18} cm^{-3}) EBL, followed by a 0.54 μm-thick p-type GaN layer (p-doping = 1×10^{18} cm^{-3}). On the p-type GaN layer, a 20 nm-thick SiO_2 is employed as the current-confined layer. Following that, a 20 nm-thick indium-tin-oxide (ITO) layer (p-doping = 1×10^{19} cm^{-3}) is employed as the current spreading layer and the diameter of the current injection aperture is designed to be 10 μm. The top metal ring contact confining the optical mode is 12 μm in diameter. Figure 1b shows the schematic diagram of the GVCSEL in this study, which is formed by replacing the 8 nm-thick $In_{0.035}Ga_{0.965}N$ LQB and 20 nm-thick $Al_{0.21}Ga_{0.79}N$ EBL in the conventional GaN-based VCSEL with a new layer. The new layer consists of a p-$Al_{0\rightarrow0.21}Ga_{1\rightarrow0.79}N$ (p-doping = 5×10^{18} cm^{-3}) layer and a p-$Al_{0.21}Ga_{0.79}N$ (p-doping = 5×10^{18} cm^{-3}) layer, where the thickness of the first layer is L nm and the thickness of the second layer is (28-L) nm. The GaN-based VCSEL samples with L equals 6 nm, 8 nm and 16 nm are named GVCSEL1, GVCSEL2 and GVCSEL3, respectively, and their parameters are given in Table 1. The conventional GaN-based VCSEL is named CVCSEL for comparison.

The physical features and optical properties of the GVCSEL and CVCSEL were investigated numerically with the Simupics3d program of Crosslight software. An important issue in simulation is the selection of proper parameters in the physical models. In this study, we used a Mg activation energy of 170 meV for GaN, which was assumed to increase by 3 meV per Al% for AlGaN [12].

The Shockley-Read-Hall (SRH) lifetime and the Auger recombination coefficient were estimated to be 100 ns and 1×10^{-34} cm^6s^{-1}, respectively [22]. The built-in polarization caused by piezoelectric polarization and spontaneous polarization was represented by fixed interface charges at every heterointerface within the device, which was calculated using the methods developed by Fiorentini et al. [23]. Here we took 50% of the theoretical value by setting the polarization screening to 0.5 [24]. We also considered the energy band offset ratio as 50:50 [25]. Other material parameters in the simulation can be found in [26].

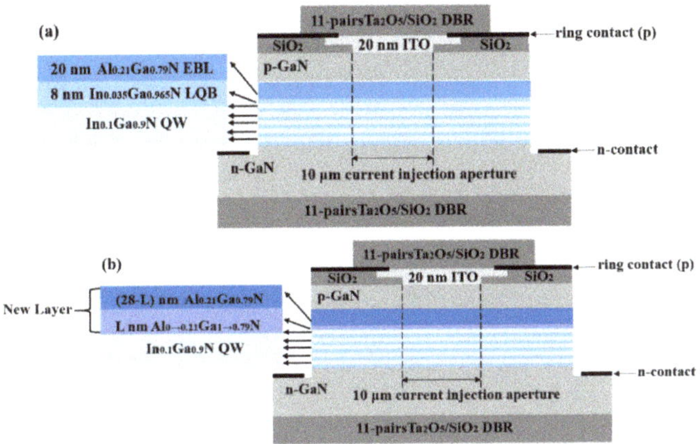

Figure 1. Schematic diagrams of the (**a**) conventional gallium nitride (GaN)-based vertical-cavity surface-emitting laser VCSEL (CVCSEL) and (**b**) proposed structure which is formed by replacing the LQB and EBL in the CVCSEL with a new layer (GVCSEL).

Table 1. The thickness of p-$Al_{0.21}Ga_{0.79}N$ layer and p-$Al_{0\to0.21}Ga_{1\to0.79}N$ layer of GVCSEL.

Sample Name	GVCSEL1	GVCSEL2	GVCSEL3
p-$Al_{0.21}Ga_{0.79}N$	22 nm	20 nm	10 nm
p-$Al_{0\to0.21}Ga_{1\to0.79}N$	6 nm	8 nm	16 nm

3. Results and Discussions

Since the output power and the threshold current are interlinked, investigating both of them is crucial to understand the performance of GaN-based VCSELs [17], we calculated the two performances of CVCSEL and GVCSEL1-GVCSEL3, as shown in Figure 2. It was found that the output power of CVCSEL was 0.179 mW and was lower than that of the GVCSELs in Figure 2a. In addition, the output power increased from 0.267 mW to 0.306 mW when the thickness of L of the p-$Al_{0\to0.21}Ga_{1\to0.79}N$ in the GVCSEL was increased from 6 nm to 8 nm, while with further increase of L from 8 nm to 16 nm, the output power decreased from 0.306 mW to 0.303 mW. Thus, the highest output power was obtained in GVCSEL2, which achieved a 70.6% increase of output power compared with the CVCSEL. Figure 2b shows that the current threshold of the four samples first decreased and then went up. The lowest threshold current was also achieved in GVCSEL2.

In order to reveal the origin of the observations in Figure 2, we calculated the distribution of the electrons and the holes and the current density flowing along the vertical direction near the active region of the four samples at an injection current of 6 mA, as shown in Figure 3. We also calculated the distribution of the radiative recombination rate as shown in Figure 4a–d. In Figure 3a, the highest electron concentration in the p-GaN layer was obtained in CVESEL, thus there was the largest electron leakage. For the GVCSELs, the electron concentration in the p-GaN layer decreased when L was increased from 6 nm to 8 nm. However, the electron concentration in the p-GaN layer increased when the thickness of L went up from 8 nm to 16 nm. Therefore, the lowest electron concentration in the p-GaN layer was obtained in GVCSEL2, which means that the largest reduction of the electron leakage was obtained in GVCSEL2. Thus, the reduction of electron leakage is one of the origins of the increase of output power and the reduction of the threshold current for CVCSEL-GVCSEL3. Figure 3b illustrates that the hole concentration in p-GaN layers of the GVCSELs was almost the same with that of the CVCSEL, which means that the hole injection is not changed in GVCSELs compared to CVCSEL. Thus, the hole injection is not the origin of the change of output power and threshold current for the samples. Figure 3c demonstrates that the electron overflow of GVCSEL was efficiently suppressed and the most effective suppression of the electron overflow was obtained in GVCSEL2. There was more uniform carrier distribution with the reduction of electron overflow. As a result, the radiative recombination rate in GVCSEL was improved, as can be seen in Figure 4a–d, and the highest one was obtained in GVCSEL2, as shown in Figure 4c. In summary, Figures 3 and 4a–d show that the increase of the output power and the decrease of the threshold current for samples of GaN-based VCSEL can be attributed to the reduced electron leakage and an increased radiative recombination rate.

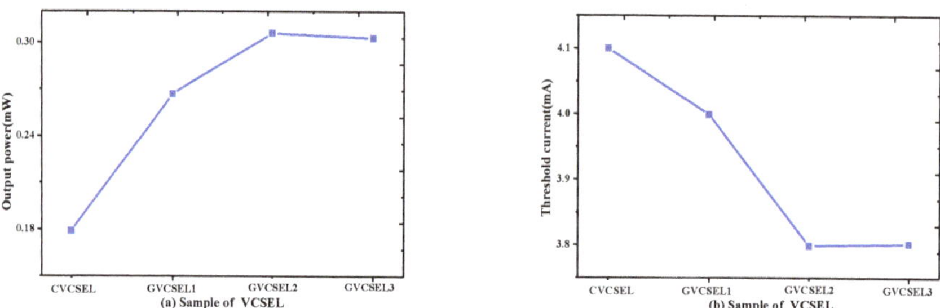

Figure 2. (**a**) Output power at the injection current of 6 mA and (**b**) threshold current of CVCSEL and GVCSEL.

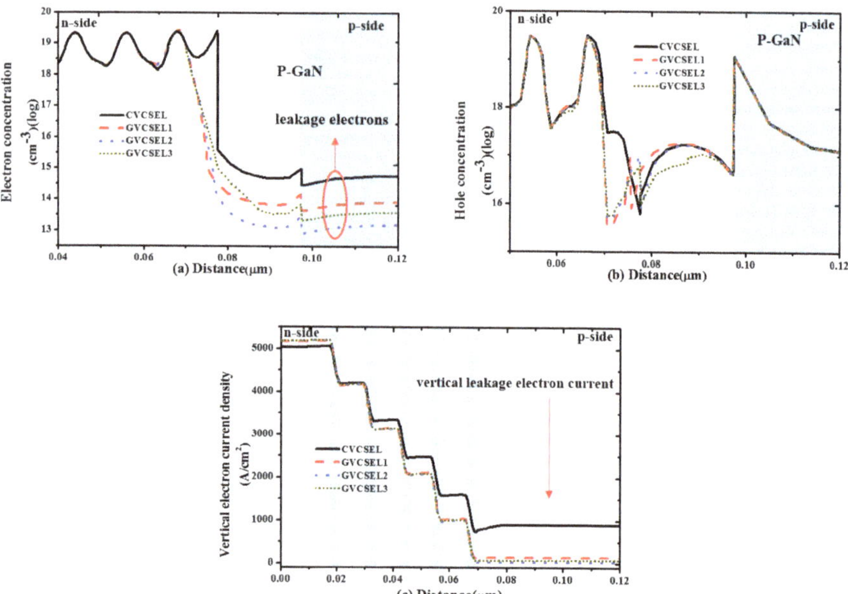

Figure 3. The distribution of (**a**) electron concentration; (**b**) hole concentration; (**c**) vertical electron current density, of GVCSEL1-3 and CVCSEL at an injection current of 6 mA.

Finally, for the purpose of investigating in-depth the origin of the reduced leaking electrons in the GVCSEL, we calculated the distribution of the energy band at an injection current of 6 mA, as plotted in Figure 4e–h. In CVCSEL, the effective barrier height of the EBL for the electrons was 223 meV, which was lower than that of GVCSEL. Therefore, in CVCSEL the electrons were easier to leak into the p-GaN layer. The increase of the electron leakage resulted in a more nonuniform carrier distribution in the active region, thus lowering the radiative recombination rate and the output power. For GVCSELs, when the thickness of L was increased from 6 nm to 8 nm, the effective barrier height increased from 263 meV to 303 meV. But the effective barrier height decreased from 303 meV to 280 meV when the thickness of L further increased from 8 nm to 16 nm. This means that the highest effective barrier height was the origin of the highest output power obtained in GVCSEL2. We can also see that the effective barrier height of EBL for the holes was about 176 meV in CVCSEL and all three GVCSELs. Therefore, the ability inject holes into the active region is almost the same in the CVCSEL and the GVCSELs, as shown in Figure 3b. In summary, the origin of the increased output power for the GVCSEL is not the rate of the injection holes but the suppressed electron leakage.

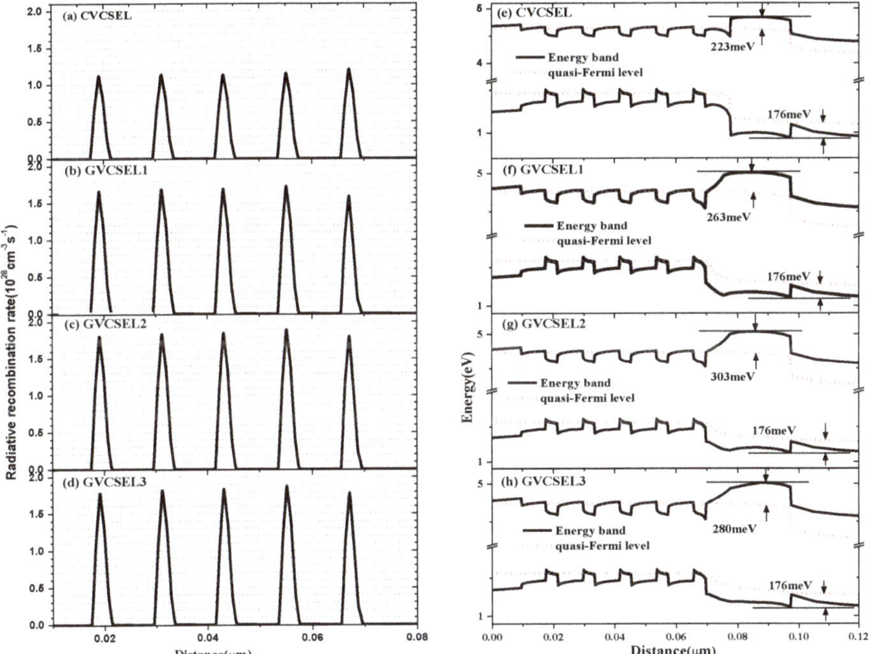

Figure 4. Distribution of (**a–d**) the radiative recombination rate and (**e–h**) the energy band of GVCSELs and CVCSEL at the injection current of 6 mA.

4. Conclusions

In our work, in order to improve the output power of GaN-based VCSEL, we proposed a structure called GVCSEL, based on the concept of modulating the carrier distribution. A new layer was used by combining the LQB and EBL in the conventional GaN-based VCSEL, which was made up of a L nm-thick composition-graded p-$Al_xGa_{1-x}N$ layer and a (28-L) nm-thick p-$Al_{0.21}Ga_{0.79}N$ layer and the Al component in the graded $Al_xGa_{1-x}N$ changed from 0 to 0.21. The thickness of L for the three GVCSEL samples was selected to be 6 nm, 8 nm and 16 nm. The numerical simulation results showed that all the GVCSELs had improved output power. The higher output power in the GVCSELs is an attribute of the decrease of the large built-in polarization field and the reduction of the free electron accumulation at the heterointerface between the LQB and the $Al_xGa_{1-x}N$ EBL. This further helped the suppression of the electron leakage and contributed to a more uniform carrier distribution in the active region, resulting in a higher radiative recombination rate in the quantum wells. What is more, this proposed layer also introduced a stronger quantum barrier which can confine the carriers in the quantum wells. In the GVCSELs, the one with a thickness of L of 8 nm obtained the highest output power, which was 70.6% stronger over that of the CVCSEL, since it had higher effective barrier height in the conduction band for electrons. Therefore, the GVCSEL output power can be accurately designed by selecting the thickness of L.

Author Contributions: Conceptualization, H.L. and J.L.; data curation, H.L.; methodology, H.L.; project administration, J.L. and M.L.; supervision, M.L.; validation, M.L.; writing original draft, H.L.

Funding: This work was supported by the Science Challenge Project (No. TZ2016003-2), National Key R&D Program of China (No. 2017YFB0403103), and National Natural Science Foundation of China (No. 61804140).

Conflicts of Interest: The authors declare no conflict of interest.

References

1. Hamaguchi, T.; Tanaka, M.; Mitomo, J.; Nakajima, H.; Ito, M.; Ohara, M.; Kobayashi, N.; Fujii, K.; Watanabe, H.; Satou, S. Lateral optical confinement of GaN-based VCSEL using an atomically smooth monolithic curved mirror. *Sci. Rep.-UK* **2018**, *8*, 10350. [CrossRef] [PubMed]
2. Hamaguchi, T.; Nakajima, H.; Fuutagawa, N. GaN-based Vertical-cavity surface-emitting lasers incorporating dielectric distributed bragg reflectors. *Appl. Sci.* **2019**, *9*, 733. [CrossRef]
3. Forman, C.A.; Lee, S.; Young, E.C.; Kearns, J.A.; Cohen, D.A.; Leonard, J.T.; Margalith, T.; DenBaars, S.P.; Nakamura, S. Continuous-wave operation of m-plane GaN-based vertical-cavity surface-emitting lasers with a tunnel junction intracavity contact. *Appl. Phys. Lett.* **2018**, *112*, 111106. [CrossRef]
4. Takeuchi, T.; Kamiyama, S.; Iwaya, M.; Akasaki, I. GaN-based vertical-cavity surface-emitting lasers with AlInN/GaN distributed Bragg reflectors. *Rep. Prog. Phys.* **2018**, *82*, 012502. [CrossRef]
5. Carlin, J.F.; Zellweger, C.; Dorsaz, J.; Nicolay, S.; Christmann, G.; Feltin, E.; Butté, R.; Grandjean, N. Progresses in III-nitride distributed Bragg reflectors and microcavities using AlInN/GaN materials. *Phys. Status Solidi B* **2005**, *242*, 2326–2344. [CrossRef]
6. Lai, Y.-Y.; Chang, T.-C.; Li, Y.-C.; Lu, T.-C.; Wang, S.-C. Electrically pumped III-N microcavity light emitters incorporating an oxide confinement aperture. *Nanoscale Res. Lett.* **2017**, *12*, 15. [CrossRef]
7. Higuchi, Y.; Omae, K.; Matsumura, H.; Mukai, T. Room-temperature CW lasing of a GaN-based vertical-cavity surface-emitting laser by current injection. *Appl. Phys. Express* **2008**, *1*, 121102. [CrossRef]
8. Forman, C.A.; Lee, S.; Young, E.C.; Kearns, J.A.; Cohen, D.A.; Leonard, J.T.; Margalith, T.; DenBaars, S.P.; Nakamura, S. Continuous-wave operation of nonpolar GaN-based vertical-cavity surface-emitting lasers. In Proceedings of the Gallium Nitride Materials and Devices XIII, International Society for Optics and Photonics, San Francisco, CA, USA, 23 February 2018; p. 105321C.
9. Wu, J.-Z.; Long, H.; Shi, X.-L.; Ying, L.-Y.; Zheng, Z.-W.; Zhang, B.-P. Reduction of lasing threshold of gan-based vertical-cavity surface-emitting lasers by using short cavity lengths. *IEEE Trans. Electron Dev.* **2018**, *65*, 2504–2508. [CrossRef]
10. Hayashi, N.; Ogimoto, J.; Matsui, K.; Furuta, T.; Akagi, T.; Iwayama, S.; Takeuchi, T.; Kamiyama, S.; Iwaya, M.; Akasaki, I. A GaN-Based VCSEL with a convex structure for optical guiding. *Phys. Status Solidi A* **2018**, *215*, 1700648. [CrossRef]
11. Kuramoto, M.; Kobayashi, S.; Akagi, T.; Tazawa, K.; Tanaka, K.; Saito, T.; Takeuchi, T. Improvement of slope efficiency and output power in GaN-based VCSELs with SiO2-buried lateral index guide. In Proceedings of the 2018 IEEE International Semiconductor Laser Conference, Santa Fe, NM, USA, 16–19 September 2018; pp. 1–2.
12. Chusseau, L.; Philippe, F.; Viktorovitch, P.; Letartre, X. Mode competition in a dual-mode quantum-dot semiconductor microlaser. *Phys. Rev. A* **2013**, *88*, 015803. [CrossRef]
13. Canet-Ferrer, J.; Martínez, L.J.; Prieto, I.; Alén, B.; Muñoz-Matutano, G.; Fuster, D.; González, Y.; Dotor, M.L.; González, L.; Postigo, P.A. Purcell effect in photonic crystal microcavities embedding InAs/InP quantum wires. *Opt. Express.* **2012**, *20*, 7901–7914. [CrossRef] [PubMed]

14. Baba, T.; Sano, D. Low-threshold lasing and Purcell effect in microdisk lasers at room temperature. *IEEE J. Quantum Elect.* **2003**, *9*, 1340–1346. [CrossRef]
15. Haglund, Å.; Hashemi, E.; Bengtsson, J.; Gustavsson, J.; Stattin, M.; Calciati, M.; Goano, M. Progress and challenges in electrically pumped GaN-based VCSELs. In Proceedings of the Semiconductor Lasers and Laser Dynamics VII, International Society for Optics and Photonics Europe, Brussels, Belgium, 28 April 2016; p. 908920Y.
16. Piprek, J.; Farrell, R.; DenBaars, S.; Nakamura, S. Effects of built-in polarization on InGaN-GaN vertical-cavity surface-emitting lasers. *IEEE Photonics Technol. Lett.* **2005**, *18*, 7–9. [CrossRef]
17. Cheng, L.; Cao, C.; Ma, J.; Xu, Z.; Lan, T.; Yang, J.; Chen, H.; Yu, H.; Wu, S.; Yao, S. Suppressed polarization effect and enhanced carrier confinement in InGaN light-emitting diodes with GaN/InGaN/GaN triangular barriers. *J. Appl. Phys.* **2018**, *123*, 223104. [CrossRef]
18. Li, Z.; Lestrade, M.; Xiao, Y.; Li, Z.S. Improvement of performance in p-side down InGaN/GaN light-emitting diodes with graded electron blocking layer. *Jpn. J. Appl. Phys.* **2011**, *50*, 080212. [CrossRef]
19. Zhang, Y.; Kao, T.-T.; Liu, J.; Lochner, Z.; Kim, S.-S.; Ryou, J.-H.; Dupuis, R.D.; Shen, S.-C. Effects of a step-graded $Al_xGa_{1-x}N$ electron blocking layer in InGaN-based laser diodes. *J. Appl. Phys.* **2011**, *109*, 083115. [CrossRef]
20. Lin, B.C.; Chang, Y.A.; Chen, K.J.; Chiu, C.H.; Li, Z.Y.; Lan, Y.P.; Lin, C.C.; Lee, P.T.; Kuo, Y.K.; Shih, M.H. Design and fabrication of a InGaN vertical-cavity surface-emitting laser with a composition-graded electron-blocking layer. *Laser Phys. Lett.* **2014**, *11*, 085002. [CrossRef]
21. Piprek, J.; Li, Z.-M.; Farrell, R.; DenBaars, S.P.; Nakamura, S. Electronic properties of InGaN/GaN vertical-cavity lasers. In *Nitride Semiconductor Devices: Principles*; Piprek, J., Ed.; Wiley-VCH: New York, NY, USA, 2007; pp. 423–445.
22. Xia, C.S.; Li, Z.S.; Lu, W.; Zhang, Z.H.; Sheng, Y.; Hu, W.D.; Cheng, L.W. Efficiency enhancement of blue InGaN/GaN light-emitting diodes with an AlGaN-GaN-AlGaN electron blocking layer. *J. Appl. Phys.* **2012**, *111*, 094503. [CrossRef]
23. Fiorentini, V.; Bernardini, F.; Ambacher, O. Evidence for nonlinear macroscopic polarization in III–V nitride alloy heterostructures. *Appl. Phys. Lett.* **2002**, *80*, 1204–1206. [CrossRef]
24. Wang, C.; Ke, C.; Lee, C.; Chang, S.; Chang, W.; Li, J.; Li, Z.; Yang, H.; Kuo, H.; Lu, T. Hole injection and efficiency droop improvement in InGaN/GaN light-emitting diodes by band-engineered electron blocking layer. *Appl. Phys. Lett.* **2010**, *97*, 261103. [CrossRef]
25. Vurgaftman, I.; Meyer, J.R.; Ram-Mohan, L.R. Band parameters for III-V compound semiconductors and their alloys. *J. Appl. Phys.* **2001**, *89*, 5815–5875. [CrossRef]
26. Turin, V.O. A modified transferred-electron high-field mobility model for GaN devices simulation. *Solid State Electron.* **2005**, *49*, 1678–1682. [CrossRef]

Article

High Lateral Breakdown Voltage in Thin Channel AlGaN/GaN High Electron Mobility Transistors on AlN/Sapphire Templates

Idriss Abid [1,*], Riad Kabouche [1], Catherine Bougerol [2], Julien Pernot [2], Cedric Masante [2], Remi Comyn [3], Yvon Cordier [3] and Farid Medjdoub [1,*]

1 IEMN (Institute of Electronics, Microelectronics and Nanotechnology), Avenue Poincaré, 59650 Villeneuve d'Ascq, France; riad.kabouche@ed.univ-lille1.fr

2 CNRS-Institut Néel, University Grenoble-Alpes, 38000 Grenoble, France; catherine.bougerol@cea.fr (C.B.); julien.pernot@neel.cnrs.fr (J.P.); cedric.masante@neel.cnrs.fr (C.M.)

3 CNRS-CRHEA, University Côte d'Azur, rue Bernard Grégory, 06560 Valbonne, France; remicomyn@gmail.com (R.C.); Yvon.Cordier@crhea.cnrs.fr (Y.C.)

* Correspondence: idriss.abid@ed.univ-lille1.fr (I.A.); farid.medjdoub@univ-lille.fr (F.M.); Tel.: +33-320-19-7840 (F.M.)

Received: 15 September 2019; Accepted: 11 October 2019; Published: 12 October 2019

Abstract: In this paper, we present the fabrication and Direct Current/high voltage characterizations of AlN-based thin and thick channel AlGaN/GaN heterostructures that are regrown by molecular beam epitaxy on AlN/sapphire. A very high lateral breakdown voltage above 10 kV was observed on the thin channel structure for large contact distances. Also, the buffer assessment revealed a remarkable breakdown field of 5 MV/cm for short contact distances, which is far beyond the theoretical limit of the GaN-based material system. The potential interest of the thin channel configuration in AlN-based high electron mobility transistors is confirmed by the much lower breakdown field that is obtained on the thick channel structure. Furthermore, fabricated transistors are fully functional on both structures with low leakage current, low on-resistance, and reduced temperature dependence as measured up to 300 °C. This is attributed to the ultra-wide bandgap AlN buffer, which is extremely promising for high power, high temperature future applications.

Keywords: GaN; high-electron-mobility transistor (HEMT); ultra-wide band gap

---

1. Introduction

AlGaN/GaN high-electron-mobility transistors (HEMTs) is a promising device for high-power and high-voltage electronic applications [1–9]. GaN-based HEMTs are already commercially available for up to 650 V applications. However, they are currently restricted to below 1 kV mainly because of the total buffer thickness limitation with low bow and high crystal quality on large wafer diameters. SiC is another attractive wide bandgap for higher voltage but still has limited impact for the moment because of cost issues. In order to address the dynamic medium and high voltage markets beyond 1200 V while benefiting from low on-resistances, low leakage current, and low switching losses in a cost-effective way, a novel breakthrough in power electronics performance requires a new generation of materials. In this frame, the so-called ultra-wide-bandgap (UWBG) [10–13] materials such as AlN (6.2 eV), which have energy bandgaps that are larger than SiC and GaN, are very promising in enabling the next leap forward in power electronics. An AlN-based material system has a unique advantage due to its prominent spontaneous and piezoelectric polarization effects, but also its flexibility in inserting appropriate heterojunctions, thus dramatically broadening the device's design space. Furthermore, AlN material represents the ideal back barrier for high voltage HEMT applications due to its large electrical breakdown field combined with a high thermal conductivity [14–17]. In turn, the AlN buffer can potentially not only increase the electron confinement in a transistor channel, but can also help to boost the breakdown voltage (BV), owing to its wider bandgap, while benefiting from an enhanced thermal dissipation as compared to GaN-based devices [18–20]. In this paper, we experimentally explore two AlGaN/GaN HEMT structures using a sub-10 nm thin and 240 nm thick GaN channel that is grown on AlN/sapphire templates.

2. Material Description and Device Fabrication

Figure 1 shows a schematic cross section of the AlN-based heterostructures grown by ammonia molecular beam epitaxy (MBE). The first structure, referred to as the thin channel structure, includes a 190 nm AlN buffer that was regrown on a 6 μm thick AlN-on-sapphire commercial template followed by a 8 nm thin GaN channel, a 10 nm AlGaN barrier layer with a high aluminum content (90%), and a 5 nm in-situ SiN cap layer. The growths were performed in a Riber Compact 21T MBE system that was equipped with 80 cc crucible effusion cells that were designed to supply a uniform flux of group III elements on 50 mm wafers [21]. After outgassing under high vacuum at 500 °C, the AlN-on-sapphire templates were annealed under NH_3 (flow rate 200 sccm) at 850 °C prior to the growth of AlN buffer using the same conditions at a growth rate of 100 nm/h. Then, the temperature was reduced to 780 °C in order to grow the rest of the structure, starting with the GaN channel. Based on previous calibration samples, the growth rates of GaN and AlN were then adjusted to produce AlGaN barriers with the desired Al content. While the structure with a 240 nm GaN channel and 29% Al barrier was capped with a thin GaN layer in quite a standard procedure, which produced high performance Radio Frequency HEMTs on silicon [22], the structure with 90% Al barrier was capped with a 5 nm SiN layer in-situ grown at 700 °C using a silicon sublimation source (SUSI) from MBE Komponenten (Weil der Stadt, Germany). The crystal quality of AlN was assessed by X-ray diffraction (XRD) using omega scans. Rocking curves around the symmetric (002) reflections are mainly sensitive to screw type threading dislocations and asymmetric (101) reflections are sensitive to any type of threading dislocations, edge, mixed, and screw type. The full widths at half maximum (FWHM) of AlN (002) and AlN (101) reflection peaks were estimated below 350 arcsec and 500 arcsec, respectively, for both samples, indicating a mean threading dislocation density in the range of fewer than $10^8/cm^2$. On the other hand, the signal of GaN reflections were too weak for the correct determination of (002) and (302) peak widths for the 8 nm channel, whereas they were evaluated at 450 arcsec and 1940 arcsec for the 240 nm GaN layer, indicating a mean threading dislocation density in the range of fewer than $10^9/cm^2$. Interestingly, such values are in a similar range as 1,7 μm thick GaN layers were developed to produce high performance HEMTs on silicon [22,23]. The 2DEG properties that were obtained by the Hall Effect measurements show a sheet carrier density of 1.9×10^{13} cm^{-2} and an electron mobility of 340 $cm^2/V{\cdot}s$. The rather low mobility can be attributed to the thin channel and/or the high Al content into the barrier layer, which can still be optimized, with large room for improvement. A high resolution Transmission Electron Microscopy picture of the active layers is also shown in Figure 1. Sharp interfaces and high material quality are observed. The second structure (Figure 2), which was grown on a similar template, consists of a 240 nm thick GaN channel, a 10 nm AlGaN barrier (29% Al), and is capped with a 2 nm GaN layer. The 2DEG properties show a charge density of 1.1×10^{13} cm^{-2} and a higher electron mobility of 1500 $cm^2/V{\cdot}s$.

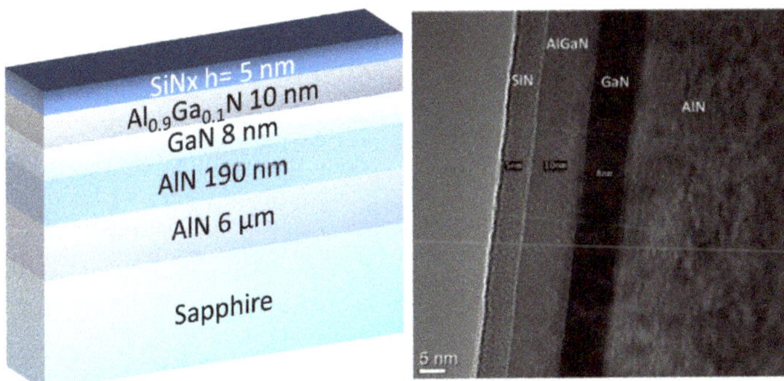

Figure 1. Schematic cross section of the thin channel structure (left) and high resolution Transmission Electron Microscopy picture of the active layers grown by molecular beam epitaxy (MBE) (right).

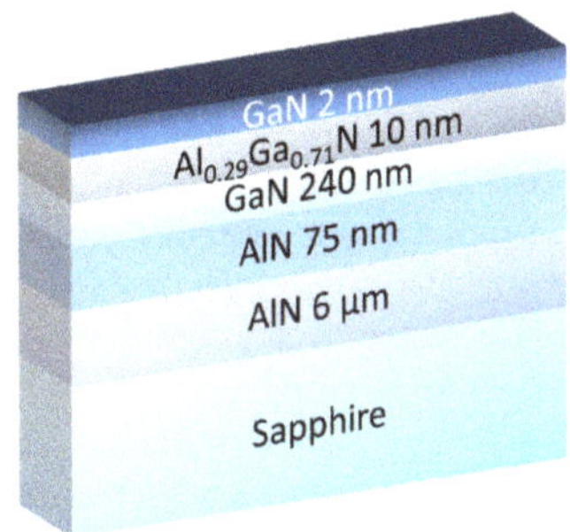

Figure 2. Schematic cross section of the thick channel structure.

For the two structures, a Ti/Al/Ni/Au metal stack was used to form the ohmic contacts on top of the barrier layers by fully etching the cap layers, using a Fluorine-based plasma for the SiN cap layer and a Chlorine-based plasma for the GaN cap layer. A rapid thermal annealing was applied at 750 °C and 875 °C on the thick and thin channel structures, respectively. Contact resistances of about 0.6 Ω·mm were obtained. Then, the device isolation was achieved by nitrogen ion implantation. Ni/Au stack was used as a gate metal and was deposited on top of the cap layers.

3. Results and Discussion

Figure 3 shows typical lateral breakdown measurements performed on isolated contacts with a 2 μm spacing for both structures while the substrate is floating. The probes are emerged in a Fluorinert solution in order to avoid electrical arcing in air under high voltage. The thin channel structure shows a lateral BV that is slightly higher than 1000 V, which corresponds to a remarkable breakdown field above 5 MV/cm. It can be pointed out that this value is far beyond that of a GaN-based material system [24]. To the best of our knowledge, this is the first high voltage demonstration on AlN-based HEMTs using a thin GaN channel. The lateral BV was measured on the thick channel structure in the same way, resulting in 270 V for the same contact distance of 2 μm (<2 MV/cm). As expected, BV is dominated by the thick GaN channel in this case and is thus similar to the best conventional GaN-based HEMTs. However, it appears that the breakdown mechanism may not be a limitation when using a sub-10 nm thickness, enabling it to benefit from the AlN bandgap for short contact distances (i.e., under high electric field).

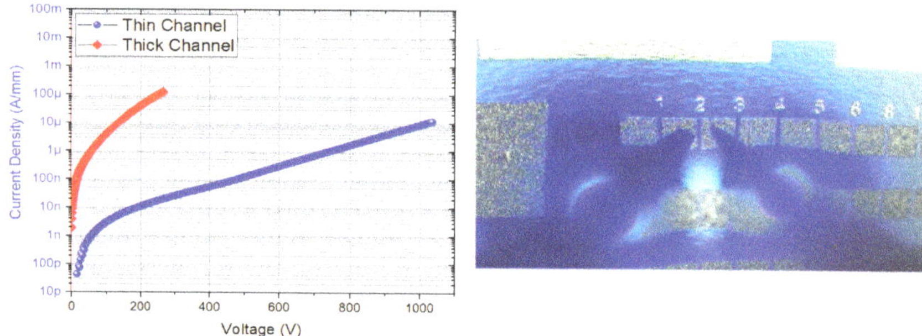

Figure 3. Lateral breakdown voltage at Room Temperature on isolated contacts with a 2 µm distance of thin and thick channel structures. The probes that were emerged in a Fluorinert solution on top of the contacts are also shown.

Figure 4 shows the lateral breakdown measurement on isolated contacts with a 96 µm spacing using a Keysight B1505A with N1268A Ultra High Voltage Expander. A significant lateral breakdown up to 10 kV (limit of the set-up) is achieved with a leakage current of 300 nA/mm on the thin channel structure. The very high blocking voltage and the low leakage current show that the heterostructure does not suffer from any parasitic conduction as full depletion down to the sapphire substrate may occur. The thick channel structure yields a value below 7000 V. On top of the much lower BV, we can also notice a higher leakage level compared to the thin channel structure. Figure 5 represents the lateral BV for various contact distances, confirming a systematically lower breakdown field for the thick channel structure. The lower breakdown field for the larger contact distances can be attributed to the regrown interface states that are activated under very high electric field. Considering that the AlN templates and the MBE regrowth are similar, it can be stated that the sub-10 nm channel thickness is beneficial for high voltage operation in this material system.

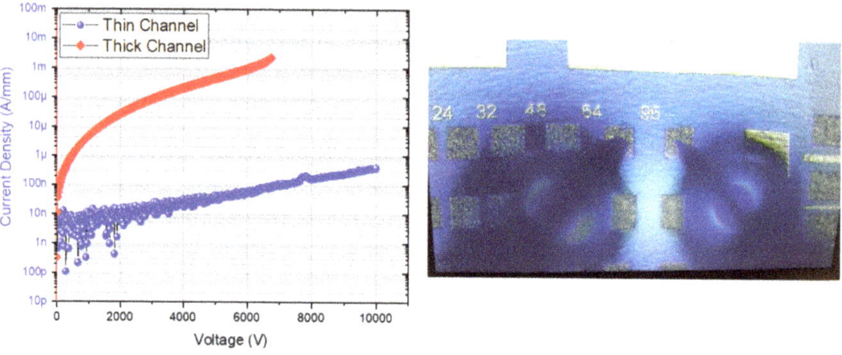

Figure 4. Lateral breakdown voltage at RT on isolated contacts with a 96 µm distance of thin and thick channel structures. The probes that were emerged in a Fluorinert solution on top of the contacts are also shown.

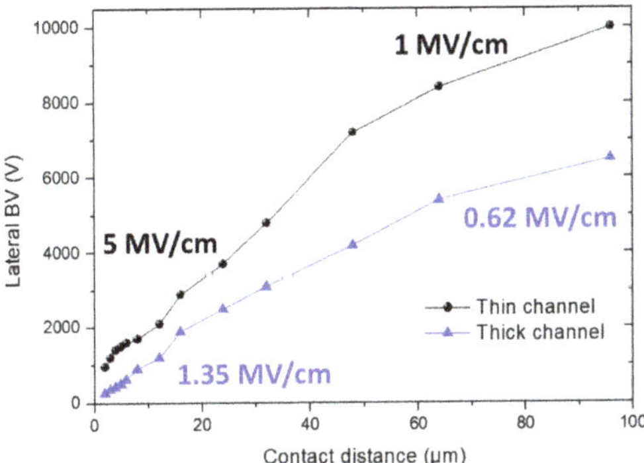

Figure 5. Lateral breakdown voltage at RT on isolated contacts with a 96 µm distance of thin and thick channel structures. The probes that were in a Fluorinert solution on top of the contacts are also shown.

DC characteristics of both structures show fully functional devices, as seen in Figure 6 for both structures. For the thin channel structure, the off-state leakage current at $V_{DS} = 4$ V is around 200 nA/mm and the static on-resistance R_{ON} scales are expected with the gate-drain distance to reach values below 15 mΩ·cm^2 for a 5 µm distance. However, despite fully functional transistors with low leakage current and low on-resistance, a rather limited BV on transistors of about 600 V was measured, which is attributed to the non-optimized SiN passivation layer. On the other hand, in agreement with the buffer characteristics, more than two orders of magnitude higher off-state leakage current is observed on thick channel transistors.

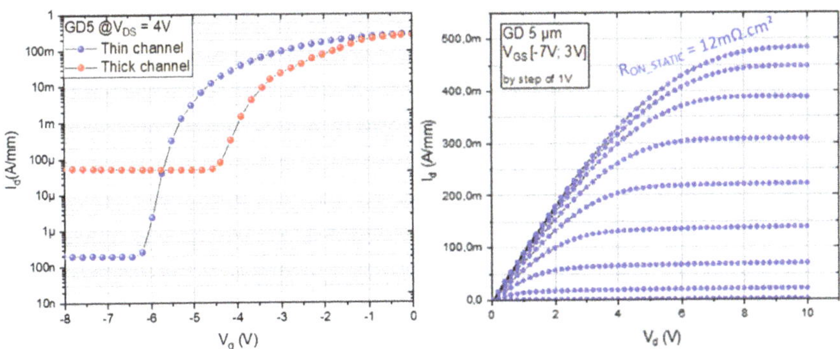

Figure 6. Transfer characteristics (left) for thin and thick channel structures and output characteristics (right) of the thin channel structure at RT.

Figure 7 depicts the transfer characteristics at various temperatures for the thin and thick channel structures at $V_{DS} = 10$ V. A low variation of the off-state leakage current and the threshold voltage (V_{TH}) is observed up to 300 °C even though the thick channel structure delivers a higher leakage current level. It can be stressed that at such a high temperature, AlGaN/GaN HEMTs using standard (Al)GaN-based buffer layers generally show V_{TH} shift and especially significant leakage current increase due to a higher charge conductivity within the buffer layers. Therefore, it appears that the use of AlN material as a buffer may enable it to push the temperature limitation of GaN-based transistors. Furthermore, as expected, the on-state current density drops with the temperature increase due to the decrease of the electron mobility, which is induced by increasing phonon scattering [25]. However, the current density degradation is lower in the case of the thin channel. This indicates that the channel temperature for the thin channel devices is lower under the same bias conditions. This effect is probably due to the lower dislocation density at the GaN/AlN interface in the case of the thin layer, in turn reflecting the benefit of the high AlN thermal conductivity [26,27]. Further experiments such as Infrared/Raman thermography would be needed in order to fully confirm the thermal stability improvement.

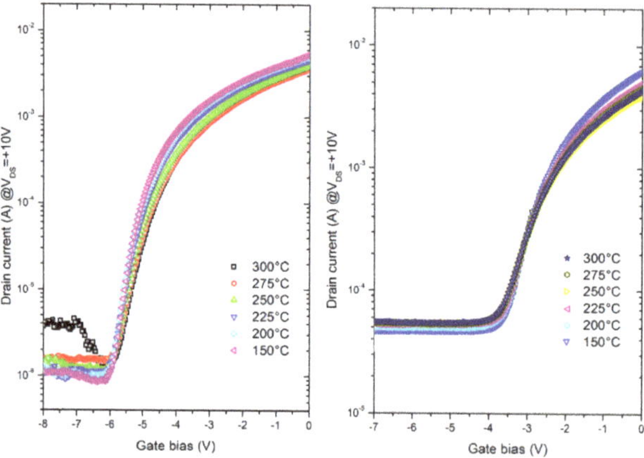

Figure 7. Transfer characteristics at V_{DS} = +10 V as a function of temperature (from 150 °C to 300 °C) of the thin channel structure (left) and the thick channel structure (right).

4. Conclusions

This work presents the fabrication and characterization of thin and thick channel AlGaN/GaN HEMTs grown on AlN/sapphire templates. Lateral buffer breakdown voltage assessment of the thin channel reveals a remarkable breakdown field of 5 MV/cm for short contact distances, which is far beyond the theoretical limits of a GaN-based material system. Furthermore, fabricated transistors are fully functional with low leakage current and low on-resistance. The use of a sub-10 nm ultrathin GaN channel may not be a limiting factor for the breakdown mechanism of transistors. Temperature measurements up to 300 °C show that the AlN buffer enables increased temperature stability of GaN-based transistors. When combined with a thin channel, the related transistors may deliver lower channel temperatures as compared to devices with thicker channels. Proper passivation and associated processing will allow us to take advantage of the properties offered by AlN-based devices, providing both low resistances and high voltages well above 1 kV.

Author Contributions: Device design, fabrication and characterization, I.A., R.K. and F.M., TEM images, C.B., Temperature measurements, J.P., C.M., R.C., Growth, Y.C., Writing-Original Draft Preparation, all co-authors.

Funding: This work is supported by the French RENATECH network and the French National grant ANR-17-CE05-00131 within the project called BREAKUP and ANR-11-LABX-0014 representing GaNeX: the "Investissements d'Avenir".

Conflicts of Interest: The authors declare no conflict of interest.

References

1. Saito, W. Reliability of GaN-HEMTs for high-voltage switching applications. In Proceedings of the International Reliability Physics Symposium, Monterey, CA, USA, 10–14 April 2011.
2. Dogmus, E.; Zegaoui, M.; Medjdoub, F. GaN-on-silicon high-electron-mobility transistor technology with ultra-low leakage up to 3000 V using local substrate removal and AlN ultra-wide bandgap. *Appl. Phys. Express* **2018**, *11*, 034102. [CrossRef]
3. Mishra, U.K.; Parikh, P.; Wu, Y.F. AlGaN/GaN HEMTs: An overview of device operation and applications. *Proc. IEEE* **2002**, *90*, 1022–1031. [CrossRef]
4. Roccaforte, F.; Greco, G.; Fiorenza, P.; Iucolano, F. An Overview of Normally-Off GaN-Based High Electron Mobility Transistors. *Materials* **2019**, *12*, 1599. [CrossRef] [PubMed]

5. Baliga, B.J. Gallium nitride devices for power electronic applications. *Semicond. Sci. Technol.* **2013**, *28*, 074011. [CrossRef]
6. Pavlidis, G.; Kim, S.H.; Abid, I.; Zegaoui, M.; Medjdoub, F.; Graham, S. The Effects of AlN and Copper Back Side Deposition on the Performance of Etched Back GaN/Si HEMTs. *IEEE Electron Device Lett.* **2019**, *40*, 1060–1063. [CrossRef]
7. Ikeda, N.; Niiyama, Y.; Kambayashi, H.; Sato, Y.; Nomura, T.; Kato, S.; Yoshida, S. GaN Power Transistors on Si Substrates for Switching Applications. *Proc. IEEE* **2010**, *98*, 1151–1161. [CrossRef]
8. Kambayashi, H.; Satoh, Y.; Ootomo, S.; Kokawa, T.; Nomura, T.; Kato, S.; Chow, T.S. Over 100 A operation normally-off AlGaN/GaN hybrid MOS-HFET on Si substrate with high-breakdown voltage. *Solid-State Electron.* **2010**, *54*, 660–664. [CrossRef]
9. Herbecq, N.; Roch-Jeune, I.; Rolland, N.; Visalli, D.; Derluyn, J.; Degroote, S.; Germain, M.; Medjdoub, F. 1900 V, 1.6 mΩ cm2 AlN/GaN-on-Si power devices realized by local substrate removal. *Appl. Phys. Express* **2014**, *7*, 034103. [CrossRef]
10. Tsao, J.Y.; Chowdhury, S.; Hollis, M.A.; Jena, D.; Johnson, N.M.; Jones, K.A.; Kaplar, R.J.; Rajan, S.; Van de Walle, C.G.; Bellotti, E.; et al. Ultrawide-Bandgap Semiconductors: Research Opportunities and Challenges. *Adv. Electron. Mater.* **2018**, *1*, 1600501. [CrossRef]
11. Kaplar, R.J.; Allerman, A.A.; Armstrong, A.M.; Crawford, M.H.; Dickerson, J.R.; Fischer, A.J.; Baca, A.G.; Douglas, E.A. Review-Ultra-Wide-Bandgap AlGaN Power Electronic Devices. *ECS J. Solid State Sci. Technol.* **2017**, *6*, Q3061. [CrossRef]
12. Razzak, T.; Xue, H.; Xia, Z.; Hwang, S.; Khan, A.; Lu, W.; Rajan, S. Ultra-wide band gap materials for high frequency applications. In Proceedings of the IEEE MTT-S International Microwave Workshop Series on Advanced Materials and Processes for RF and THz Applications (IMWS-AMP), Ann Arbor, MI, USA, 16–18 July 2018.
13. Anderson, T.J.; Hite, J.K.; Ren, F. Ultra-Wide Bandgap Materials and Device. *ECS J. Solid State Sci. Technol.* **2017**, *6*, Y1. [CrossRef]
14. Shealy, J.R.; Kaper, V.; Tilak, V.; Prunty, T.; Smart, J.A.; Green, B.; Eastman, L.F. An AlGaN/GaN high-electron-mobility transistor with an AlN sub-buffer layer. *J. Phys. Condens. Matter* **2002**, *14*, 3499. [CrossRef]
15. Yafune, N.; Hashimoto, S.; Akita, K.; Yamamoto, Y.; Tokuda, H.; Kuzuhara, M. AlN/AlGaN HEMTs on AlN substrate for stable high-temperature operation. *Electron. Lett.* **2014**, *50*, 211–212. [CrossRef]
16. Wu, Y.F.; Kapolnek, D.; Ibbetson, J.P.; Parikh, P.; Keller, B.P.; Mishra, U.K. Very-high power density AlGaN/GaN HEMTs. *IEEE Trans. Electron Devices* **2001**, *48*, 586–590.
17. Kinoshita, T.; Nagashima, T.; Obata, T.; Takashima, S.; Yamamoto, R.; Togashi, R.; Kumagai, Y.; Schlesser, R.; Collazo, R.; Koukitu, A.; et al. Fabrication of vertical Schottky barrier diodes on n-type freestanding AlN substrates grown by hydride vapor phase epitaxy. *Appl. Phys. Exp.* **2015**, *8*, 061003. [CrossRef]
18. Cahill, D.G. Analysis of heat flow in layered structures for time-domain thermoreflectance. *Rev. Sci. Instrum.* **2004**, *75*, 5119–5122. [CrossRef]
19. Choi, S.R.; Kim, D.; Choa, S.H.; Lee, S.H.; Kim, J.K. Thermal Conductivity of AlN and SiC Thin Films. *Int. J. Thermophys.* **2006**, *27*, 896–905. [CrossRef]
20. Strite, S.; Morkoç, H. GaN, AlN, and InN: A review. *J. Vac. Sci. Technol. B* **1992**, *10*, 1237. [CrossRef]
21. Cordier, Y.; Pruvost, F.; Semond, F.; Massies, J.; Leroux, M.; Lorenzini, P.; Chaix, C. Quality and uniformity assessment of AlGaN/GaN Quantum Wells and HEMT heterostructures grown by molecular beam epitaxy with ammonia source. *Phys. Stat. Sol. C* **2006**, *3*, 2325–2328. [CrossRef]
22. Cordier, Y. Al(Ga)N/GaN high electron mobility transistors on silicon. *Phys. Status Solidi A* **2015**, *212*, 1049–1058. [CrossRef]
23. Baron, N.; Cordier, Y.; Chenot, S.; Vennégués, P.; Tottereau, O.; Leroux, M.; Semond, F.; Massies, J. The critical role of growth temperature on the structural and electrical properties of AlGaN/GaN high electron mobility transistor heterostructures grown on Si (111). *J. Appl. Phys.* **2009**, *105*, 033701. [CrossRef]
24. Rowena, I.B.; Selvaraj, S.L.; Egawa, T. Buffer thickness contribution to suppress vertical leakage current with high breakdown field (2.3 MV/cm) for GaN on Si. *IEEE Electron Device Lett.* **2011**, *32*, 1534–1536. [CrossRef]
25. Khan, M.N.; Ahmed, U.F.; Ahmed, M.M.; Rehman, S. An improved model to assess temperature-dependent DC characteristics of submicron GaN HEMTs. *J. Comput. Electron.* **2018**, *17*, 653–662. [CrossRef]

26. Razeeb, K.M.; Dalton, E.; Cross, G.L.; Robinson, A.J. Present and future thermal interface materials for electronic devices. *Int. Mater. Rev.* **2017**, *63*, 1–21. [CrossRef]
27. Cho, J.; Bozorg-Grayeli, E.; Altman, D.H.; Asheghi, M.; Goodson, K.E. Low Thermal Resistances at GaN–SiC Interfaces for HEMT Technology. *IEEE Trans. Electron Devices Lett.* **2012**, *33*, 378–380. [CrossRef]

Review

Vertical GaN-on-GaN Schottky Diodes as α-Particle Radiation Sensors

Abhinay Sandupatla [1,*], Subramaniam Arulkumaran [2,3], Ng Geok Ing [1,2,*], Shugo Nitta [3], John Kennedy [4] and Hiroshi Amano [3]

1 School of Electrical and Electronics Engineering, Nanyang Technological University, Singapore 639798, Singapore
2 Temasek Laboratories in Nanyang Technological University, Research Techno Plaza, 50 Nanyang Drive, Singapore 639798, Singapore; Subramaniam@ntu.edu.sg
3 Center for Integrated Research of Future Electronics (CIRFE), IMaSS, Nagoya University, Nagoya 464-8603, Japan; nitta@nagoya-u.jp (S.N.); amano@nuee.nagoya-u.ac.jp (H.A.)
4 National Isotope Center, GNS Science, Lower Hutt 5010, New Zealand; J.Kennedy@gns.cri.nz
* Correspondence: abhinay001@e.ntu.edu.sg (A.S.); eging@ntu.edu.sg (N.G.I.)

Received: 16 April 2020; Accepted: 8 May 2020; Published: 20 May 2020

Abstract: Among the different semiconductors, GaN provides advantages over Si, SiC and GaAs in radiation hardness, resulting in researchers exploring the development of GaN-based radiation sensors to be used in particle physics, astronomic and nuclear science applications. Several reports have demonstrated the usefulness of GaN as an α-particle detector. Work in developing GaN-based radiation sensors are still evolving and GaN sensors have successfully detected α-particles, neutrons, ultraviolet rays, x-rays, electrons and γ-rays. This review elaborates on the design of a good radiation detector along with the state-of-the-art α-particle detectors using GaN. Successful improvement in the growth of GaN drift layers (DL) with 2 order of magnitude lower in charge carrier density (CCD) ($7.6 \times 10^{14}/cm^3$) on low threading dislocation density ($3.1 \times 10^6/cm^2$) hydride vapor phase epitaxy (HVPE) grown free-standing GaN substrate, which helped ~3 orders of magnitude lower reverse leakage current (I_R) with 3-times increase of reverse breakdown voltages. The highest reverse breakdown voltage of −2400 V was also realized from Schottky barrier diodes (SBDs) on a free-standing GaN substrate with 30 μm DL. The formation of thick depletion width (DW) with low CCD resulted in improving high-energy (5.48 MeV) α-particle detection with the charge collection efficiency (CCE) of 62% even at lower bias voltages (−20 V). The detectors also detected 5.48 MeV α-particle with CCE of 100% from SBDs with 30-μm DL at −750 V.

Keywords: GaN-on-GaN; schottky barrier diodes; high-energy α-particle detection; low voltage; thick depletion width detectors

---

1. Introduction

Nuclear reactions normally emit different high-energy particles such as α-particle, β-particles, neutrons, γ-radiation and x-rays. As each of these emitted particles interacts with matter differently, a study into α-particle detection is highly important [1]. Due to the high mass and density of an α-particle, the distance traveled by α-particle is limited to only a few centimeters during which it loses all its energy along the path. In general, α-particles traverse only a few microns in any solid before losing all its energy. The energy transferred from the α-particle gets converted into heat. This distance traversed by α-particle plays an important role as only the atoms in this area can interact with the α-particle.

1.1. Gaseous Ionization Detectors

Ionization detectors are used to detect ionizing radiation like α-particle and β-particle. Figure 1 shows a typical schematic of an ionization chamber in which an external voltage is applied to keep the conditions in ionization region. A basic ionization detector consists of a chamber filled with a suitable gaseous medium (see Figure 1). Ionization detector is dependent on the effect of a charged particle passing through the gaseous medium. The gaseous medium should have the following qualities:

- chemically stable and inert
- low ionization energy realizes maximum ionization of the medium
- low sensitivity to radiation damage to realize a longer lifetime of the detector.

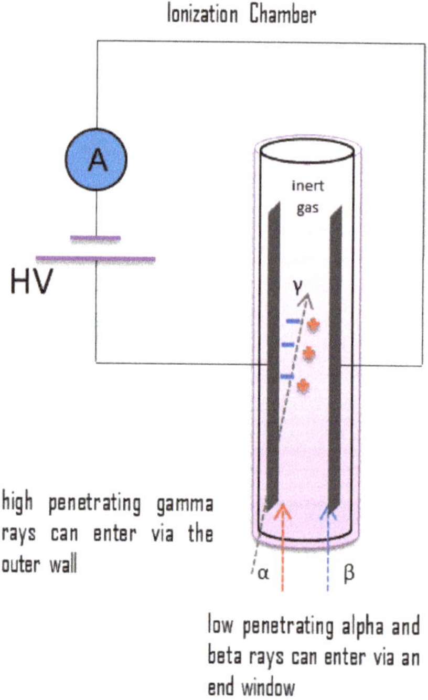

Figure 1. Schematic set-up of an Ionization detector [2,3].

Typically noble gases like helium (He) and argon (Ar) are used in nuclear power plants to measure α-particles, β-particles and γ-rays.

The ionization chamber has two electrodes across which a very large voltage (>1 kV) is held. When ionizing radiation enters the ionization chamber it generates electron-ion pairs, whose behavior is dependent on the external electric field. Under the high electric field, the generated electron-ion pairs move towards opposite electrodes as the extremely high electric field prevents their recombination.

Ionization chambers are preferred for high radiation dose rates as they do not have any dead time as these detectors have no inherent amplification of the signal. The absence of an amplification component enables the use of ionization chamber immediately after large current detection. In addition, the absence of amplification also helps to provide excellent energy resolution as amplification increases electronic noise.

Although Ionization chambers have many advantages discussed above, the high voltage requirement and the required directionality of incident α-particles restricts its use. The use of a gas chamber increases the fragility of the equipment and reduces portability of the detector. Further details on the principle and method of operation of ionization detectors can be found in Lamarsh, J.R. et al. [1], Burn, R.R. et al. [4] and Rossi, B. et al. [5].

1.2. Scintillation Detector

Scintillation is a flash of light observed when a transparent material interacts with a charged particle. By detecting the flashes of light produced by a scintillator using a photodetector detection of radiation is possible (see Figure 2). A scintillation detector mainly consists of two key elements.

- Scintillator—it generates photons in response to incident radiation.
- Photodetector—a sensitive photodetector converts the incident light into an electrical signal.

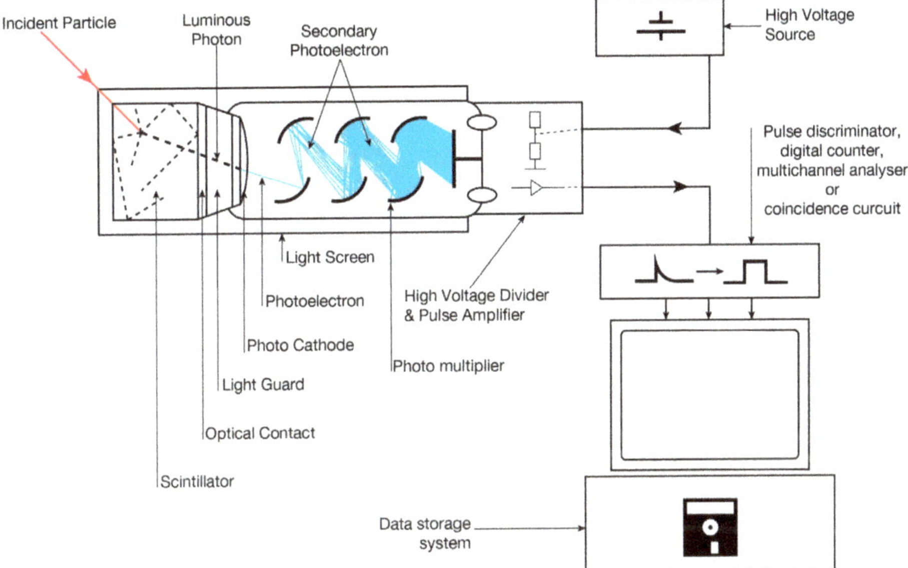

Figure 2. Schematic set-up of a Scintillation detector [3].

The basic operating principle of the scintillation detector involves the conversion of incident radiation energy to optical energy by a scintillator, which produces flashes of varying intensity. The intensity of the optical energy generated is dependent on the energy of the incident radiation (see Figure 2).

Scintillation detectors are highly beneficial for its high efficiency, precision and counting rates.

The intensity of flashes generated by the scintillator and the output voltage is directly proportional to the energy of the incident particle. Therefore scintillators can be used for the determination of the number of incident α-particles and their energy.

The use of Scintillation based detector involves multiple energy conversions resulting in conversion losses and introduction of noise. Along with multiple energy conversions presence of amplifiers in the circuit results in a significant dead time which reduces the maximum dosage of α-particles which can be detected. The amplifier also introduces electronic noises resulting in poorer energy resolution. Apart from the disadvantages due to the setup, scintillation based detectors generate 1/10 the intensity of light due to the incidence of α-particles when compared electrons of the same energy. This reduced sensitivity is due to α-particles being heavier in comparison to electrons. Inorganic crystals like ZnS are generally used in the fabrication of scintillators [1,2]. Further details on the principle and method of operation of scintillation detectors can be found in Teo, W.R. et al. [6] and Knoll, G. et al. [7].

1.3. Solid-State Semiconductor Detectors

Solid-state detectors are made of semiconductors and operate by generating current on interaction with ionizing radiation. The interaction of semiconductor material with ionizing radiation like α-particle results in the excitation of an electron. This excited electron moves out of its energy level creating electron-hole (e-h) pairs. The energy of the incident radiation particle is utilized to generate multiple e–h pairs, hence higher the incident energy higher the e-h pair generation. Figure 3 shows a schematic of a semi-insulating GaAs α-particle detector.

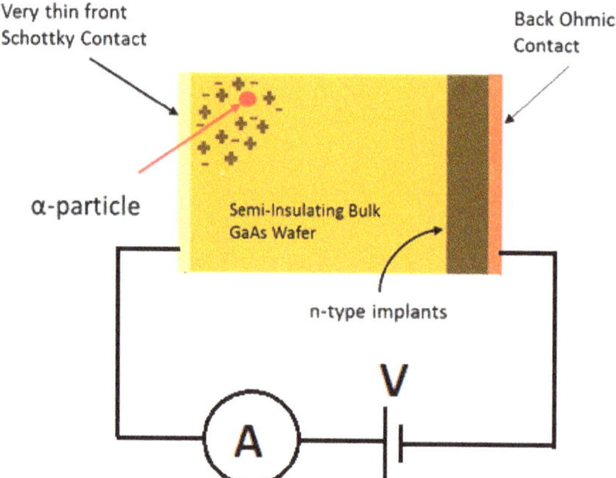

Figure 3. Schematic diagram of GaAs Schottky Barrier Diodes for α-particle detection.

Operating principles of Semiconductor radiation detectors:

- Ionizing radiation enters the depletion width (DW) of Schottky Barrier Didoes.
- Radiation passing through the DW generates e-h pairs. The number of e-h pairs generated depends on the incident energy of the particle.
- The external electric field forces e-h pairs to traverse to the electrodes and result in a pulse which can be recorded by an external circuit. The ratio of detected energy to incident energy has been termed as charge collection efficiency (CCE).
- Generated pulse carries information about the energy of the incident radiation and the number of such pulses gives the intensity of the radiation.

Based on the principles of operation of semiconductor detectors, for their successful operation, the following considerations are important. These considerations can be classified into material characteristics and device characteristics.

- Material Characteristics—Material characteristics like the displacement energy (E_d) and e-h pair creation energy of the semiconductor play a vital role. While E_d determines the lifetime of the detector, e-h pair creation energy determines the sensitivity of the detector.
- Device characteristics—Electrical characteristics of the fabricated device like the generated DW and leakage current determine the maximum detectable energy incident on the detector and the sensitivity of the detector.

Properties of Selected Semiconductors

While electrical properties like E_d and e-h pair creation energy of the semiconductors play a vital role in determining the detector performance, other properties like thermal conductivity, bandgap, breakdown strength and so forth, also play a role. Bandgap defines the lowest detectable energy, thermal conductivity regulates the maximum operable temperature range and breakdown strength determines the maximum DW, which can be generated for any material. A summary of important material characteristics has been listed in Table 1 below.

Table 1. Material characteristics of different semiconductors used for the α-particle detection along with the best-reported detector performances.

Properties	Si	GaAs	SiC	GaN	Diamond
Band gap (eV)	1.12	1.4	3.3	3.4	6
Electron Mobility (cm^2/V·s)	1450	8500	800–1000	1000	1800–2200
Sat. elec. Drift velocity ($\times 10^7$ cm/s)	1.0	1.2	2	2	2.7
Breakdown Strength (MV/cm)	0.5	0.4	2.2	3.3	10
e-h pair creation (eV)	3.6	4.3	7.8	8.9	13
Displacement Energy, E_d (eV) [8]	13	10	C-20 Si-35 [9]	Ga-45 N-109 [10]	35 [11]
Bias Voltage Required to achieve 100% CCE (V)	24 [12]	150 [13]	200 [14]	300 [15]	15 [16]
Detected Spectral Energy Resolution (%)	0.23	15	0.25	1.29	0.35

Si-based radiation detectors—Si has a decently high E_d of 13 eV accompanied by a well-developed device fabrication technology [8]. While high E_d results in a good life-time of the detector, the established fabrication technology is important to fabricate detectors with wide DW and low leakage currents. Diodes are made from narrow strips of Si (~100 μm), which are then reverse biased to generate a thick DW. As α-particles pass through this DW, they cause small ionization currents that are detected and measured. Arranging multiple such thin detectors in an array can provide an accurate picture of the α-particle distribution in a measurement setup. Such strip detectors are widely used in the Inner Tracking System (ITS) of A Large Ion Collider Experiment (ALICE) [3]. The matured Si technologies have resulted in the development of Si detectors with the lowest energy resolution (0.23%) [12] which is very advantageous in differentiating various spectra of α-particles based on their energies.

Similarly, other semiconductors like GaAs were also explored as an alternative to Si-based detectors. Although GaAs has lower E_d at 10 eV [8], GaAs is a direct energy bandgap semiconductor that allows direct transition of electrons from the valence band to conduction band without any change in their momentum. Electrons in the direct conduction band valley experience very high mobility (~7000 cm^2/V·s) which helps the device function at lower voltages. The high electron mobility in GaAs accompanied with developed growth process development of GaAs have led to the development of Semi-insulating GaAs based α-particle detectors. The low doping density helps generate a thick DW at even at low voltages resulting in decently good energy resolution (0.89%) [17]. The primary drawback of both GaAs and Si-based detectors is that they have low E_d resulting in a lower lifetime of the detector.

Although SiC, GaN and Diamond are not as developed in terms of growth and fabrication, their material characteristics exhibit their immense potential as radiation detectors due to their higher E_d. Among the semiconductors listed in the Table 1, diamond shows the best material characteristics for radiation detection [16], which has resulted in the fabrication of α-particle detectors with a high energy resolution of 0.35% while detecting 5.48 MeV α-particles with 100% CCE at 15 V. Despite the superior material characteristics, the difficulties involved with the growth of a single crystalline diamond accompanied by the cost involved restricts the usage of single crystalline diamond for radiation detection. While SiC based α-particle detector have performed exceedingly well in low voltages with good energy resolution (0.25%), the E_d of SiC is still lower than GaN making GaN a better choice for fabricating α-particle detectors.

2. GaN α-Particle Detector

Among the III-V semiconductors, Gallium Nitride (GaN) emerged as the best semiconductor materials for lighting [18–20], electronic [21,22] and sensing applications [23,24] due to their superior inherent material properties such as a high direct bandgap, critical electric field, electron and saturation velocity in comparison with other popular semiconductors. High energy bandgap accompanied with large theoretical E_d (109 eV for N and 45 eV for Ga) [25] and high thermal stability (melting point 3500 K at a 9 GPa pressure [26]) has also resulted in GaN being used for radiation detection applications [27]. Compared with semiconductor materials like Si and GaAs, GaN can operate at higher temperatures for a longer time. A review article by Sellin, P.J. in 2006 compared different wide bandgap semiconductors in high radiation environments and concluded that GaN was a promising candidate for α-particle detection despite GaN being relatively immature as a semiconductor [28].

The first group to report an α-particle detector fabricated on GaN employed a 2–2.5 μm thick GaN layer grown by metalorganic chemical vapor deposition (MOCVD) on a sapphire substrate. While these detectors performed reasonably well, their performance was highly limited due to the thin DW and high leakage currents in the devices. The absence of free-standing GaN substrate resulted in hetero-epitaxial GaN which used sapphire, Si or SiC as its substrates. The high lattice mismatch between epitaxial GaN with its substrate has resulted in high threading dislocation density (TDD). High TDD increases the reverse leakage current (J_R), which is detrimental to the α-particle detector performance.

With the improvement in GaN growth technology researchers developed free-standing GaN with low TDD and thereby GaN-on-GaN wafers. This led to the development of Schottky barrier diodes (SBD) which have thick epitaxial layer and low TDD. Zhao et al. has previously reported the effects of reduced TDD by comparing SBD characteristics of GaN on sapphire and GaN-on-GaN SBDs. Use of GaN substrate has helped reduce TDD by ~3 orders of magnitude (see Table 2) thereby reduce leakage current by more than ~6 orders of magnitude [29].

Table 2. Lowest reported threading dislocation density (TDD) in GaN drift layers on different substrates.

Parameter	Si	Sapphire	SiC	GaN
Lattice mismatch	−17%	−33%	3.5%	0
Thermal Mismatch	116%	−23%	24%	0
TDD (/cm^2)	~10^9	~10^9	~10^7	~10^4

In addition to the semiconductor material properties, the electrical properties of DW also affect the performance of the detector. Devices like p-n diode, pin diode and Schottky diode structures can be used to generate a DW. Detector performance widely depends on the device fabricated.

2.1. p-n Diodes

Sugiura, M. et al. has recently reported for the first time a p-n diode-based α-particle detector. Figure 4 shows the schematic of GaN p-n diode used for α-particle detection [30]. These p-n diodes exhibited a mobility/life-time product of 4.6×10^{-5} cm^2/V which is lower than the values reported for CdTe (~10^{-3} cm^2/V) and TlBr (~10^{-3} cm^2/V). From the values of mobility/life-time product M. Sugiura et al. also concluded that GaN is a suitable material for radiation sensing applications.

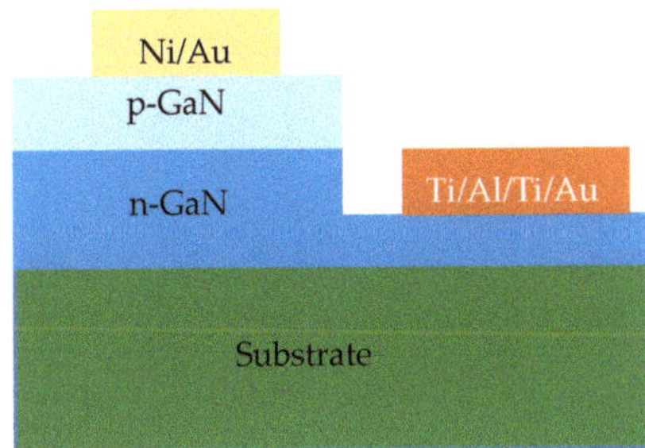

Figure 4. Cross-sectional schematic of GaN p-n diodes for α-particle detectors.

2.2. PIN Diodes

Wang, G. et al. has reported a PIN diode based α-particle detector. The use of an 8 μm intrinsic GaN layer between p-GaN and n-GaN increases the thickness of formed DW, which in turn helps to detect higher energy particles with improved sensitivities [31].

Figure 5 shows the diode schematic of the PIN α-particle detector used to detect 700 keV α-particles with a CCE of 80% and an energy resolution of 50%.

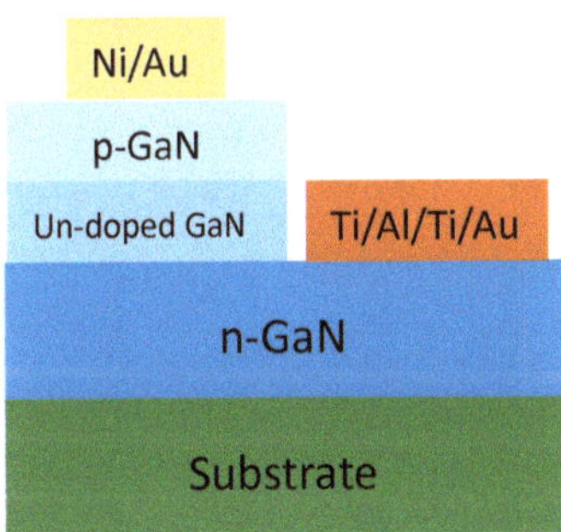

Figure 5. Cross-sectional schematic of GaN PIN diode structure for α-particle detector.

Authors also predict the use of a thicker intrinsic layer may increase detectable energy and reduce energy resolution.

2.3. Schottky Barrier Diodes

Unlike the p-n diodes and PIN diodes, SBDs have been the most popular GaN device for radiation detection. Multiple research groups have fabricated different structures at different stages of the development of GaN growth technologies. The various schematics of different structures of GaN SBD based α-particle detectors have been reviewed by Wang, J. et al. [32].

The best performing α-particle detector structure is shown in Figure 6. This kind of structure is called a sandwich structure. The use of a thick GaN layer sandwiched between both electrodes helps generate a thick DW hence detect higher energies of α-particles. This structure was first employed by Lee et al. [33] reported the first implementation of a sandwich structure, in which the GaN layer had unintentional H5 traps in the top 30 μm of the active area. These H5 traps resulted in reducing the charge carrier density (CCD) resulting in the detection of 5.1 MeV α-particle energy with 90% CCE. The sandwich structure was also fabricated by Mulligan et al. [34] but the detector had a very high CCD ($10^{16}/cm^3$) resulting in the formation of a thin DW and detection of only 325 keV α-particles. Most recently, Xu, Q. et al. has reported α-particle detector based on a sandwich structure that can detect 5.48 MeV α-particles with 100% CCE at −550 V [35], which is the highest detected α-particle energy.

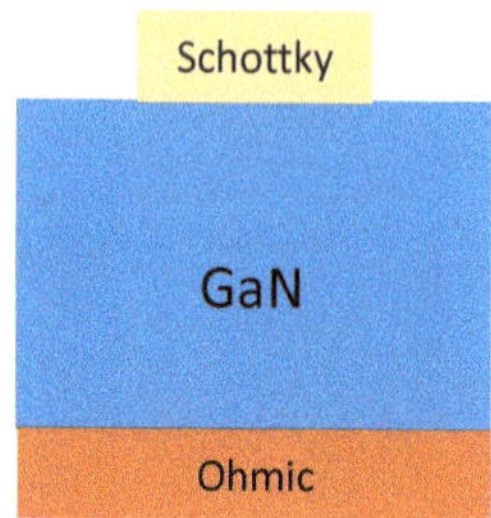

Figure 6. Sandwich structure of alpha particle detectors.

In comparison to the thin film structures, sandwich structures with free-standing GaN substrates have lower TDD. surface morphology Thicker DW helps to detect higher energy particles. Other than TDD, CCD also plays an important role in the generation of a thick DW. Higher CCD reduces DW, hence it is mandatory to reduce the CCD of the detector to detect higher energies. The use of a bulk GaN substrate with low CCD increases the resistance of the detector thereby increasing the voltage required to function at full potential. Table 3 lists all reported GaN α-particle detectors including their structure, detected energy and CCE.

Table 3. The state-of-the-art GaN-based α-particle detectors.

Affiliation	Type of Detector	DLT (μm)	Reverse Bias (V)	Source Energy (MeV)	Det. Energy (MeV)	CCE
Vilnius Univ. [36]	DSC	2–2.5	28	5.48	0.55	92%
Institute of Rare Metals [37]	Mesa	3	–	5.157	1.129	100%
Univ. of Glasgow [38]	Mesa	2.5	16	5.48	0.477	94%
		12	150	5.48	1.2	53%
Ohio State Univ. [34]	Sandwich	bulk	20	5.48	0.325	6%
Ohio State Univ. [35]	Sandwich	bulk	550	5.48	5.48	100%
Chonbuk National Univ. [33]	Sandwich	30/bulk	120	5.1	4.59	90%

From Table 3, it can be observed that higher energy has been detected by sandwich structures but they require high reverse voltage bias conditions for successful operation. To overcome the high-voltage requirement of a sandwich structure, a low CCD epitaxial layer on highly doped GaN substrate could be used [15]. The low CCD epitaxial layer helps to operate the detector at lower voltages however, highly doped substrate helps to form a low resistance Ohmic contacts.

3. Design Considerations and Material Characteristics

In order to improve the low voltage functionality of a GaN-based α-particle detector, an epitaxial layer whose thickness is corresponding to the target detectable α-particle is required. The required thickness of DW can be simulated using stopping range of ions in matter (SRIM) [39].

From the simulation results shown in Figure 7, 14.58 μm was determined to be the minimum DW required to detect 5.48 MeV α-particle energy generated from a ^{241}Am source. In order to generate a 14.58 um DW SBDs with 15 μm and 30 μm drift layer thickness with very low CCD need to be fabricated.

Material characteristics of the GaN DL like crystalline quality, threading dislocation density (TDD), surface morphology and CCD of GaN DL play an important role in determining the detector performance. Use of thick GaN substrate to grow the 15 μm and 30 μm GaN DL has ensured the high crystalline quality of the DL which was measured by 2 theta-omega scan using XRD. The full-wave half maximum (FWHM) was measured at 108.4 arc.sec and 260.6 arc.sec in 002 and 102 orientations, respectively (see Figure 8a and Table 4). These values of measured FWHM are lower than the maximum reported values of 310 arc.sec (002) and 350 arc.sec (102) [40]. The use of a GaN substrate has also reduced the TDD in DL generated due to lattice mismatch between the substrate and the DL. A TDD of

$3.6 \times 10^6/cm^2$ was measured using multiphoton excitation photoluminescence microscopy (MPPL) [41] which was similar to the TDD of the GaN substrate (see Figure 8b and Table 4). Polishing of the DL has reduced the rms roughness of the surface of the DL 0.206 nm (see Figure 8c and Table 4). While to reduce the unintentional n-type CCD and increase DW p-type dopant (Mg) was doped in the drift layer [42,43]. The presence of Mg and its concentration in the DL was extracted from Secondary Ions Mass Spectroscopy (SIMS) analysis. The reduced CCD of $7.6 \times 10^{14}/cm^3$ was measured from the elemental concentrations and verified Hall measurements (see Figure 8d and Table 4).

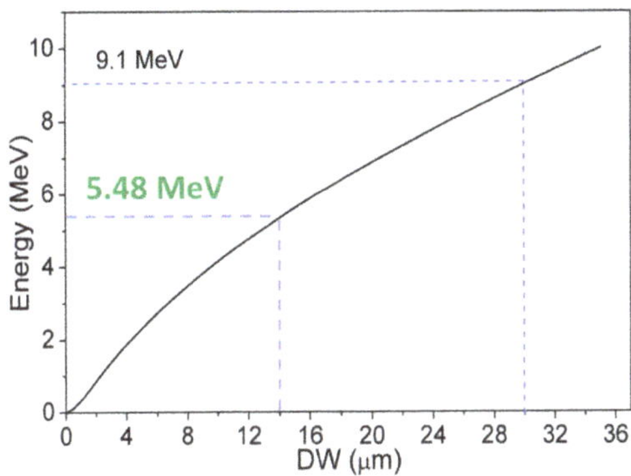

Figure 7. α-particle range in GaN calculated by stopping range of ions in matter (SRIM).

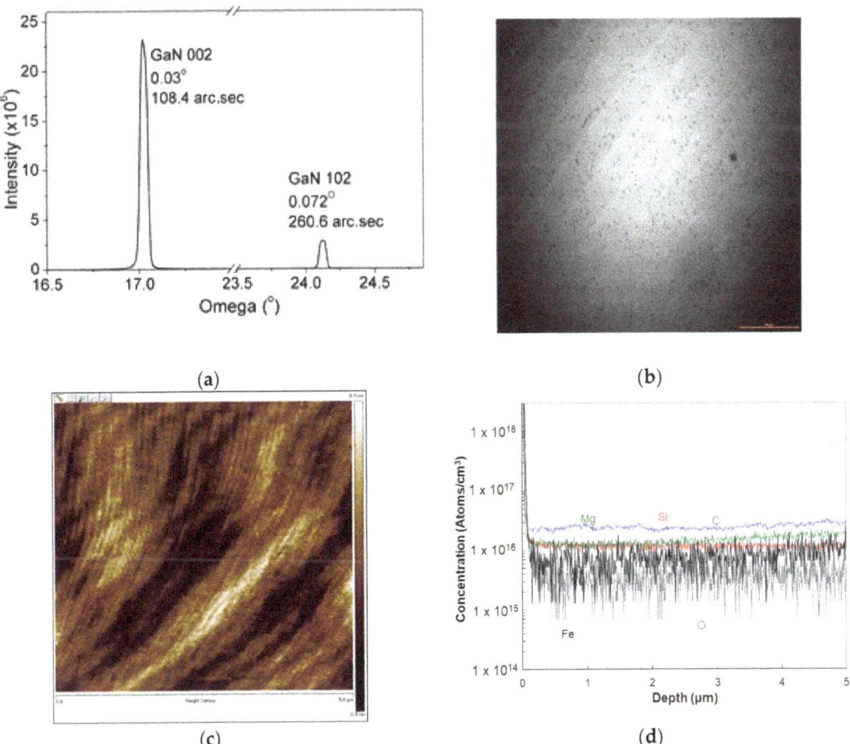

Figure 8. (**a**) X-ray diffraction (XRD), (**b**) TDD (MPPL), (**c**) Atomic force microscopy (AFM) and (**d**) SIMS Characteristics of the wafer.

Table 4. Material properties of MOCVD grown GaN drift layers (DL)s such as 2 theta-omega scan, root mean square (RMS) surface roughness, TDD, elemental concentrations and charge carrier density (CCD) measured by SIMS (Si limit 1.0×10^{14} and Mg limit 2.0×10^{14}) and Hall.

	XRD (2 theta-omega scan)		TDD	RMS	Si conc.	Mg conc.	CCD = N_D-N_A (/cm^3)	
DLT (µm)	002 (arc.sec)	102 (arc.sec)	(MPPL) ($\times 10^6$/cm^2)	Roughness (AFM) (nm)	(/cm^3) (N_D)	(/cm^3) (N_A)	SIMS	Hall
15	108.4	260.6	3.3	0.206	1.5×10^{16}	1.5×10^{16}	7.6×10^{14}	7.5×10^{14}
30	130	236	4.2	0.210	6×10^{15}	1×10^{15}	5×10^{15}	5.2×10^{15}

4. Detector Fabrication

The SBD fabrication started with a complete cleaning of the wafer with piranha solution (H_2SO_4:H_2O_2 = 4:1) and organic cleaning (acetone and isopropanol) followed by dipping the wafer in buffered oxide etchant (BOE) for 2 min for the formation of an excellent metal-semiconductor interface [44]. After the preparation of the surface, the ohmic contact was formed by depositing Ti/Al/Ni/Au (20/120/40/50 nm) at the bottom of the wafer (N-face) of the wafer using e-beam, followed by rapid thermal annealing at 775 °C for 30 s in N_2 ambience. Ti acts as the first layer of Ohmic stack which forms a low-resistance contact, as Ti helps in the generation of N-vacancies after annealing, which increases CCD and promotes tunneling [45]. The second layer deposited was Al which is used to absorb excessive Ti material [44], while Ni is used as a barrier metal, which confines the downward diffusion of the fourth layer (Au) [46]. The top layer of Au protects layers below from oxidization [47]. Multiple SBDs of varying sizes were then fabricated by depositing Ni/Au (50/1000 nm) on the Ga-face of the wafer. Ni was selected to be the first layer due to the difference in work functions of GaN (4.2 eV) and Ni (5.04 eV) [48], which helps to form good Schottky contact (see Figure 9).

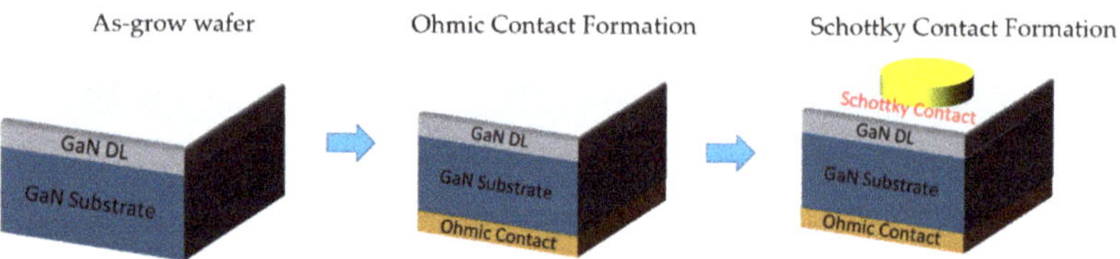

Figure 9. Fabrication of 1 mm GaN Schottky barrier diodes (SBD).

5. Electrical Characterization of SBD

To understand the effects of Mg-compensation on the performance of the SBDs, the electrical characteristics of an Mg-compensated SBDs with 2 different DL thicknesses (15 µm and 30 µm).

5.1. I-V Characterization

Figure 10a and b show both the reverse and forward current characteristics of SBDs with 15 µm and 30 µm. It can be observed in Figure 10a that both SBDs exhibit similar I_R at −20 V. Figure 10b shows the forward characteristics in which we observe a slight decrease in forward saturation current (I_{sat}), the drop observed could be due to increase in DL thickness. Increased series resistance due to an increase in DL is the primary cause of the decrease in I_{sat}.

From the measured forward I-V characteristics, Ideality Factor (n) and Barrier Height (Φ_B) were extracted using Equations (1) and (2) [2].

$$\Phi_B = KTqlnI_{sat}AA*T^2 \tag{1}$$

$$I = (eqVnKT/-1), \tag{2}$$

where I_{sat} is the forward saturation current, A is the SBD contact area, K is the Boltzmann's constant and A* is the Richardson's constant with a theoretical value of 26.9 A/cm^2.K^2.

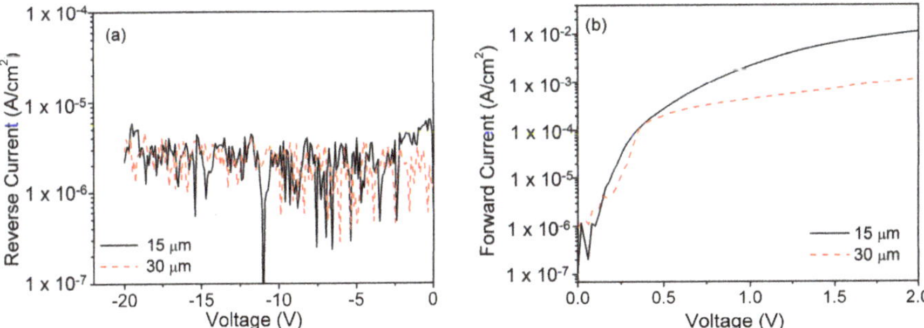

Figure 10. (**a**) Reverse and (**b**) Forward *I-V* characteristics of SBDs with 15 μm and 30 μm GaN DL.

Average ideality factors of 1.03 and 1.05 were extracted from 10 SBDs with compensated and conventional DL, respectively. The near-unity ideality factor signifies an excellent metal-semiconductor interface at the Schottky-semiconductor contact. The use of the Piranha solution followed by organic cleaning and dipping in BOE for 2 min has resulted in the formation of an excellent metal-semiconductor interface. The similarity of values in n among the SBDs with both 15 μm and 30 μm DLs indicates the thickness of DLs does not play any role in the determination of n [49]. Similarly, the extracted Φ_B for both SBDs of 0.81 eV (15 μm) and 0.78 eV (30 μm) SBDs are close to each other and similar to other reported Φ_B for Ni-based Schottky contacts (see Table 5). From the comparison, the extracted n and Φ_B were found to be within the reported range of 1.01 to 1.4 for the ideality factor and 0.74 eV to 1.1 eV for barrier height.

Table 5. List of state-of-the-art SBDs fabricated on GaN-on-GaN wafers.

Affiliation	DL (μm)	CCD (/cm^3)	Barrier Height (eV)	Ideality Factor
NTU [49]	15	7.6×10^{14}	0.81	1.03
	30	3×10^{15}	0.78	1.3
Army Res. Labs Maryland [50,51]	6.6	3×10^{16}	0.75	1.05
	2	2×10^{16}	0.79	1.1
AIST Japan [52]	–	5×10^{16}	0.74	1.09
Toyoda Gosei Co. [53]	10	2.2×10^{16}	1.01	1.01
University of Fukui [54]	12	10^{16}	1.05	1.03
Univ. of Notre Dame [29]	0.3	3×10^{16}	1.1	1.4

5.2. Capacitance–Voltage (C–V) Characteristics

C–V measurements were performed to extract the DW of the SBDs. No significant variation in capacitance was observed in a voltage range of −20 V to 5 V (see Figure 11), which signifies the complete depletion of the DL [55,56].

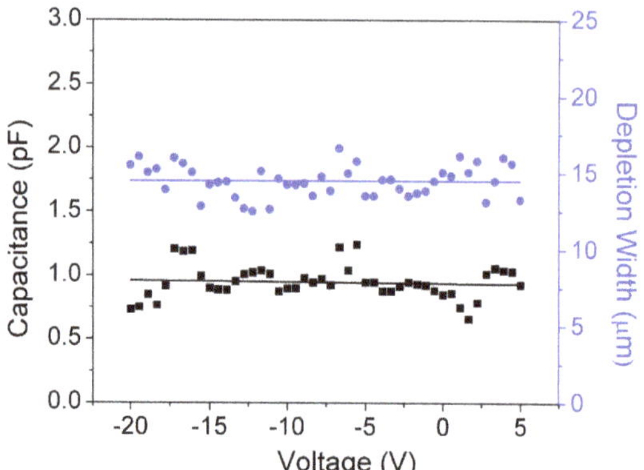

Figure 11. Variation of capacitance and DW with voltage of 0.5 mm diameter GaN SBDs with 15 μm.

DW can be extracted from the C–V characteristics using Equation (3):

$$C = \varepsilon_0 \varepsilon_r (A/DW). \tag{3}$$

A uniform DW of ~15 μm was measured at all voltages (−20 V to 5 V), which implies the total depletion of DL even at 0 V.

5.3. Reverse Conduction Mechanism of SBD

Reverse Conduction mechanism (CM) helps to understand the physical constituents leading to the reverse leakage current (J_R) Thermionic Emission (TE) is present if all electrons traverse over the barrier and Thermionic Field Emission (TFE) is the dominant CM if electrons tunneling through the barrier. The study of CM at elevated temperatures is worthy and important to understand the overall performance of the fabricated SBDs. Moreover, high voltages are required to generate a thick DW which will help in the detection of higher energies with improved sensitivity and higher CCE [15,35]. CM of J_R was extracted by comparing measured J_R with theoretically calculated J_R using equations for TE and TFE. Further details on the extraction of CM have been reported [43,57]. Figure 12 shows the measured J_R at different temperatures over a wide voltage range for SBDs with (a) 15 μm DL and (b) 30 μm DL. I-V-T characteristics of both devices have been divided into 3 zones depending on the observed CM.

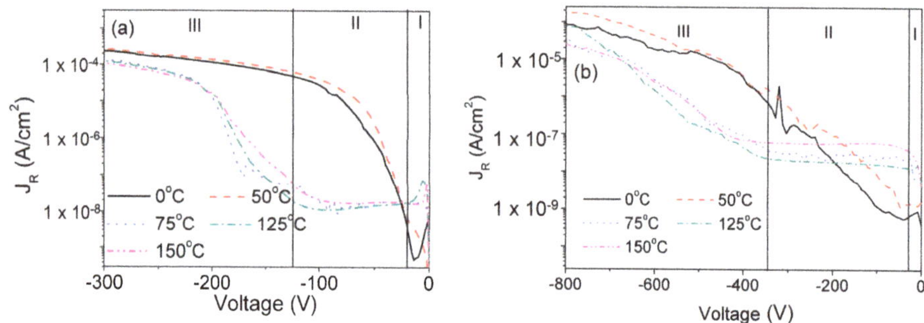

Figure 12. I-V-T characteristics of vertical GaN SBDs with (**a**) 15 μm DL, (**b**) 30 μm DL. Adapted from [58] A. Sandupatla et al. 2020 Appl. Phys. Express in press https://doi.org/10.35848/1882-0786/ab93a0. Copyright [2020] by Japanese Society of Applied Physics.

Various changes in CM were observed with a change in both temperature and voltage, which have been shown in Figure 13. To understand the physical significance of the change in CM The activation energy was also extracted in all voltage zones [55,56]. The extracted E_a of 0.4 eV corresponds to the presence of Mg ions in the DL [59,60]. The activation of N-vacancies [61] (E_a = −1.67 eV) with an increase in temperature resulted in the trapping of tunneling electrons and changing the CM from TFE to TE in Zone-II of both SBDs. Similarly, activation of C-traps (E_a = 0.69 eV) released electrons into the depletion region when a high reverse voltage was supplied. The release of these electrons increased the probability of tunneling through the DW changing the CM to TFE. This change of CM at elevated temperatures can be used in aid of the design of high breakdown voltage SBDs for high-power switching and high-energy radiation sensing applications.

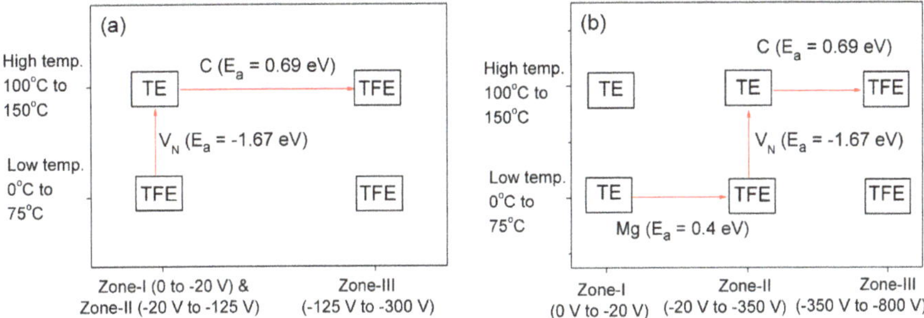

Figure 13. Change of CM with voltage zones and temperature ranges in SBDs with (**a**) 15 μm DL and (**b**) 30 μm DL. Adapted from [58] A. Sandupatla et al. 2020 Appl. Phys. Exp. https://doi.org/10.35848/1882-0786/ab93a0. Copyright [2020] by Japanese Society of Applied Physics.

5.4. Breakdown Voltage of SBD

The maximum voltage and power handling capability of SBDs are determined by its breakdown voltage characteristics. High voltages are essential to generate thick DW which is a primary requirement for an improved radiation detector performance. GaN has a high bandgap and high electric field strength, which makes it an optimum material to fabricate devices with high V_{BD}. For breakdown characterization, SBDs were exposed to increasing voltages until it reaches the set compliance of 1 A/cm^2 or when it reaches the catastrophic failure of the device. The SBDs were also dipped in Flourinert FC-40 prior to the measurements to insulate the SBDs from atmospheric flashover [62]. Figure 14 shows the semi-log breakdown characteristics of fabricated vertical SBDs with both 15 μm and 30 μm DLs. Thicker DW in SBDs with 30 μm DLs results in larger V_{BD}.

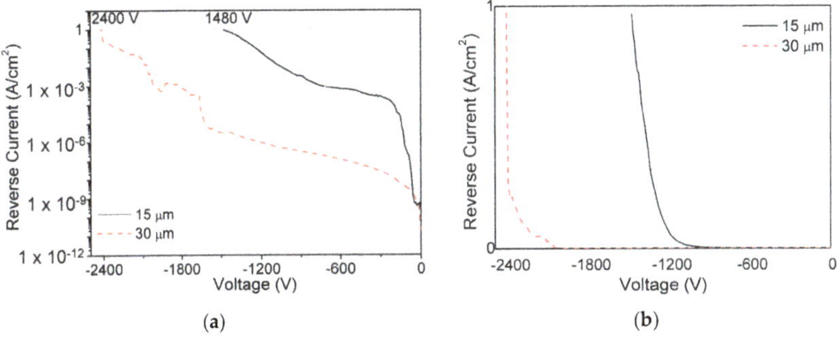

Figure 14. Reverse breakdown voltage characteristics of SBDs with 15 μm and 30 μm in (**a**) log and (**b**) linear scale.

Figure 15 shows the plot of V_{BD} vs. DLT for the state-of-the-art reported vertical SBDs [63–68]. Saitoh et al. realized a V_{BD} of 1100 V for the SBDs with DLT of 5 μm after using a field plate (FP) [63]. Shibata, D. et al. has reported the use of junction barrier Schottky (JBS) with p-type termination on SBDs with 13 μm thick DL to measure a V_{BD} value of 1600 V [64]. The improvement in V_{BD} was reported to be due to the reduction of CCD in the MOCVD grown GaN DLs [43,58].

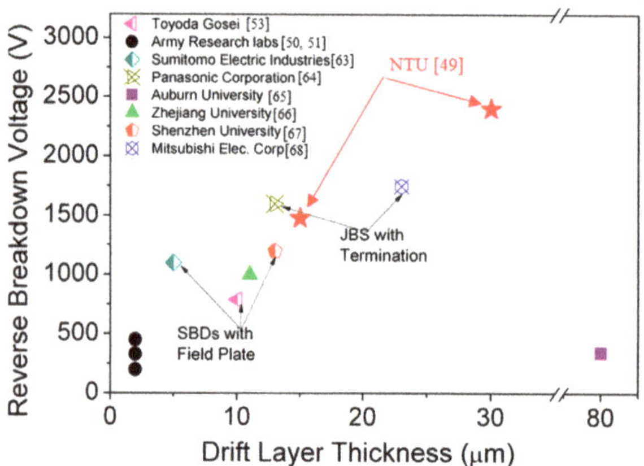

Figure 15. Benchmarking of measured reverse breakdown voltages of vertical GaN SBDs with state-of-the-art results.

6. Measurement Setup for α-Particle Detection

After electrical characterization, both wafers were diced into individual detectors and packaged onto a dual inline package (DIP) with silver paste for the ground contact (cathode) and wire-bonding for the Schottky contact (anode) (see Figure 16a). 5.48 MeV α-particles were generated from ^{241}Am source with an active area of 7 mm^2, which was placed at 8 mm from the detector (as shown in Figure 16b). Radionuclides of ^{241}Am were deposited onto a stainless-steel disc of 16 mm diameter, which was held in place by a plastic holder. The change in the current flowing through the circuit due to interaction with an α-particle was amplified by passing through pre-amplifier, amplifier and signal processing circuit. A Si surface detector from ORTEC was used as a reference, along with an ORTEC-671 amplifier for energy calibration.

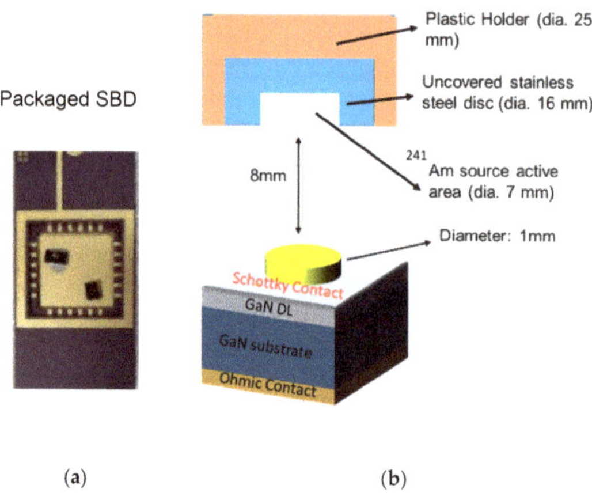

Figure 16. (**a**) Packaged Device and (**b**) Schematic drawing of Source-Detector measurement setup (not to scale).

6.1. Detection of α-Particle Spectra

The performance of any α-particle detectors is defined by its CCE. CCE is defined as the ratio of energy detected and the energy incident on the detector. The acquired data needs to be calibrated using a standard Si detector as a reference [69,70]. The fabricated GaN detectors require higher energy for the generation of e-h pairs in comparison with the reference Si detectors. The final detected energy is described by the following Equation (4) [34]:

$$E = E_0 + W_{GaN}/W_{Si} \times k \times \text{Channel}, \tag{4}$$

where E is the energy absorbed, E_0 is the loss in energy at the metal-semiconductor interface, which can be estimated from Transport of Ions in Matter (TRIM) simulations; k is a calibration factor of the reference detector; W_{GaN} is 8.9 eV and W_{Si} is 3.6 eV.

6.1.1. Low Voltage α-Particle Detection

For the detection of high energy α-particles, researchers have increased the DW of the detectors by fabricating them on GaN substrates. These detectors have generated 27 μm of the depleted region at very high voltages (−550 V) [34]. The requirement of high voltages in the generation of a thick DW increases the detector complexity and size, which severely affects its portability.

The α-particle energy spectra obtained from the 15 μm detectors under low-bias conditions (−20 to −80 V) are shown in Figure 17a. Figure 17b compares the variation of CCE with the voltages of different reported detectors (sandwich structures). Detectors with compensated DL exhibited lower variation in CCE (7%) in comparison to reports using a bulk GaN-based sandwich detector. The observation of lower variation in CCE was reported to be due to formation of a thick DW.

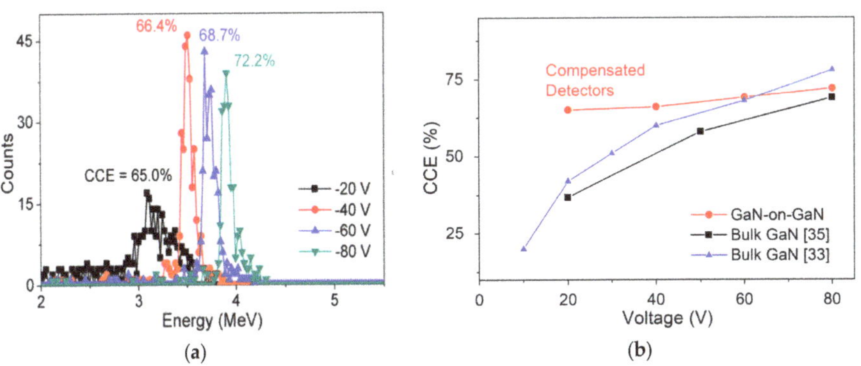

Figure 17. (**a**) Acquired α-particle spectra of compensated detectors for different voltages (−20 V to −80 V) and (**b**) Comparison of variation in CCE with voltages (−20 V to −80 V) for state-of-the-art α-particle detectors.

6.1.2. High Voltage α-Particle Detection

Although researchers have increased the energy detected by using bulk GaN-based sandwich structures, the complexity of generating a thick DW only two research groups have successfully developed α-particle detectors capable of detecting 5.48 MeV energy generated from ^{241}Am source. CCE of the compensated detectors improved from 72% at −80 V to 96.7% at −300 V (see Figure 18a) due to the increase in DW. The high-voltage performance of the compensated detectors requires 250 V lower bias conditions in comparison to the detector fabricated by Q. Xu et al. (see Figure 18b).

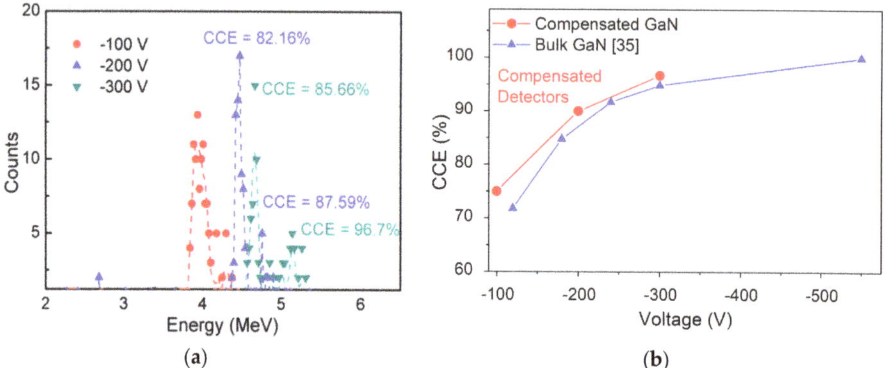

Figure 18. (**a**) α-particle spectra of compensated GaN detectors for different applied voltages (−100 V to −300 V) and (**b**) Comparison of variation in CCE with voltages (−100 V to −550 V) for state-of-the-art α-particle detectors.

Figure 19 shows the α-particle energy spectra obtained by SBD detectors with 30 μm DL at different voltages (−400 V to −750 V). An increase in applied bias conditions increases the detected energy increasing CCE. CCE of 100% in the detection of 5.48 MeV α-particle was obtained at −750 V. The high CCE obtained ensures complete energy detection from incident charged α-particle.

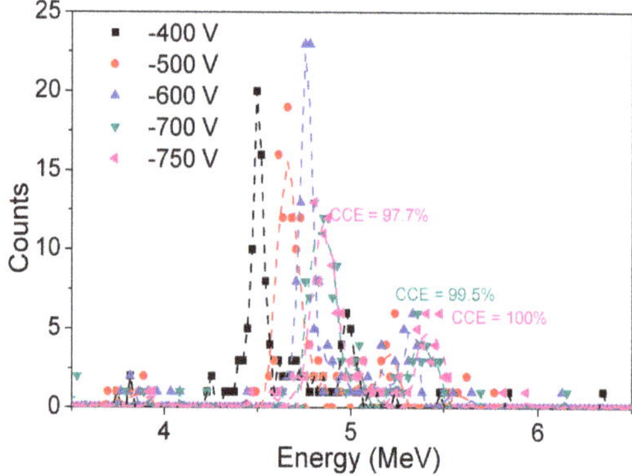

Figure 19. Acquired α-particle energy spectra of GaN SBDs at different voltages in vacuum.

6.1.3. Variation in α-Particle Spectra-Air vs. Vacuum (SBDs with 15 μm DL)

Figure 20 shows the comparison of the energy spectrum of GaN detectors biased at −100 V measured in a vacuum and in air reported for compensated detectors. 7% reduction in CCE was reported due to the presence of air. In vacuum, the complete energy of an α-particle is transferred to the detector, resulting in the detection of higher energies. While in air energy of α-particles is lost due to scattering. This loss in α-particle energy lowers detected CCE.

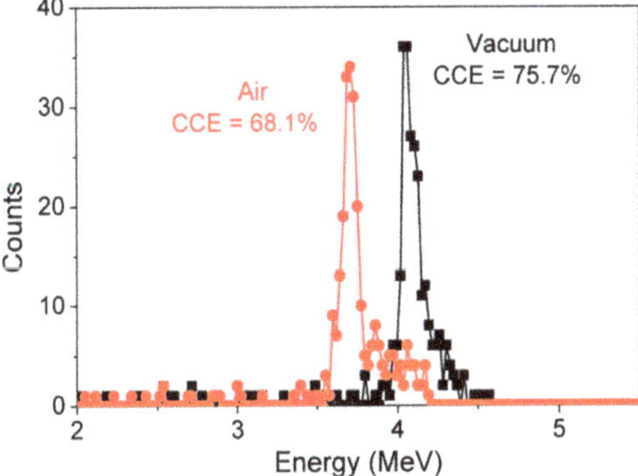

Figure 20. Acquired α-particle energy spectra of GaN SBDs at −100 V under air and in a vacuum.

6.2. Benchmarking

Figure 21 compares the performance of various reported GaN-based α-particle detectors as a function of detected energies. About 30% higher CCE was reported for the compensated 15 µm α-particle detectors at −20 V. In addition, compensated detectors also exhibit 96.7% CCE at −300 V, which is 250 V in comparison to other published literature. These promising results pave the way for compensated α-particle detectors to achieve high CCE with low operating voltage.

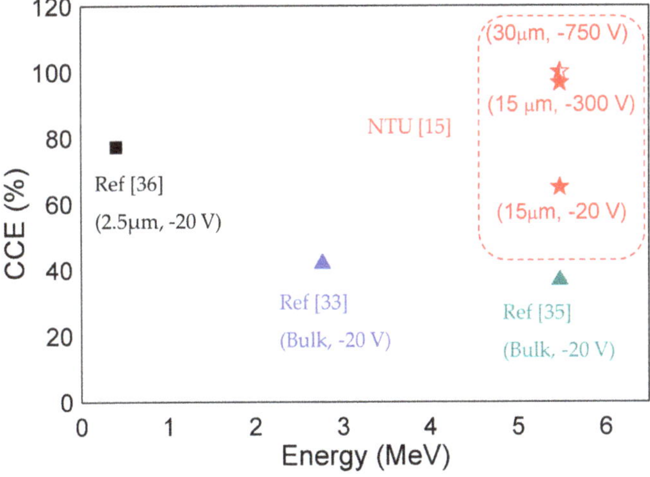

Figure 21. Benchmarking of extracted CCE of compensated detectors with epitaxial-grown GaN detectors (squares) and bulk GaN detectors (triangles) at low voltages.

7. Summary

GaN SBDs have demonstrated great potential in radiation detection on the virtue of its high displacement energy, wide bandgap and critical electric field strength. Conventionally, GaN particle detectors employ either a thin GaN epitaxial layer on the hetero-epitaxial substrate or thick free-standing GaN substrate to fabricate a radiation detector. While the thin epi-layer detector worked at low voltages (−28 V) with high CCE, they are only able to detect very low energies (< 1 MeV). The defects present in GaN such as high TDD and unintentional doping, have been the major constraints in terms of improving the detector performance. With improvement in growth technologies, free-standing GaN detectors with low TDD and doping densities have been fabricated which has increased the detected energy (5.48 MeV) but they require very high voltages (−550 V) to detect these energies with 100% CCE.

High performing GaN α-particle detector can be achieved by designing a thin compensated GaN DL on conducting GaN substrate. Compensation of thin DL by doping p-type (Mg) ions in DL reduces the CCD generated by unintentional n-type doping. The increase in DW can reduce the J_R and increased the V_{BD} without affecting the ideality factor and barrier height of the SBDs. A ~3 times improvement in V_{BD} from 430 V to 1480 V due to compensation has helped realize SBDs with the highest reported V_{BD} at 2400 V (SBDs with 30 μm DL) without any additional termination or field plates.

The improvement in reverse characteristics have helped in improvement of the performance of the detector by improving the sensitivity and reducing the voltage requirement of the detector. These compensated GaN-on-GaN SBDs exhibit a 30% increase in CCE (65%) at low voltages (−20 V) in comparison to previously reported GaN α-particle detectors. High CCE of 96.7% was also measured at −300 V, which is 250 V lower than the previously reported bias requirement. The improved performance in α-particle detection is due to the formation of thicker DW at low voltages. The spectral resolution of 71 keV is also 30% better than the previous reports. Presence of air results in scattering of α-particle incident on the detector, which in turn reduced its efficiency by ~7%. While detectors with thicker DW do require a higher voltage (750 V) to detect α-particles with 100% CCE, they can also detect higher energies (9.1 MeV).

Author Contributions: A.S. along with S.A. and N.G.I. have designed the experiments; Simulations and fabrication of devices was completed by A.S.; Result analysis was completed by A.S., S.A. and N.G.I.; J.K. performed radiation detection testing and analysis.; S.N. and H.A. have grown the GaN material for fabrication of devices; A.S. wrote the paper, which was reviewed by all authors. All authors have read and agreed to the published version of the manuscript.

Funding: This research received no external funding

Conflicts of Interest: The authors declare no conflict of interest.

References

1. Lamarsh, J.R.; Baratta, A.J. *Introduction to Nuclear Engineering*; Prentice-Hall: Upper Saddle River, NJ, USA, 2001.
2. Stacey, W.M. *Nuclear Reactor Physics*; John Wiley & Sons: Hoboken, NJ, USA, 2001.
3. Connor, N. What-Is-Detection-of-Alpha-Radiation-Alpha-Particle-Detector-Definition. Available online: https://www.radiation-dosimetry.org/what-is-detection-of-alpha-radiation-alpha-particle-detector-definition/ (accessed on 14 December 2019).
4. Burn, R.R. *Introduction to Nuclear Reactor Operation*; Detroit-Edison: Detroit, MI, USA, 1988.
5. Rossi, B.; Staub, A.H.H. *Ionization Chambers and Counters*; McGraw-Hill Book Company Inc: New York, NY, USA, 1949.
6. Teo, W.R. *Techniques for Nuclear and Particle Physics Experiments*; Springer: Berlin/Heidelberg, Germany, 1993.
7. Knoll, G. *Radiation Detection and Measurement*; John Wiley and Sons Inc.: Ann Arbor, MI, USA, 2010.
8. Ionascut-Nedelcescu, A.; Carlone, C.; Houdayer, A.; Bardeleben, H.J.V.; Cantin, J.L.; Raymond, A.S. Radiation hardness of gallium nitride. *IEEE Trans. Nucl. Sci.* **2002**, *49*, 2733–2738. [CrossRef]
9. Devanathan, R.; Weber, W. Displacement energy surface in 3C and 6H SiC. *J. Nucl. Mater.* **2000**, *278*, 258–265. [CrossRef]
10. Nord, J.; Nordlund, K.; Keinonen, J. Molecular dynamics study of damage accumulation in GaN during ion beam irradiation. *Phys. Rev. B* **2003**, *68*, 184104. [CrossRef]
11. Bourgoin, J.C.; Massarani, B. Threshold energy for atomic displacement in diamond. *Phys. Rev. B* **1976**, *14*, 3690. [CrossRef]
12. Ören, Ü.; Nilsson, J.; Herrnsdorf, L.; Rääf, C.L.; Mattsson, S. Silicon Diode as an Alpha Particle Detector and Spectrometer for Direct Field Measurements. *Radiat. Prot. Dosim.* **2016**, *170*, 247–251. [CrossRef]
13. Kang, S.M.; Ha, J.H.; Park, S.-H.; Kim, H.S.; Lee, N.H.; Kim, Y.K. Radiation Response of a Semi-insulating GaAs Semiconductor Detector for Charged Particle at Variable Operating Temperature. *Prog. Nucl. Sci. Technol.* **2011**, *1*, 282–284. [CrossRef]

14. Zatko, B.; Dubecky, F.; Sagatova, A.; Sedlačová, K.; Ryc, L. High resolution alpha particle detectors based on 4H-SiC epitaxial layer. *J. Instrum.* **2015**, *10*, C04009. [CrossRef]
15. Sandupatla, A.; Arulkumaran, S.; Ranjan, K.; Ng, G.I.; Murmu, P.P.; Kennedy, J.; Nitta, S.; Honda, Y.; Deki, M.; Amano, H.; et al. Low Voltage High-Energy α-Particle Detectors by GaN-on-GaN Schottky Diodes with Record-High Charge Collection Efficiency. *Sensors* **2019**, *19*, 5107. [CrossRef]
16. Wang, L.; Lou, Y.; Su, Q.; Shi, W.; Xia, Y. CVD diamond alpha-particle detectors with different electrode geometry. *Opt. Express* **2005**, *13*, 8612–8617. [CrossRef]
17. Chmill, V. *Radiation Tests of Semiconductor Detectors*; Kungl Tekniska H'Ogskolan Fysiska Institutionen: Stockholm, Sweden, 2006.
18. Meneghini, M.; Trevisanello, L.; Meneghesso, G.; Zanoni, E. A Review on the Reliability of GaN-Based LEDs. *IEEE Trans. Device Mater. Reliab.* **2008**, *8*, 323–331. [CrossRef]
19. Nakamura, S.; Mukai, T.; Senoh, M. Candela-class high-brightness InGaN/AlGaN double-heterostructure blue-light-emitting diodes. *Appl. Phys. Lett.* **1994**, *64*, 1687–1689. [CrossRef]
20. Trivellin, N.; Meneghini, M.; Zanoni, E.; Orita, K.; Yuri, M.; Meneghesso, G. A review on the reliability of GaN-based laser diodes. In Proceedings of the 2010 IEEE International Reliability Physics Symposium (IRPS), Anaheim, CA, USA, 2–6 May 2010.
21. Del Alamo, J.; Joh, J.; Del Alamo, J.A. GaN HEMT reliability. *Microelectron. Reliab.* **2009**, *49*, 1200–1206. [CrossRef]
22. Ranjan, K.; Arulkumaran, S.; Ing, N.G.; Sandupatla, A. Investigation of Self-Heating Effect on DC and RF Performances in AlGaN/GaN HEMTs on CVD-Diamond. *IEEE J. Electron Devices Soc.* **2019**, *7*, 1264–1269. [CrossRef]
23. Sun, X.; Li, D.; Li, Z.; Song, H.; Jiang, H.; Chen, Y.; Miao, G.; Zhang, Z. High spectral response of self-driven GaN-based detectors by controlling the contact barrier height. *Nat. Sci. Rep.* **2015**, *5*, 16819. [CrossRef]
24. Alifragis, Y.; Roussos, G.; Pantazis, A.K.; Konstantinidis, G.; Chaniotakis, N. Free-standing gallium nitride membrane-based sensor for the impedimetric detection of alcohols. *J. Appl. Phys.* **2016**, *119*, 74502. [CrossRef]
25. Nord, J.; Nordlund, K.; Keinonen, J.; Albe, K. Molecular dynamics study of defect formation in GaN cascades. *Nucl. Instrum. Methods Phys. Res. Sect. B Beam Interact. Mater. At.* **2003**, *202*, 93–99. [CrossRef]
26. Porowski, S.; Sadovyi, B.; Gierlotka, S.; Rzoska, S.J.; Grzegory, I.; Petrusha, I.; Turkevich, V.; Stratiichuk, D. The challenge of decomposition and melting of gallium nitride under high pressure and high temperature. *J. Phys. Chem. Solids* **2015**, *85*, 138–143. [CrossRef]
27. Sellin, P.; Vaitkus, J. New materials for radiation hard semiconductor dectectors. *Nucl. Instrum. Methods Phys. Res. Sect. A Accel. Spectrometers Detect. Assoc. Equip.* **2006**, *557*, 479–489. [CrossRef]
28. Weaver, B.D.; Anderson, T.J.; Koehler, A.D.; Greenlee, J.D.; Hite, J.K.; Shahin, D.I.; Kub, F.J.; Hobart, K.D. On the Radiation Tolerance of AlGaN/GaN HEMTs. *ECS J. Solid State Sci. Tech.* **2016**, *5*, Q208–Q212. [CrossRef]
29. Zhao, P.; Verma, A.; Verma, J.; Jena, D. *Comparison of Schottky Diodes on Bulk GaN Substrates & GaN-on-Sapphire*; CS Mantech: New Orleans, LA, USA, 2013.
30. Sugiura, M.; Kushimoto, M.; Mitsunari, T.; Yamashita, K.; Honda, Y.; Amano, H.; Inoue, Y.; Mimura, H.; Aoki, T.; Nakano, T. Study of radiation detection properties of GaN pn diode. *Jpn. J. Appl. Phys.* **2016**, *55*, 05FJ02. [CrossRef]
31. Wang, G.; Fu, K.; Yao, C.-S.; Su, D.; Zhang, G.-G.; Wang, J.-Y.; Lu, M. GaN-based PIN alpha particle detectors. *Nucl. Instrum. Methods Phys. Res. Sect. A Accel. Spectrometers Detect. Assoc. Equip.* **2012**, *663*, 10–13. [CrossRef]
32. Wang, J.; Mulligan, P.L.; Brillson, L.J.; Cao, L.R. Review of using gallium nitride for ionizing radiation detection. *Appl. Phys. Rev.* **2015**, *2*, 031102. [CrossRef]
33. Lee, I.H.; Polyakov, A.; Smirnov, N.B.; Govorkov, A.V.; Kozhukhova, E.A.; Zaletin, V.M.; Gazizov, I.M.; Kolin, N.G.; Pearton, S.J. Electrical properties and radiation detector performance of free-standing bulk n-GaN. *J. Vac. Sci. Technol. B* **2012**, *30*, 21205. [CrossRef]
34. Mulligan, P.L.; Wang, J.; Cao, L.R. Evaluation of freestanding GaN as an alpha and neutron detector. *Nucl. Instrum. Methods Phys. Res. Sect. A Accel. Spectrometers Detect. Assoc. Equip.* **2013**, *719*, 13–16. [CrossRef]
35. Xu, Q.; Mulligan, P.L.; Wang, J.; Chuirazzi, W.; Cao, L. Bulk GaN alpha-particle detector with large depletion region and improved energy resolution. *Nucl. Instrum. Methods Phys. Res. Sect. A Accel. Spectrometers Detect. Assoc. Equip.* **2017**, *849*, 11–15. [CrossRef]

36. Vaitkus, J.V.; Cunningham, W.; Gaubas, E.; Rahman, M.; Sakai, S.; Smith, K.; Wang, T. Semi-insulating GaN and its evaluation for α particle detection. *Nucl. Instrum. Methods Phys. Res. Sect. A Accel. Spectrometers Detect. Assoc. Equip.* **2003**, *509*, 60–64. [CrossRef]
37. Min, L.; Guo-Guang, Z.; Kai, F.; Guo-Hao, Y. Gallium Nitride Room Temperature α Particle Detectors. *Chin. Phys. Lett.* **2010**, *27*, 052901. [CrossRef]
38. Polyakov, A.; Smirnov, N.B.; Govorkov, A.V.; Markov, A.V.; Kozhukhova, E.A.; Gazizov, I.M.; Kolin, N.G.; Merkurisov, D.I.; Boiko, V.M.; Korulin, A.V.; et al. Alpha particle detection with GaN Schottky diodes. *J. Appl. Phys.* **2009**, *106*, 103708. [CrossRef]
39. Ziegler, J.F.; Ziegler, M.; Biersack, J. SRIM—The stopping and range of ions in matter (2010). *Nucl. Instrum. Methods Phys. Res. Sect. B Beam Interact. Mater. At.* **2010**, *268*, 1818–1823. [CrossRef]
40. Shinoda, H.; Mutsukura, N. Structural properties of GaN layers grown on Al_2O_3 (0001) and GaN/Al_2O_3 template by reactive radio-frequency magnetron sputter epitaxy. *Vacuum* **2016**, *125*, 133–140. [CrossRef]
41. Tanikawa, T.; Ohnishi, K.; Kanoh, M.; Mukai, T.; Matsuoka, T. Three-dimensional imaging of threading dislocations in GaN crystals using two-photon excitation photoluminescence. *Appl. Phys. Express* **2018**, *11*, 031004. [CrossRef]
42. Sandupatla, A.; Subramaniam, A.; Ng, G.I.; Ranjan, K.; Deki, M.; Nitta, S.; Honda, Y.; Amano, H. Enhanced breakdown voltage in vertical Schottky diodes on compensated GaN drift layer grown on free standing GaN. In Proceedings of the 13th International Conference of Nitride Semiconductors (ICNS-13), Bellevue, DC, USA, 7–12 July 2019.
43. Sandupatla, A.; Subramaniam, A.; Ng, G.I.; Ranjan, K.; Deki, M.; Nitta, S.; Honda, Y.; Amano, H. Improved breakdown voltage in vertical GaN Schottky barrier diodes on free-standing GaN with Mg-compensated drift layer. *Jpn. J. Appl. Phys.* **2019**, *59*, 010906. [CrossRef]
44. Hwang, Y.-H.; Ahn, S.; Dong, C.; Zhu, W.; Kim, B.-J.; Ren, F.; Lind, A.G.; Jones, K.S.; Pearton, S.J.; Kravchenko, I. Effect of Buffer Oxide Etchant (BOE) on Ti/Al/Ni/Au Ohmic Contacts for AlGaN/GaN Based HEMT. *ECS Meet. Abstr.* **2015**, *69*, 111–118. [CrossRef]
45. Wang, L.; Mohammed, F.M.; Adesida, I. Differences in the reaction kinetics and contact formation mechanisms of annealed Ti/Al/Mo/Au Ohmic contacts on n-GaN and AlGaN/GaN epilayers. *J. Appl. Phys.* **2007**, *101*, 13702. [CrossRef]
46. Feng, Q.; Li, M.; Hao, Y.; Ni, Y.; Zhang, C. The improvement of ohmic contact of Ti/Al/Ni/Au to AlGaN/GaN HEMT by multi step annealing method. *Solid State Electron.* **2009**, *53*, 955–958. [CrossRef]
47. Yang, L. *Planar and Non Gold Metal Stacks Processes and Conduction Mechanism of AlGaN/GaN HEMTs on Si*; Nanyang Technological University: Singapore, 2016.
48. Huang, Y.; Huang, Z.; Zhong, Z.; Yang, X.; Hong, Q.; Wang, H.; Huang, S.; Gao, N.; Chen, X.; Cai, D.; et al. Highly transparent light emitting diodes on graphene encapsulated Cu nanowires network. *Sci. Rep.* **2018**, *8*, 13721. [CrossRef] [PubMed]
49. Sandupatla, A.; Arulkumaran, S.; Ng, G.I.; Ranjan, K.; Deki, M.; Nitta, S.; Honda, Y.; Amano, H. GaN drift-layer thickness effects in vertical Schottky barrier diodes on free-standing HVPE GaN substrates. *AIP Adv.* **2019**, *9*, 045007. [CrossRef]
50. Tompkins, R.P.; Khan, M.R.; Green, R.; Jones, K.A.; Leach, J.H. IVT measurements of GaN power Schottky diodes with drift layers grown by HVPE on HVPE GaN substrates. *J. Mater. Sci. Mater. Electron.* **2016**, *27*, 6108–6114. [CrossRef]
51. Tompkins, R.P.; Smith, J.R.; Kirchner, K.W.; Jones, K.A.; Leach, J.H.; Udwary, K.; Preble, E.; Suvarna, P.; Leathersich, J.; Shahedipour-Sandvik, F. GaN Power Schottky Diodes with Drift Layers Grown on Four Substrates. *J. Electron. Mater.* **2014**, *43*, 850–856. [CrossRef]
52. Yamada, H.; Chonan, H.; Takahashi, T.; Shimizu, M. Electrical properties of Ni/n-GaN Schottky diodes on freestanding m-plane GaN substrates. *Appl. Phys. Express* **2017**, *10*, 41001. [CrossRef]
53. Tanaka, N.; Hasegawa, K.; Yasunishi, K.; Murakami, N.; Oka, T. 50 A vertical GaN Schottky barrier diode on a free-standing GaN substrate with blocking voltage of 790 V. *Appl. Phys. Express* **2015**, *8*, 71001. [CrossRef]
54. Tanaka, T.; Kaneda, N.; Mishima, T.; Kihara, Y.; Aoki, T.; Shiojima, K. Roles of lightly doped carbon in the drift layers of vertical n-GaN Schottky diode structures on freestanding GaN substrates. *Jpn. J. Appl. Phys.* **2015**, *54*, 41002. [CrossRef]

55. Grant, J.; Bates, R.; Cunningham, W.; Blue, A.; Melone, J.; McEwan, F.; Vaitkus, J.; Gaubas, E.; O'Shea, V. GaN as a radiation hard particle detector. *Nucl. Instrum. Methods Phys. Res. Sect. A Accel. Spectrometers Detect. Assoc. Equip.* **2007**, *576*, 60–65. [CrossRef]
56. Wang, J. *Evaluation of GaN as a Radiation Detection Material*; The Ohio State University: Columbus, OH, USA, 2012.
57. Abhinay, S.; Arulkumaran, S.; Ng, G.; Ranjan, K.; Deki, M.; Nitta, S.; Honda, Y.; Amano, H. Effects of Drift Layer Thicknesses in Reverse Conduction Mechanism on Vertical GaN-on-GaN SBDs grown by MOCVD. In Proceedings of the Electron Devices Technology and Manufacturing Conference (EDTM), Singapore, 13–15 March 2019.
58. Sandupatla, A.; Subramaniam, A.; Ng, G.I.; Ranjan, K.; Deki, M.; Nitta, S.; Honda, Y.; Amano, H. Change of high-voltage conduction mechanism in vertical GaN-on-GaN Schottky diodes at elevated temperatures. *Appl. Phys. Exp.* **2020**. [CrossRef]
59. Pearton, S.J. *GaN and Related Materials II*; Gordon Breach and Science Publication: Gainesville, Florida, 2000.
60. Ko, K.; Lee, K.; So, B.; Heo, C.; Lee, K.; Kwak, T.; Han, S.W.; Cha, H.Y.; Nam, O. Mg-compensation effect in GaN buffer layer for AlGaN/GaN high-electron-mobility transistors grown on 4H-SiC substrate. *Jpn. J. Appl. Phys.* **2016**, *56*, 015502. [CrossRef]
61. Xie, Z.; Sui, Y.; Buckeridge, J.; Catlow, C.R.A.; Keal, T.W.; Sherwood, P.; Walsh, A.; Farrow, M.R.; Scanlon, D.O.; Woodley, S.M.; et al. Donor and acceptor characteristics of native point defects in GaN. *arXiv* **2018**, arXiv:1803.06273. [CrossRef]
62. Arulkumaran, S.; Vicknesh, S.; Ing, N.G.; Selvaraj, S.L.; Egawa, T. Improved Power Device Figure-of-Merit (4.0×10^8 V^2 Ω^{-1} cm^{-2}) in AlGaN/GaN High-Electron-Mobility Transistors on High-Resistivity 4-in. Si. *Appl. Phys. Express* **2011**, *4*, 084101. [CrossRef]
63. Saitoh, Y.; Sumiyoshi, K.; Okada, M.; Horii, T.; Miyazaki, T.; Shiomi, H.; Ueno, M.; Katayama, K.; Kiyama, M.; Nakamura, T. Extremely Low On-Resistance and High Breakdown Voltage Observed in Vertical GaN Schottky Barrier Diodes with High-Mobility Drift Layers on Low-Dislocation-Density GaN Substrates. *Appl. Phys. Express* **2010**, *3*, 081001. [CrossRef]
64. Shibata, D.; Kajitani, R.; Handa, H.; Shiozaki, N.; Ujita, S.; Ogawa, M.; Tanaka, K.; Tamura, S.; Hatsuda, T.; Ishida, M.; et al. Vertical GaN-based power devices on bulk GaN substrates for future power switching systems. In Proceedings of the SPIE OPTO, San Francisco, CA, USA, 11 April 2018.
65. Wang, Y.; Xu, H.; Alur, S.; Sharma, Y.; Tong, F.; Gartland, P.; Issacs-Smith, T.; Ahyi, C.; Williams, J.; Park, M.; et al. Electrical characteristics of the vertical GaN rectifiers fabricated on bulk GaN wafer. *Phys. Status Solidi C* **2011**, *8*, 2430–2432. [CrossRef]
66. Yang, S.; Han, S.; Li, R.; Sheng, K. 1 kV/1.3 $m\Omega\cdot cm^2$ vertical GaN-on-GaN Schottky barrier diodes with high switching performance. In Proceedings of the International Symposium on Power Semiconductor Devices, Chicago, IL, USA, 13–17 May 2018.
67. Liu, X.; Liu, Q.; Li, C.; Wang, J.; Yu, W.; Xu, K.; Ao, J.-P. 1.2 kV GaN Schottky barrier diodes on free-standing GaN wafer using a CMOS-compatible contact material. *Jpn. J. Appl. Phys.* **2017**, *56*, 26501. [CrossRef]
68. Hayashida, T.; Nanjo, T.; Furukawa, A.; Watahiki, T.; Yamamuka, M. Leakage current reduction of vertical GaN junction barrier Schottky diodes using dual-anode process. *Jpn. J. Appl. Phys.* **2018**, *57*, 040302. [CrossRef]
69. Kennedy, J.V.; Markwitz, A.; Trodahl, H.J.; Ruck, B.J.; Durbin, S.M.; Gao, W. Ion beam analysis of amorphous and nanocrystalline group III-V nitride and ZnO thin films. *J. Electron. Mater.* **2007**, *36*, 472–482. [CrossRef]
70. Murmu, P.P.; Kennedy, J.V.; Ruck, B.J.; Rubanov, S. Micostructural, electrical and magnetic properties of erbium doped zinc oxide single crystals. *Electron. Mater. Lett.* **2015**, *11*, 998–1002. [CrossRef]

MDPI
St. Alban-Anlage 66
4052 Basel
Switzerland
Tel. +41 61 683 77 34
Fax +41 61 302 89 18
www.mdpi.com

Micromachines Editorial Office
E-mail: micromachines@mdpi.com
www.mdpi.com/journal/micromachines

www.ingramcontent.com/pod-product-compliance
Lightning Source LLC
LaVergne TN
LVHW070105170726
843515LV00007B/1614

9783036505664